HAMPSHIRE BIRI 2005

Hampshire Ornithological Society

(Registered Charity no. 1042309)
www.hos.org.uk

EDITOR

ALAN COX
E-mail: alanfjcox@compuserve.com

PRODUCTION EDITOR

MIKE WALL
E-mail: mike@bike2nature.co.uk

COUNTY RECORDER

JOHN CLARK
4 Cygnet Court, Old Cove Road,
Fleet, Hampshire GU51 2RL
E-mail: johnclark@cygnetcourt.demon.co.uk

© Hampshire Ornithological Society - February 2007
ISBN 0-9509805-7-9

Cover photograph: Desert Wheatear (Richard Ford)

HOS is pleased to acknowledge the help of Hampshire County Council, Southampton City Council and Portsmouth City Council in the publication of this Report.

HOS wishes to acknowledge the continued generous support of London Camera Exchange, Winchester, one of Hampshire's leading suppliers for binoculars, telescopes, cameras and video equipment. Discounts are available to members of HOS.

London Camera Exchange (Winchester) Limited
15, The Square,
Winchester,
SO23 9ES
Tel: 01962 866203

The views expressed in this report are not necessarily those held by the Hampshire Ornithological Society

Contents

[1] All *HOSRP* and *BBRC* species accounts were written by John Clark

Chairman's Report 2005

During the time that I have been a member of HOS there have been some major changes in the status of Hampshire's birds. We take it for granted now when we visit the coast that we may well see Little Egrets, a species that was rare in Hampshire twenty years ago. Gadwall, Mediterranean Gull and Cetti's Warbler are others that have become increasingly common. In contrast, we have become resigned to the fact that many of our farmland birds are much harder to find today than they were back then. If proof was needed, HOS's 2005 Corn Bunting survey has provided it as described later in this report. Yet, despite the negative trends, there is possibly some light at the end of the tunnel for our farmland birds. Their prospects may be brighter now than they have been for decades. The change has been brought about first and foremost by the recognition that there is a real problem related to current farming practices. This occurred thanks, in large part, to survey work carried out by volunteers including many HOS members. But who would have predicted that birdwatchers could bring about changes in the EU Common Agricultural Policy? Recently, political pressure has begun to change the emphasis away from encouraging more intense farming towards improved environmental management of farmland. As described in last year's Bird Report, the new Environmental Stewardship schemes hold promise of much improved prospects for farmland birds. It's too early to say whether these will succeed. There may be surprises such as bio-fuels in the pipeline and, of course, it's not just farmland birds that are in trouble. Many of our woodland and river valley birds are declining as well. So the message for HOS is clear. If we want to have a voice in protecting our birds and their habitats, we must continue to gather the data and make it available as widely as possible.

We have also seen major changes within the Society itself; I would single out technology as the greatest factor driving these. That may come as a surprise when binoculars, the main technology involved in bird-watching, have not changed a great deal over the past 200 years. I appreciate that Zeiss, Leica and the rest can point to coated lenses, internal focussing, even image stabilisation but the fundamentals of bird-watching have not really changed very much. It's been a case of evolution rather than revolution. Where the technology has really made a difference is in communications and IT. I remember the time when the only way to find out what was happening anywhere in the country was to call Nancy's café in Cley. Nowadays we have mobile phones, pagers, e-mail and the web. These changes have certainly impacted on HOS. We still have our excellent Newsletter and the annual Bird Report but these have been supplemented by an e-mail discussion group and a website, which was revamped, thanks to Annette and Robert Clayson, in 2005. It's commonplace now to see a bird and return home to find excellent photographs already available on the web. The days when we used to cut incoming paper records into strips to glue them on to species sheets are long gone. Records are now entered into a computerised records database. Many of us are entering our records into a personal database or even entering them on line into a national database. Where is this leading? I'm not sure but I am certain that HOS needs to move with the times. We need to streamline our recording system to make it easier both to submit records and input them to the database. We have made some progress but much remains to be done.

2005 was a good year for HOS. Membership, at close to 1200, reached an all time high; the Society's financial position was strong but, most importantly, I believe the Society made a real contribution to safeguarding Hampshire's birds. In the north east, data from HOS members has helped protect the few remaining areas of heathland by getting them included in the Thames Basin Heaths Special Protection Area. Data from the 2005 Corn Bunting survey are being used to target Environmental Stewardship schemes aimed at helping Corn Bunting numbers to recover. The Society can be proud of these results but it can only continue to flourish through the ongoing efforts of many people; I want to thank all of those who have contributed to HOS in 2005. There are too many individuals involved to mention them all by name so I will point to the enormous amount of work they do as evidenced by their output which includes the Bird Report, the Newsletter, the Walks Programme, Members' Day, the surveys, the database and the Website. I will mention three members of the Management Committee who stood down in 2005 or early 2006 - Peter Dudley as HOS Secretary, Alison Wall as Membership Secretary and Richard Jacobs as Librarian. I want to thank Peter, Alison and Richard for their assistance and contributions to the Society. It is also appropriate here to express the Society's appreciation to many organisations that have supported and cooperated with HOS in 2005. These include Hampshire County Council, Portsmouth and Southampton City Councils, the RSPB, BTO, HWT, HBIC and Game Conservancy Trust.

Regrettably, I must end this report on a sad note. In 2005, HOS lost three members who each played important roles in the development of Hampshire ornithology. Peter Le Brocq, Brian Renyard and John Taverner all passed away during the year. Their obituaries will be found in this Report. As County Recorder, Editor of the Bird Report and as a great publicist and educationalist, John Taverner's contribution was particularly prominent. In memory and appreciation of John, HOS intends to create a permament memorial to him at Needs Ore.

John Eyre

Chairman Hampshire Ornithological Society (HOS)

Peter Francis Le Brocq

1922-2005

When Peter Le Brocq died on March 13th 2005, aged 82, Hampshire lost one of its most expert amateur naturalists. He was not only a capable birdwatcher, but was an excellent botanist and entomologist who possessed an encyclopaedic knowledge of these taxa, including many obscure and difficult to identify species.

He was born in Crondall, and during the Second World War did military service in the Royal Tank Corps and latterly in a top secret installation on the Isle of Wight. In the early 1950s he worked at Plant Protection, Fernhurst, concurrently with studying for a degree at Wye College, Kent. Unfortunately, he subsequently suffered from illness and was unable to sustain full time employment again. This of course gave him the opportunity to indulge his natural history interests, which he had harboured since an early age. Perusal of county bird reports for Kent, Surrey, Sussex and Hampshire over the last half century show that he was a regular contributor of records.

In Hampshire, perhaps his favourite patch was Woolmer Pond and Forest, which he watched over a period of over 50 years from the early 1950s until his death. He was strongly influenced by his good friend Ewart Jones, who died in 1998, and they both became founder members of the 'Portsmouth Group' and were regular visitors to the Selsey Bill area, where sea watching was a particular passion. In 1954, Peter discovered Honey Buzzards in the New Forest, and this led to their subsequent study by other Portsmouth Group members.

Peter exhibited a certain obsession about optics; at one time he owned up to 12 pairs of binoculars! He particularly favoured large binoculars, including a pair of 15 x 60 Barr and Strouds (the famous Lebrocqulars), which he invariably rested on a walking stick. He much preferred binoculars to telescopes, which he didn't get on with.

I was introduced to Peter by Ewart Jones in the early 1970s, and was immediately struck by his enthusiasm and intensity for all matters natural history. In his later years, he became obsessive in his study of certain species – particularly nesting Hobbies, roosting Hen Harriers and Great Grey Shrikes. He would often return to a location evening after evening to confirm a particular detail about a bird's plumage or its behaviour.

Peter enjoyed a certain notoriety amongst the birding fraternity. He was a well-practised conversationalist, both in the field and over the telephone; he even labelled himself the telephone terrorist! There are many stories, perhaps somewhat apocryphal, from various birders of the lengthy phone calls with Peter of up to four hours duration, with Peter doing virtually all the talking! Nonetheless, despite his intensity at times, he was a wonderful naturalist and birder who was always willing to share his knowledge with others. In an age of conformity and political correctness, his eccentricity was most refreshing! He is sadly missed.

John Clark - with thanks to Dave Billett, Gilbert Rowland and Mick Scott

John Heugh Taverner
1929-2005

Forgive me if I digress a little. Do you remember when you started secondary school after going up in the world from primary? The day seems filled with lessons and homework and cramming a lot into a small brain. For me there was a bright light amongst all this French, Maths and History, a teacher who did Nature Study lessons. This was just an extra subject for interest and whilst not everyone was particularly interested, I was able to do something I really wanted to. That teacher was John, and not only were there periods spent learning about migration, flight and other topics but he actually took us into the field to see exciting birds, in new places. We went to the Needs Ore and the New Forest and stayed at St Catherine's Point on the southern tip of the Isle of Wight in spring (my first trip away from home), Cley, the Camargue and even Cape Clear in Ireland during autumn, and all that whilst still at school. I doubt whether many teachers would do that today, but it instilled in me a lifelong interest, for visiting distant places and looking at and recording birds. His generous nurturing, sense of humour and expertise made these experiences enjoyable and memorable. He was an excellent teacher and rarely suffered from the problems of badly behaved students, who were well aware of the consequences of 'trying it on'! After I left school, John remained a good friend for the rest of his life.

John's family came from Bournemouth, which was then part of Hampshire and his father instilled a lifetime fascination with birds when he took him to see the nesting auks and seabirds on the Purbeck cliffs. He returned many times to the area which was one of his all time favourite spots in the world. He studied for an honours degree in Geography and Economics at Southampton University and spent a lot of time playing football. He started teaching in Bournemouth, but soon moved to Peter Symond's Grammar School in Winchester, where he stayed till his retirement.

John become a member of the Hampshire Field Club, Ornithological Section and was closely involved with the running of the section and its Chairman Edwin Cohen. He assisted recording and co-authored the *Revised List of Hampshire and Isle of Wight Birds* that Cohen was preparing before he died. He then co-edited the *Hampshire Bird Report* for many years and eventually became the sole editor for many more years, putting some 21 years into the task. He helped steer the ever growing Ornithological Section in 1978, to the thriving and independent Hampshire Ornithological Society, that we know today. When he relinquished editorship of the *Bird Report*, he didn't rest on his laurels. His widening interest into other branches of the natural history, especially butterflies and became editor of the *Hampshire Butterfly Report* and subsequently editor of *The Butterflies of Hampshire*, then *The Dragonflies of Hampshire*. At this time he continued his passion for photographing wildlife[1], leading foreign birdwatching trips, teaching evening classes and writing engaging columns in the local press (*Hampshire* magazine, *The Echo* and *Hampshire Chronicle*) about local Hampshire wildlife, every month for some 35 years. He also submitted a number of scientific notes and papers to British Birds, mainly about Eiders and Spotted Redshanks, Black-headed and Mediterranean Gulls at his

[1] Photograph shows such a close-up photo-session at Eric Ashby's home in the New Forest,

chosen 'patch' Needs Ore, that he visited regularly right up to his last years. His final contribution to his studies at Needs Ore, were two booklets: *A Colony of Seabirds*, about the gull and tern colony and *The Welcome Immigrants*, the story of the arrival of the Mediterranean Gulls to breed. John was part of the revival of Hampshire ornithology which slowly gained momentum after the Second World War. He was particularly impressed with the careful studies of shorebirds made at Farlington Marshes by the 'Portsmouth Group' and realised that they were some of the leading lights in the country.

John was born on May 3rd,1929 and died on December 3rd, 2005. In his lifetime he carved himself an unassailable position in the ranks of Hampshire ornithologists and ornithology. He is survived by his wife Pat, children Terri, Philip and Stephen.

David Thelwell

The following appreciation is from another former pupil, now a teacher in Devon:

I knew John Taverner from my earliest days of birding. I used to visit Needs Ore and the Gins every Sunday morning with two friends. John would collect us in his car and we would spend the morning at the reserve. All the time developing our skills and field craft under his gentle tutorage. Our birding horizons were quickly widened with visits into the New Forest and then the almost annual spring migration to St. Catherine's Point on the Isle of Wight, staying in the old oil store and monitoring bird movements. To this day I can still picture getting up in the middle of the night to experience "a fall at the light". Catching the birds as they landed, dazed on the parapet, we would lower them down the side of the lighthouse in cotton bags, so that they could be identified, measured and ringed at first light.

John it was, who took us on our first 'twitches'. To Durlston Head to see the Brown Thrasher in 1966 and then to Chew Lake in 1967 for a Wilson's Phalarope. He began to look even further afield for us. Trips were organised to Cley, where we met the famous Richard Richardson and then to Cape Clear for seabird passage. He helped us to enjoy the experiences of foreign birding with visits to the Camargue and the Coto Doñana.

It's only now in later life, when you look back, that you can really begin to appreciate the impact he had on my generation of birdwatchers. Many of us are still actively involved in birding today. Friendships forged then have never been lost. The thrill and excitement of birding and the willingness to share experiences, that John instilled in us as teenagers, are still strong today. A lasting tribute to this engaging, sincere, humorous man, is surely that through his ability to communicate his love for a hobby to a new generation, he has helped ensure that future generations will have the opportunities to experience the wonders of the natural world that helped shape our lives.

David Fieldsend

The following appreciation is from Commander A.Y. Norris RN, FRGS and was written following David Thelwell's earlier obituary published in Kingfisher No 109:

David Thelwell has written a fine and comprehensive obituary which described the life and accomplishments of the late John Taverner, a naturalist *par excellence* and a superb stalwart of HOS. Obituaries are the obverse of the coin. I hope that I may not appear too presumptuous in attempting a personal appreciation of this remarkable man, perhaps, the reverse of that same coin.

I met John Taverner for the first time in 1963 at the AGM of the Ornithological Section of the Hampshire Field Club in circumstances which were to be testing that evening for his resources of tact and patience. I was aware that for some years, Edwin Cohen and Dr Canning Suffern had been engaged in ornithological hostilities by letter. Cohen, the Recorder, had disputed somewhat testily numbers of Suffern observations from Titchfield

Haven. Suffern's reactions were less than appropriate but as Richard Leach and Norman Pratt comment in "Birds of Hampshire" one-upmanship among bird enthusiasts in Hampshire was something of a practice in those days, Suffern and Cohen no less. The 'Doc' had some determination for a public face-to-face with Cohen; the opportunities of the AGM and free transport in my car to Southampton were too good for him to miss. That evening John Taverner in urbane fashion, with nicely concealed objectivity 'disarmed' Suffern; my companion was reduced to uncharacteristic peacefulness while 'Any Other Business' became a lively discussion of some merit on the direction, some would say politics, of bird studies in Hampshire. For my elderly mentor and great friend, the occasionally irascible Suffern, John was thereafter regarded sadly as being a mite too close to the County Recorder but the hostilities of the epistolary war were thankfully over if not, much reduced. Just a few years ago, John and I were picking over the events in the history of County bird politics and he lamented that he never got to know the 'Doc' as he might have done and that the events of the AGM of 1963 probably had much to do with that result. Nevertheless, this combination of personal qualities and well-informed objectivity, evinced so magisterially at that meeting back in 1963, have been for me the hall-marks of John Taverner's performance in the conduct of all his duties, professional and otherwise, as they unfolded over the years. Much later, John demonstrated such courtesy and good manners in apologising to Canning Suffern for the undue rejection of an important record made in 1948, accepting that record and publishing the apology in one of his books.

For me, John Taverner's brief yet kindly written contributions down the years to the *Southern Evening Echo* and to the *Hampshire Chronicle* were worthy models of how natural history should be introduced to the un-initiated and particularly the young. To practised naturalists, there was always something new to absorb from these elegantly written and very special notes.

We all know that John was essentially the example of the consummate enthusiast. For me, clear as this characteristic was, I will remember him especially for his accompanying smile through those big glasses and his laconic sense of humour as he was speaking of things. His *curriculum vitae* tells of his National Service in the Royal Artillery. CVs are meant to be serious stuff but John recorded his military achievement as 'Malta - Lance Bombadier'! It would be a dullard who on reading this part of his CV did not smile; I am sure that he was a very good lance corporal. A HOS field meeting under his leadership was arranged at Needs Ore for late June, one year. Things transpired to be very quiet, the tide was out, there were a few Crows, one or two Oystercatchers, small numbers of gulls and numbers of Canada and Greylag geese were unusually subdued. John recognised that the tour wasn't going to offer much ornithologically for his party of birders and he engaged in an artful description of the invertebrates and plants spliced with a number of anecdotes which had everyone smiling for much of the visit, meanwhile his lovely grin was never far away. It was a subtle performance from which I, for one, benefited and which I recall with pleasure.

As an officer in the Profession of Arms, leadership for me has always been about example and the ability of those in authority to inspire others to do their best for the general good. In volunteer groups, such an effect is not easy to obtain. Yet in all his years of office, John did and I suspect that few realised that he was leading us with wisdom and a clear sense of the aim. John had a core of steel and like all the best leaders, he rarely had cause to show it such was his gift of inspiration.

I recall John at Needs Ore with much pleasure. Here, not surprisingly, he always seemed to be at his best no matter whether he was involved with observing, conducting others around the Reserve or doing special things like counting nests in the gull colony or ternery. He wrote or collaborated with the writing of some fine books but for me and appropriately, his little book *A Colony of Seabirds* reflects all that is best about John Taverner. That book is a gem written by a marvellous naturalist and a lovely man. Back in the 1950s, Canning Suffem encouraged me to read J K Stanford's *A Bewilderment of Birds* and over the decades I have thought that it was the most pleasurable book I ever read to reflect an authors deep love of birds. Now I have a second such book which also reflects that deep love for and joy of being with birds. I am glad that it was written by John about his special place on the north shore of The Solent. He was such a remarkable man whose long and valued industry has served the cause of conservation and biodiversity in Hampshire so well and I for one, am missing him deeply.

Tony Norris

Bryan William Renyard
1937-2004

Bryan was an early member of the 'Portsmouth Group' first appearing at Farlington Marshes in 1952 while still at Portsmouth Grammar School where he was a contemporary of Colin Tubbs. At about that time he was diagnosed as a Type-One diabetic. Despite this, he lived a remarkably active life until a few years before his death. He cycled everywhere in this locality, for fifty years using the same carefully maintained Raleigh Sports bicycle. He worked as a scientific officer for the Ministry of Defence Central Metallurgical Laboratories where work involved studies on corrosion and antifouling measures on warships.

From the very beginning Bryan was a meticulous observer and careful recorder. He was particularly, though not exclusively, interested in estuarine species and was a life-long member of the Wildlife and Wetland Trust (formerly the Wildfowl Trust), RSPB, BTO, the County Wildlife Trust from its inception as well as the Sussex counterpart. He was an active member of the Portsmouth Natural History Society, now defunct, being for some years on the committee and its Treasurer. As a very regular and frequent visitor to Farlington Marshes it was appropriate that he was a member of the management committee from 1962 when it became a reserve.

As part of the 'Portsmouth Group' team who since 1952 counted wildfowl and waders in Langstone, Bryan concentrated on the Harbour mouth and the eastern shore i.e. the west Hayling side. In the 1970s with the dramatic increase in numbers of Brent geese and their change of habit from being entirely intertidal to largely grass grazers, he had to locate the areas on Portsea Island which they favoured such as playing fields and undeveloped strips. This was no easy task as flocks could be fragmented onto small areas, were liable to disturbance and the only practical way to cover the whole length of the Island was by bicycle.

In 1967 Bryan took over as county organiser for wildfowl counts which he did until 1974 during which period he expanded the areas covered to include some previously uncounted waters. He corresponded regularly with counters, keeping them up to date and supplying them with relevant Wildfowl Trust literature. He was extremely well read and was a mine of information. It was not unknown for him to reel-off from memory the life history of a Brent goose whose ring-number he had just read.

In company with John Taverner, Michael Bryant and other friends, Bryan visited most European countries, Tunisia and the USA, as well as being well travelled in Britain. He was keenly interested in classical music, having a large record collection and was a regular concert-goer. Bryan died on 12th December 2004.

David Billett

2007-2011 BIRD ATLAS

The next Atlas Project starts in November 2007. During the period 2007 to 2011, HOS expects to update all of our information on bird distributions, in both breeding season and winter. The project will involve the whole of Britain and Ireland, but in Hampshire we will be making a special effort, and plan on obtaining bird data from each of the county's 1,033 2km x 2km squares (tetrads). This will of course only be achieved with the help of all HOS members.

Why do we need another Atlas? There have been big changes in the distribution of many species since the last atlas period of 1988-91. A quick glance at the previous county distribution maps incorporated in Birds of Hampshire (Clark & Eyre 1993) will soon convince you that an update is needed. Then if you compare the national Breeding Atlas of 1972 (Sharrock 1976) with the maps produced in 1993 (Gibbons et al 1993), you'll see that many changes, sometimes unexpected, occur over a twenty-year period - And the only Atlas survey of wintering birds was undertaken way back in 1981-1984 (Lack 1986). So, now is the time for all the bird distribution maps, summer and winter, to be revised.

What will be involved? The survey will use broadly similar methods in summer and winter, and will provide opportunities for different levels of involvement and skill. Here are some of the key points:

- Winter - November to February: four winters from 2007/08 to 2010/11.

- Breeding Season - April to July: four seasons from 2008 to 2011.

- Fieldwork will involve a combination of 'Timed Tetrad Visits' and 'Roving Recorder Visits' – the first providing vital information on relative abundance, and the two together providing the total species list and evidence of breeding.

- Timed Tetrad Visits will be for two hours, during which all the birds seen and heard will be counted. There will need to be two such visits (an early one and a late one) in each season (winter and summer), but each tetrad need only be surveyed in one year. In Hampshire, Timed Tetrad Visits will be undertaken in every tetrad (25 in each 10-km square, and 1,033 in the county).

- Roving Recorder Visits will simply provide records of species with the aim of amassing comprehensive species lists for each tetrad.

Much work remains to be done on the detail. At the time of writing, the British Trust for Ornithology (BTO) are working on the design of forms and instructions, along with online forms and online applications that will help observers and organisers in the input and management of data. Locally, we plan on making full use of the national forms and systems, and hope to avoid the complexities of additional forms or processes.

Where will it lead? At the national level, the BTO are envisaging publication of a revised Atlas book in 2013. Within the county, current thoughts are that there will be a revised Birds of Hampshire book with atlas maps for both winter and summer. This may depend on developments in technology, and on members' expectations and preferences by 2013. Also of course, it will all depend on finance.

So how can you help with the Atlas? You will appreciate by now, that the absolute minimum level of coverage is for the Timed Tetrad Visits. This will require at least 8,000 hours of fieldwork by Hampshire birders(2 hours per visit x 2 visits per season x 2 seasons x 1000 tetrads), and of course there will also need to be lots of additional hours looking for the more elusive species and obtaining evidence of breeding. Sounds like quite a lot, but if all of HOS's 1000 plus members were to get involved and make timed visits to just one tetrad , then the timed visits could be completed in the first year, and there would be three more years to concentrate on extra species! That said, it would obviously be no good if everyone were covering the same tetrads. So we are already arranging and agreeing with observers, who will be covering exactly which areas. If you can help, and especially if you would be willing to cover any of the farmland areas of the county, then please e-mail me. We need a minimum of 200 volunteers arranged very soon so that we can make a good start in November 2007. This will be the major bird survey of the next 20 years and we feel you will want to be part of it.

References:

Clark, J.M. and Eyre, J.A. 1993, *Birds of Hampshire*. Basingstoke: HOS.

Gibbons, D.W., Reid, J.B. & Chapman, R.A. 1993, *The New Atlas of Breeding Birds in Britain and Ireland: 1988-1991*. London: Poyser.

Lack, P. 1986, *The Atlas of Wintering Birds in Britain and Ireland*. Staffordshire: Poyser.

Sharrock, J.T.R. 1976, *The Atlas of Breeding Birds in Britain and Ireland: 1972*. Tring: The British Trust for Ornithology.

Glynne Evans, Waverley, Station Road, Chilbolton, Stockbridge, Hampshire SO20 6AL
hantsbto@hotmail.com

EDITORIAL

This is the 49th consecutive annual report on Hampshire's birds. Firstly, I would like to take the opportunity on behalf of the Society to recognise the achievements of Mike Wall who has acted as production editor for this and the four preceding reports: unfortunately, due to unforeseen circumstances, Mike has regrettably and after a lot of thought decided to step down from this role.

Collating the artwork, photographs and text into a single document which our printers can work from without further adjustments is a professional task which Mike has undertaken voluntarily for the Society; and we are indebted to him both for the timely way this has been achieved and for the considerable costs he has saved the Society. The Society would like to offer its very best wishes to you Mike for your future activities in natural history, wherever your interests take you.

In the last few years we have seen a surge forward in bird photography – so much so that virtually every county rarity/scarcity is recorded by an enthusiastic band of "digiscopers" or, more recently, digital single lens reflex camera (D-SLR) users. This year's report moves to a larger format to improve reproduction of both photographic images and artwork. In order to assist with photograph selection we were fortunate recently to have Richard Ford join the editorial team as Photographic Editor.

The next few pages list the observers who contributed to a total of just over 39,000 records: those ringers engaged in programmes during the year and all survey workers. An impressive total of 503 people contributed with their fieldwork to the making of The Report; it is dedicated to you all.

As The Report will show, 2005 was an important year for both surveys and finding rare birds in Hampshire. I hope the span of articles reflects this and my thanks to all the authors who are listed on the contents page. I thank all systematic list writers for their contributions; their names too are all listed on the contents page against the species they wrote up. Systematic list writers completed their work in a most timely manner between early July and mid August. It takes several weeks to check on the accuracy of records and fill in any gaps that we detect; this year has been very demanding in this respect. Once more John Clark and Jason Crook have provided expert advice and unflagging assistance in editing. This year too John has drafted the HOSRP/BBRC assessed species and prepared the tables for the waterbirds covered by the WeBS counts. Pauline Cox has again undertaken the demanding role of proof reader.

As I see the writing of the report reach its conclusion, with such wonderful observations described, I still have regrets. It is incomplete - look at Annex 2 of the systematic list for confirmation of one aspect of this. I would urge all members to act on our behalf to ensure we receive important county records promptly and comprehensive in detail - not just their own observations but to encourage others to submit their records too. The bird news services are excellent for communicating new arrivals, but we are seeing an increasing number of potential records lost through failure to notify the County Recorder. In this respect, please look at the revised notes for submissions of records later in this report and available also on the web-site. Take the trouble to check the threshold numbers for submitting records; some species are becoming scarcer so we have lowered the thresholds. If my message here is heeded next year's report should be better still.

Alan Cox

Editor Hampshire Bird Report

All editorial correspondence should be sent to: **alanfjcox@compuserve.com** or otherwise to Alan Cox, Crosswater Cottage, Lake Lane, Dockenfield, FARNHAM, GU10 4JB.

The Hampshire Bird Report Team

Production Editor: Mike Wall

Assistant Editors (systematic list review): John Clark, Jason Crook

Photographic Editor: Richard Ford

Artwork: Dan & Rosemary Powell, David Thelwell & Marc Moody

Photographers: Duncan Bell, Jason Crook, Alan Cox, Mike Duffy, Richard Ford, Trevor Hewson, Nigel Jones, Nick Montegriffo, Peter Raby, George Spraggs, Simon Stirrup, Marcus Ward, Simon Woolley and Russell Wynn

Article reviewers: John Eyre & Glynne Evans

Draft corrections & proof reading: Pauline Cox

Distribution: Margaret Boswell

HOS Database

Keith Betton, John Clark, Jason Crook, Mark Edgeller and Matthew Shaft

HOS Records Panel

John Clark (Hampshire Recorder), Keith Betton, Jason Crook, Mark Edgeller, Simon Ingram, Marc Moody & David Thelwell

BTO Surveys & WeBS

Glynne Evans (BTO Coordinator), John Clark, Brian Sharkey, John Shillitoe, & Keith Wills

Table of Abbreviations used throughout the Report:

BBRC	British Birds Rarities Committee	HOSRP	HOS Records Panel
BBS	Breeding Bird Survey	LNR	Local Nature Reserve
BoH	Birds of Hampshire[1]	m.o.	many observers
BOU	British Ornithologists' Union	NF	The New Forest
BTO	British Trust for Ornithology	NNR	National Nature Reserve
CBC	Common Bird Census	RT	Rubbish Tip
CB	Cress Bed	RVS	River Valley Survey
CP	Country Park	SF	Sewage Farm/Works
GCT	Game Conservancy Trust	SL	Systematic List
GC	Golf Course	SP	Sand Pit
GP	Gravel Pit(s)	WBBS	Waterways Breeding Bird Survey
H	Heronry Census	WeBS	The Wetland Bird Survey
HBR	Hampshire Bird Report		

[1] Clark, J.M., & Eyre, J.A., 1993. *Birds of Hampshire.* HOS.

LIST OF OBSERVERS AND CONTRIBUTORS

The following made contributions to the 2005 Hampshire Bird Report as shown through submission of their observations to the County Recorder (indicated by 'Re'), as active ringers (see Key to ringers) and as volunteers participating in BTO and HOS surveys (see Key to surveys). We apologise for any inadvertent errors or omissions.

Key to ringers: Licensed ringers who contributed records during 2005 are indicated by their licence category (either 'A', 'C' or training TR) and if a member of a recognised ringing group this is shown as follows: FRG – Farlington, IRG – Itchen. Other contributors to the Ringing Report – RR.

Key to surveys:
At	Winter Atlas Pilot Survey 2005-06 (to be reported HBR2006)
Bb	BTO Breeding Bird Survey or Waterways BBS- 2005
CB	HOS Corn Bunting Survey - 2005
TO	BTO Tawny Owl Survey – 2005
Wb	WeBS Wetland Bird Survey – 2005
Wg	Winter Gull BTO Survey additional coastal sites - 2005
Ws	BTO Scarce Woodland Bird Survey - 2005

Adams MC Re
Alexander G A Re
Allen C Re
Andrew F Re
Andrews JK Re
Andrews M Re
Archer BM Re
Arnold J C Re
Baker MJ Re Wb
Baldwin J Re
Ball DJ Re Wb
Banks R Re
Barbagallo P Bb
Barker DJ Re
Barnes E Re
Barrett G&E Re
Bates A Re
Bates CM Re
Batho GS Re Bb Ws
Beckett P Re
Bell DA A-FRG Re
Bell NMG Re
Bennett EJB C Re
Bennett G Bb
Bentall S Re
Betton KF Re At CB
Bill DI Re Wb
Billett DF Re
Billett RA Re
Blakeley AF Re
Blunden AC C
Bond A Re
Borwick RM Re
Boswell L&P Re CB Ws
Boswell S Re
Bray NP C
Brazier A TO
Brett EC A
Brickwood M Wb
Briggs KB Re Wb Bb
Broadhurst A Bb TO
Brook R Re
Brown IH Re
Brown RI Re

Browning AA C
Burton G Re
Butler TJ Re
Butterworth AMB Re Bb Ws
Calderwood G Re
Calderwood I Re Wb CB
Carpenter RJ Re CB
Carpenter TF Re
Carr P A Re TO
Carter CI Re
Casson J Re
Chapleo C Re
Chapman JW Re Wb
Chapman RA Re Wb
Chappell L Re
Chawner J Re
Cheese TE Re
Cheke RA A
Christie DA Re
Clare BA Re
Clark A Wb
Clark JM Re Wb CB
Clark MRR Re
Clay GH Re
Cleave AJ Re
Clements J Re At
Cloyne JM Re Wb
Cockburn C Re Wb
Codlin TD A
Colcomb K Bb
Cole G At
Colin-Stokes R RR
Collins AR Re
Collins CB Re
Collins M Re
Collman JR Re CB Ws
Combridge P Wb
Compton DK Re
Cook R Re
Cooke RE Re Wb
Cooper JH At
Cooper Mtn Re
Cooper P Re
Cooper PF Re

Copsey S Re
Coulton M Re
Coumbe MJ Re Wb Bb CB
Coward KT Re
Cowx AJ Re Bb
Cox AFJ&PR Re Bb At TO
Cox IN Bb At CB Wg
Cozens B Re
Cripps J Re
Crisp K Re CB
Croger RE C-FRG
Crook J Re Wb
Crowley PJ Re
Cummins D Re
Curson S Re
Curtis C CB
Cuthbert CR A-IRG
Cutts M Re
Darvill G Re CB TO Wg Ws
Davidson P Re
Davies K TR
Dawson A&C Re Wb
De Potier A Re
de Retuerto MA C Re At
De Vries P Re
Dedman JW Re Wb Bb
Dewhurst J TR-FRG
Dicker DS A
Dicker G Re
Dicks DEJ Re Wb CB
Dixon M Re CB
Dodsworth P Bb At
Donegan E Re
Doran TMJ Re Wb
Downey B Re
Drewett A At
Drewitt E Re
Dudley B A
Duffin BS A Re Wb TO
Duffin TM Re
Duffy MD Re
Durnell PR Re Wb
Edgeller ML Re Wb
Edmunds H Re

Ellery T *Re*
Ellis D *Re*
Evans AD *A*
Evans DG *Re Bb*
Evans GC&SL *Re Bb At CB TO Wb Ws*
Evans J *A*
Evans J *CB*
Evans M *Re*
Eyre JA *Re Bb At CB Ws*
Facer RD *Re*
Faherty M *Re*
Faithfull J *Re Wb*
Farwell G *Re*
Feare CJ *Bb*
Fellows BJ *Re Wb*
Fletcher M R *RR*
Folkes P&A *Wg*
Ford RE *Re TO*
Friend BJ *Re*
Fry D *Wb*
Fulton A *Re*
Gammage PA *Re*
Garner M *Re*
Gibbons MJ *Re*
Giddens GS *A Re*
Gilbert S *Wb*
Gilchrist WLRE *Bb*
Gillingham M *Re Wb*
Gleed-Owen C *Re*
Goater B *Re*
Godden NR *Re*
Godesen S *Re*
Good C *Re*
Goodridge J *Re*
Goodspeed JR *Bb TO*
Gowen J *Wb*
Gray J *Re*
Green AE *Bb CB*
Green G *Re*
Greensmith A *Re*
Greensmith JD *Re*
Grist M *Re Ws*
Groves R *Re*
Gumn DG *Re Wb*
Gutteridge AC *Re*
Hale APS *Re*
Hall JA *Re*
Hallett R *Re*
Hallier RG *Re*
Hampton M *Re*
Hart R *Re*
Hartill S *Re*
Hawtree J *Re*
Hay MJW *Re Ws*
Hayden A *Re Wb*
Hazel AJ *C*
Heath BA *Re*
Hedley B *Re*
Hellyer R *Re*
Hemsley M *Re*
Hewson T *Re*
Hibberd JW *AP*
Hibberd P *Ws*
Hill J *Re*
Hilton JI *Re*
Hoare P *Re*
Hobby P *Re Wb TO*
Hobson J *Wb*
Hold AJ *Re Bb At Ws*

Holland D *Re*
Holland MG *CP*
Hollins JRW *Re Wb Bb*
Holt D *Re*
Hoodless AN *TR*
Horacek-Davis G *Re*
Houghton D *Re*
Howell R *Re*
Hughes RM *Re*
Hull N&J *Re*
Humphrey S *Re*
Humphrys N *Re*
Hunt PR *Bb*
Hutchins PE *Re CB*
Huxley GH *Re Wb At CB*
Ingram S *Re Wb*
Irwin RN *Re*
Ivon-Jones B *Re*
Jackson C *Re*
Jackson M *Re*
Jacobs RJK *Re*
James AR *Re*
James DBL *Re CB*
James LC *TR*
James RMR *Re*
Janaway D *Re*
Jennings F *Re*
Jennings TJ *Re*
Jeske E *Re*
Johnson AC *Re*
Johnson K *Re*
Johnston S *Re*
Jones A *Re*
Jones C *Re Ws*
Jones G *Re*
Jones JR *Re Ws*
Jones NR *A*
Jones S *Re*
Julian I *Re*
Keane PS *Bb*
Keen SG *Re*
Keil IJ *Re*
Kelson D *Re*
Kent HN *Bb At Ws*
Kilby M *Re Wb*
King M *Re*
King RA *Re*
King SS *Re Wb*
Kinross P *TO*
Kirkman DS *Re*
Lachlan C *Bb*
Lambden M *Re*
Lambert DR *TO*
Lander C&P *Re*
Lane K *Re*
Lane S *A*
Langford M *Wb*
Lankester SR&S *Re Wb Bb Ws*
Larter M *Re Ws*
Lavelle MA *Ws*
Lawman TA *Re Wb*
Lawn MR *Re*
Layfield AR *Bb Ws*
Leach BJ *Re CB*
Leeke-Bennett L *Re*
Legge WGD *Re*
Lemon R *Re*
Lenney M *Re*
Lester A *Re*

Levell J *Re*
Levett RK *Re Wb CB*
Lewis Adrian *CB*
Lewis Alan *Re*
Lewis J *Re Wb*
Lilley HA *Re*
Lilley M *Re*
Lister D *TO*
Litjens M *Re Wb At CB Ws*
Little JA *Re TO*
Littleboy N *Re Wb*
Locke A *Re Bb TO*
Lord B *Wb*
Lord P *Re*
Love M *Wb*
Lowings V *Re*
Lymbery PJ *C-FRG Re*
Mainstone C *Re*
Mansfield SJ *Re At CB*
Marchant RH *Re Bb At CB TO*
Marjeram W *Re Wg*
Marston PC *Re At*
Martin A *Re At CB*
Martin CGA *Re*
Martin KP *C*
Matthews PJ *Re*
Maundrell AJ *Re Bb At TO*
Maycock KW *Re*
Middlecote B *Re*
Millington SJ *Re*
Mills D *Re*
Minns D *At*
Minns D *CB*
Mist J *Re*
Mitchell D&M *Re Bb Ws*
Montegriffo NJ *Re Ws*
Moody MP *Re*
Moon J&J *Re*
Moseley JC *Re Bb*
Mould-Ryan RB *Re At CB*
Munday E *Re*
Munts D *Re*
Nash P *Re*
Newman MAH *Re Bb Ws*
Nicholson JB *Re*
Nicoll N *Re*
Norris AY *Bb At*
Northwood C *Re*
Norton JA *Re Wb*
Nundy JA *Wb*
Oakes S *Re*
Offer DC *A*
Offer DC *TO*
Oram MA *Re*
Oram P *Re*
Orchard P *Re*
Orchard-Webb M *Re*
Osborne G *Re CB*
Owen R *Re*
Page AG *C Re*
Pain J *A Wb*
Painter MG *Re*
Palmer MJ *Re*
Parfitt A *Wb*
Parkin LT *TR Re*
Parminter TJ *Re Wb*
Parsons AJ *Re*
Peace ND *Re Wb Bb*
Pearce DF *Re*

Pearce JV *Re*
Pearce RK *Re CB Wg Ws*
Pearson DJ *Re Wb*
Peters BF *Re*
Peters JN *Bb At TO Ws*
Phillips M *Wb*
Pibworth I *Re Wb*
Piggott SS *Re*
Pinchen BJ *Re Wb*
Pike L *Wb*
Pink A *Re*
Pitt MJ *Re Wb CB*
Pitt R *Re*
Pleasance J *Re*
Pointer RG *Bb At*
Polley AJ *Re Wb*
Potts PM *A-FRG Re*
Powell A *Re*
Powell D *Re*
Pratt EM *Re*
Preston R *Re*
Priest SN *Re Bb*
Pugh T *Re*
Pyrah R *Re*
Quinn PR *Re*
Raby PN *Re Wb*
Radden D *Re CB*
Rafter M *Re CB*
Ralphs IL *Re*
Ransom DA *Re Wb*
Raymond CJ *Re*
Raynor EM *Re*
Raynor P *Re CB*
Read MJ *Re*
Redway JA *TR Re*
Reeves DJ *Re*
Rhodes AS *Re Wb*
Rich G *Re*
Rickwood BA *Re Wb*
Righelato R *Ws*
Rix JB *Re*
Roberts BJ *Re Bb At CB TO Ws*
Roberts ET *A-FRG Bb*
Roberts GCM *A-FRG*
Robertson D *Re CB*
Robertson M *Re*
Robinson S *Re*
Rogers G&L *Re CB*
Rolfe MD *Re TO*
Rosenvinge HP *Re*
Roskell AJ *Re*
Ross JG *Re Wb*
Rowe J *Re*
Rowland GJS *Re Wb CB*
Ruscoe SR *Re*

Russell D *Re CB*
Sayer K *Re*
Schmedlin RJ *Re*
Scouller A *Re*
Seargent R *Re*
Severs R *Re*
Shaft M *Re Bb CB Ws*
Sharkey B *Re Bb TO Wb Ws*
Sharp G *Re*
Sheldrake P *Re*
Shillitoe JRD *Re Wb Bb At CB TO*
Ship R *Re Wb*
Simcox WF *A-IRG*
Simpson A&I *Re*
Smale J *Re*
Smallbone AC *Re Wb Bb*
Smart ADG *Wb*
Smith CJ *Bb*
Smith L *Re*
Smith M *Re*
Smith PM *Re*
Smith SJ *Re*
Souter R *At*
Southam M *Re*
Sporne L *Re*
Sporne SH *Re*
Spraggs GSA *Re*
Spring-Smyth J *Re*
Stephenson GC *Re CB Ws*
Sterry P *Re*
Stevens D *Re*
Stewart AH *Wb*
Still R *Re*
Stouse K *Re CB*
Strangeman PJ *Re Wb Bb Ws*
Sullivan AP *Re*
Summerhayes JLV *Re*
Swallow J *Re*
Tamblyn J *Re*
Taylor DH *Re*
Taylor J *Re*
Taylor S *Re*
Terry MH *Re*
Thelwell DA *Re Wb Bb CB*
Thomas PE *Re*
Thompson PGL *Bb At CB TO*
Thornton GA *Bb*
Thurston MH *Re*
Timlick T *Re Wb*
Tizzard PJ *Re*
Treacher D *Re TO*
Tubbs JM *Re*
Turner JT *Re*
Turner KA *Re*
Turner R *Re CB*

Twine P *Re Wb*
Unsworth DJ *Re Wb At*
Uphill N *Re*
Veall RM *Re*
Venables HJ *Re Bb At*
Verity J *Re*
Vine G *Re*
Vokes K *Re*
Walker TH *A Re At*
Wall MJ *Re Wb*
Wallen A *Re*
Wallis MR *TO*
Walmsley AP *Bb TO Ws*
Ward M&Z *Re Bb CB Ws*
Waters WE *Re*
Watson RF *Re Bb TO*
Watts IR *Re Wb*
Wearing MF *Re*
Webb B *Re*
Webb RM *Re Bb*
Webster BC *Re*
Webster P *Re*
Welch AJ *C Re*
Wheatley D *Wb*
Wildish MF *Re Ws*
Willard C *CB*
Willetts G *Re*
Williams G *Re*
Williamson DAJ *Re*
Williamson MJ *Re Bb CB*
Wills KB *Re Wb At CB TO Ws*
Wilson C *Re*
Wilson DC *Re*
Wilson G *Re*
Winspear RJ *Bb*
Winsper JL *Bb*
Winter PD *Re Wb Bb*
Wiseman EJ *Re Wb Bb TO Ws*
Wood JKR&J *Re Bb CB*
Wood S *Re Bb*
Woodley B *Ws*
Woods A *Re*
Woods M&R *Re*
Woolfries SA *Re CB TO Ws*
Woolley SK *Re Bb At TO*
Worgan A *Re*
Wright AM *Re*
Wright SJ *Re Wb CB*
Wynde AR *Re Bb*
Wynn RB *Re*
Yates SR *Re Bb*
Yelland D *Bb CB*
Young CL *Re TO*
Young K *Re*

We would also like to thank the following organisations who provided information to the County Recorder or Duncan Bell as coordinator of the ringing report.

Birdguides Information Service
Farlington Ringing Group
Itchen Valley Ringing Group
RSPB Stone Curlew team

Birdline South East
Forestry Commission (NF raptor data)
New Forest Ornithological Club
Titchfield Haven Web site

REVIEW OF 2005

Alan Cox

Eider passing The Needles   *© Russell Wynn*

It was an exceptional year by any standard – more species were recorded in Britain and Ireland in one calendar year than ever before;, this was reflected in Hampshire with a county year list of 265 species accepted, with one or two more still subject to record assessment. The year list includes two new to the county avifauna and some extreme Siberian, continental and Nearctic vagrants (throughout *BBRC* accepted rarities appear in **bold/underlined** and *HOSRP* acceptances in **bold**). Other headlines were the largest invasion of Waxwings ever recorded in the county; a three-figure count for both breeding pairs and fledged young of Mediterranean Gull and a seemingly remorseless rise in the Little Egret population, including a single roost count one short of 250. On the debit side was the plight of the county's farmland birds. The HOS organised Corn Bunting survey graphically illustrated the speedy decline in the population of this species. Tree Sparrows, which had clung on to a site in the north-east of the county for a decade, were sadly absent in the last winter period. Perhaps, through funded conservation measures arising from new initiatives in farmland management, we will be in time to save Yellowhammer, Skylark and other formerly much commoner species; there is no room here for complacency.

January A wild month, mostly mild but very disturbed windy weather at the start and end, rather more settled mid month. East and west of The Solent there were sightings of Common Scoters, Eiders, Long-tailed Ducks, Shags, Gannets, Kittiwakes (81 at Hurst on 10th), Guillemots and Razorbills and perhaps, more unexpectedly, over-wintering Little Stints and Black Redstarts. By the month's end all three regular diver species had been seen from Hurst Castle, with nine Red-throated, four Black-throated and a Great Northern;

in addition a party of five Scaup was the only record in month. There was welcome news of a second small flock of up to seven Purple Sandpipers that appear to have established Barton on Sea as a winter quarter, in addition to the regulars at Southsea Castle. Between Langstone Harbour and Hayling Island all five European grebes were represented from mid month: one/two Red-necked off Black Point/Sandy Point from 13th, up to seven Slavonian there and 14 Black-necked typically at the north-east end of Langstone Harbour. There were also nine over-wintering Sandwich Terns in that area and two Little Gulls mid-month. At Sandy Point it was possible to see over-wintering Firecrest and Chiffchaff as well but, more reliably for the latter, it paid to visit a sewage works - there were six at Chilbolton, eight at Barton Stacey and up to 14 at Eastleigh (also two long-staying Firecrests there). As these records show there was much to be found away from the coast; in fact the most reported species in month, with 168 mostly inland records, was Waxwing. In the Farnborough area the Waxwing numbers increased throughout the month: 44 at Cove on 2nd; 105 at Church Crookham on 22nd and 184 at Fleet on 30th. There were also two other three-figure counts at Calmore and Alton. Surprisingly, Jack Snipe are more numerous at Bourley Heath in the north-east than any coastal site - a year maximum of 17 there compared to a total of just 13 from all other sites. There were at least four Bitterns in the county and five Great Grey Shrikes, with three wintering in the New Forest, another at Ludshott Common and one briefly at Hannington. There was also a count of 26 Woodlarks at Hundred Acres, Wickham – the largest flock seen all year.

In the Avon valley the flock of 12 White-fronted Geese first seen in December was relocated from 9th-21st, but none was seen thereafter all year. Nearby, at Blashford Lakes, 149 Shoveler were registered during the WeBS count - the highest count of the year – one of three species for which the site has national importance as a winter refuge. At Needs Ore there were 3815 Lapwing – a site maximum for the year; in addition, a flock of 15 **Tundra Bean Geese** seen earlier in the winter was relocated. The Dark-bellied Brent Geese county total of 13,736 on 15th/16th was the highest since December 2002. This was somewhat surprising because just 13% of the flocks were first-winters; this low productivity was confirmed in coastal counties from Hampshire to Lincolnshire, which in winter hold close to 50% of the small world population of 215,000. These data were gathered both as part of WeBS and the Goose & Swan Monitoring Programme organised by the Wetlands & Wildfowl Trust.

Amongst the Dark-bellied Brents in Langstone Harbour was a long-staying family party of an adult and two first-winter Pale-bellied Brents, almost certainly from the Nearctic population. Two **Black Brants** were also wintering, one in Langstone Harbour and the other in the Gosport area. Other returning over wintering vagrants were the **Great White Egret** at Ibsley Water, Blashford Lakes, a second-winter **Iceland Gull** and an adult **Ring-billed Gull** both at Gosport. A **Pallas's Warbler** briefly illuminated a Lymington garden on 15th and nearby an adult female Marsh Harrier was seen on 21st - the only winter record. A **Grey Phalarope** at Ibsley Water on 22nd was an unusual discovery in winter. There were 433 Redwing at Black Dam, Basingstoke, on 30th - the largest flock of the early year. A total of 168 species was recorded during the month.

February High pressure centred to the west of Ireland dominated the first ten days, keeping it mostly dry and fairly mild. The Avon valley Bewick's Swan group rose to 12, including two first-winters. On 2nd a third **Black Brant** was found at Black Point, Hayling Island where it was regularly seen thereafter. On the north of the island a flock of at least 5000 Golden Plover was on the mudflats at Northney. This is the largest ever Hampshire count; there was also another major flock of 1550 at Keyhaven/Pennington Marshes. With mild weather across Britain, it seemed likely that this influx was of continental origin. Field characters to distinguish the European race *apricaria* from the more numerous Holarctic *altifrons* race in winter plumage have yet to be described, but it is possible that both forms winter in the county. There were 11 Slavonian Grebes in Hayling Bay – the highest site count of the year. A WeBS county total of 2099 Grey Plovers [4.0%][1] was the highest of the year and 700 Fieldfare at Long Sutton was the largest flock of the early year.

The only Smew recorded in the early year was present for one day at Ibsley Water and the Eider flock at Hill Head was at a year maximum of 97. Over-wintering, typically "garden-feeder", Blackcaps totalled 58 in January/February and were widespread. More unusual was a Turtle Dove present all winter at Southampton and last seen on 6th: it was the fifth ever overwintering record. There were two reported Hen Harrier roosts in the county; one in the New Forest peaked at six and one on the Wealden heaths held three; at least five seen elsewhere were additional to these. The largest flock of Waxwing ever recorded in Hampshire was at Maybush, Southampton on 13th - it was established through photographic analysis that at least 352 were present.

The February county WeBS total for Redshank was 1769 [15.3%][1]; other *Tringa* wader count maxima included: nine Spotted Redshanks and eight Greenshanks on Keyhaven/Pennington Marshes and three Green Sandpipers at Blashford Lakes. Other wader early year maxima site counts were: 14 Avocets in Langstone Harbour; four Common Sandpipers at Riverside Park, Southampton; 172 Snipe at Woolmer Pond; two over-wintering Whimbrel on Bury Marshes; 711 Bar-tailed Godwits at Northney, Hayling Island and 340 Black-tailed Godwits in Portsmouth Harbour. The *Calidris* sandpiper numbers were dominated by a count of 28,239 Dunlin in Langstone Harbour [19.6%, **2.2%**] - the highest for over a decade; 1100 Knot were also there; 255 Sanderling at Sandy Point and just three Ruff, now a very scarce winterer, at Keyhaven/Pennington Marshes. Wildfowl counts included 720 Gadwall [4.2%, **1.2%**] at Blashford Lakes and 401 Pintail in the west Solent [1.4%].

A cold front mid month saw Arctic air flood across southern England, with some snow showers into the county and several nights of sub-zero temperatures to the month's end. On 15th a first-winter **Glaucous Gull** was found feeding with other gulls in the inclosures at Marwell Zoo. At a site in the centre of the county a Short-eared Owl roost count peaked at nine. The first two Velvet Scoters of the winter were off Sandy Point on 18th. The colder weather triggered some wildfowl movements and the expanded Teal flock at Bramshill in the north-east briefly included a male **Green-winged Teal** on 20th. A Scandinavian Rock Pipit was detected at Farlington Marshes and remained there for four weeks. The first of four Long-eared Owls found in the year was at Lasham on 26th. A total of 167 species was seen in February (year-to-date total 175).

March A cold and wintry start with winds blowing in from the north/north-east. A male Red-crested Pochard on the Hamble estuary on 5th remained to summer on The Solent. Its arrival date was good for an overshooting continental vagrant (breeding as close as central France); nevertheless, with resident feral populations in neighbouring counties, it may just as easily have been of feral origin. The first unequivocal migrant, a White Wagtail, did arrive on 5th at Appleshaw, Andover, and was one of 47 in early spring. A Goosander roost at Ibsley Water reached an impressive 60 on the evening of 6th; given these numbers, it was surprising there was no subsequent evidence of breeding. At the end of the first week, the first Fulmar of the year flew east past Sandy Point.

Temperatures slowly recovered to near average from 9th. With the milder weather the Bewick's Swans departed the Avon valley and a relocating Scaup was on Yateley GP from 12th-14th. There were 18 Black-necked Grebes in Langstone Harbour – a year maximum. A Ring-necked Parakeet was at Locks Heath and 350 Bramblings were in Alice Holt Forest (highest count in early year). A mild and damp Atlantic south-westerly air stream set in, boosting the arrival of early migrants, but with low-lying coastal fog on many days, migration watching was hampered. An adult Spoonbill was at Titchfield Haven (the first of

---

[1] Square bracketed data are % of national population estimates in normal font and international (normally European) estimates in bold.

ten in spring). The first Swallow at Testwood Lakes on 13th was nine days earlier than the 35-year average. The following day first arrivals included a Wheatear at Bighton Lane CB and a Stone-curlew back on its breeding ground in the north-west of the county. There were then new arrivals almost daily. The first returning Sandwich Tern was followed by Little Ringed Plover; a Tree Pipit at Barton on Sea GC on 16th shared the earliest date recorded with 1992 and 2003. The departure of Brent Geese on 17th from the eastern harbours gathered pace when 500 left high to the east as the first Sand Martin, Yellow Wagtail, a very early Sedge and Willow Warbler arrived in the county. An Osprey at Tanners Lane, Lymington on 19th was the first of 28 in spring and a Marsh Harrier, the following day at Itchen Valley CP, the first of 20. A minor fall of 42 Wheatears spanned the coast on 20th. The first Common Tern, Redstart, Pied Flycatcher, House Martin and Cuckoo all arrived from 23rd-28th. During this period a total of 1105 Meadow Pipits were logged moving north over Hurst beach. There was also evidence of a passage of Water Pipits boosting overwintering numbers: the totals of nine at Keyhaven and 29 at Lower Test Marshes were year maxima.

A **Bluethroat** frequented a garden in Liss Forest for several days and became a news item with a photograph in the local newspaper. A first-winter **Caspian Gull** at Black Point on 30th was one of only two records that were submitted in year with adequate descriptions. On the last day of the month a small fall of ten Willow Warblers were nearby on Sinah Common. A total of 188 species was seen in March (year-to-date total 198).

April A ridge of high pressure affected the southern half of Britain, overnight fog continued to plague the coast. Four male Garganeys put down on Keyhaven Marsh squabbling for the attention of one female and another pair was nearby. The first Hobbies were over Bishops Waltham Moors and Bolderwood, NF on 2nd/3rd; other first arrivals were Ring Ouzel and Reed Warbler, both in the north-east; a Whitethroat at Sandy Point; an early Little Tern at Normandy and the last sighting of Great Grey Shrike in the New Forest. A good easterly passage on 3rd off Hurst included: 117 Brent Geese, 39 Eiders, 55 Common Scoters, three Red-throated, a Black-throated and two Great Northern Divers, a Little Gull and two Razorbills. A weather front pushed south-eastwards on 4th, introducing a more mobile westerly flow. The first Arctic Skua was in Hayling Bay and a Nightingale sang near Ashlett Creek. A further weather front pushed eastwards during 7th, introducing a colder north/north-westerly air flow as it cleared. Last departures included a Slavonian Grebe from Hill Head and a Jack Snipe from Lower Test Marshes. In the colder weather a Hoopoe made a stopover at Cams Bay, Portsmouth Harbour, and the following morning a male chiffchaff was singing at Longparish. The chiffchaff seemed to share some of the characters of Iberian Chiffchaff but its identity was not confirmed; however, there was no disputing the first Wood Warbler to return to the New Forest. More settled conditions now prevailed as a ridge of high pressure extended eastwards, across the southern half of Britain, clearing skies. This triggered the departure of the last Fieldfare and the arrival of the first Garden and Grasshopper Warblers – the latter one of 12 in spring – followed by Swift and Lesser Whitethroat. As noted nationally, it had been a terrible first half to April for hirundine passage, but it was finally relieved by movements of 500 Swallows over Pennington Marsh on 16th. The last Redwing and Water Pipit then departed, but there were still 140 Waxwings at Basingstoke.

A period of disturbed weather then caused a lull in migration, but dry sunny weather arrived on 22nd and winds went round to the south-east, improving sea-watching conditions. At Sandy Point passage included: 50 Common Scoters, 82 Whimbrels, four Arctic Skuas, 39 Little Gulls, 300 Sandwich and 185 Common Terns; this contrasted with movements at Hurst Beach which included: 139 Common Scoters, a Black-throated Diver, 139 Whimbrels, two Arctic Skuas, five Little Gulls, 76 Sandwich and 11 Arctic Terns (first arrivals). Prevailing south-easterly winds and showers and bands of rain from 23rd-27th maintained the coastal passage. On 23rd movements east off Hurst included: 23 Arctic and a Pomarine Skua; two Black, 325 Common and three Arctic Terns; 11 Little Gulls and

125 Whimbrel. Inland a Black Tern was at Fleet Pond. The next day a party of 11 Pomarine Skuas settled on the sea off Hurst - the best count from any of the English Channel watchpoints; the accompanying cast included six Manx Shearwaters and seven Arctic Skuas. Even well inside The Solent, at Stokes Bay, the early morning passage included 29 Common Scoters, four Arctic Skuas, 71 Sandwich, 131 Common Terns and 17 Yellow Wagtails. Nearby a Red Kite was over Titchfield Common, potentially another continental immigrant, as was more certainly the first Whinchat at Lepe. At Sandy Point on 25th/26th, there were 370 Swallows and 53 House Martins north and a Serin also over together with two Nightingales, three Ring Ouzels and a Grasshopper Warbler on the reserve. Other influxes included: a first-day Turtle Dove, 110 Swifts and 32 Yellow Wagtails at Farlington Marshes; 100 Swifts at Titchfield Haven; 200 Sand Martins at

Testwood Lakes and nine Tree Pipits at Hurst Castle. After a scarce passage previously in the month, a fall of 60 Wheatears across Keyhaven/Pennington Marshes provided half the county total on 24th/25th. After three previous days of strong easterly passage of seabirds, the best in spring was on 26th. Peak counts past either Hurst or Sandy Point included: five Red-throated, four Black-throated and three Great Northern Divers; 13 Fulmars; three Manx Shearwaters; 90 Gannets; 50 Bar-tailed Godwits; nine Arctic, one Pomarine and eight Great Skuas; 18 Little Gulls; 700 Common and 26 Arctic Terns. Next day the first Spotted Flycatcher was in the north-east, the first Curlew Sandpiper arrived at Titchfield Haven for a five-day stay and more briefly a **White Stork** circled there and left east. Lingering rain cleared early on 28th but sea fog persisted on the coast through 29th/30th. The month ended with a Spotted Crake heard calling at a suitable breeding site. A total of 204 species was seen in April (year-to-date total 223).

Tree Pipit   © *Nigel Jones*

May A complex area of low pressure to the west of Britain dominated the weather and brought a change to a continental airflow. The exciting discovery of a **Collared Pratincole** at Farlington Marshes set the adrenalin running. It had been five years since the last performed for the masses not too distantly at Pagham, West Sussex, but this latest individual departed within forty minutes to frustrate a Peregrine Falcon's interest in exotic food and the birding fraternity. It was the 59th record for Britain but only the sixth in the last ten years and the third record for the county, after a gap of 19 years. The appearance of a Blue-headed Wagtail was scant consolation for those that hurried to the site. More expectedly, returning Nightjars were heard for their first evening at Bourley North and in the New Forest. On 2nd a pair of Montagu's Harrier arrived in off the sea at Hurst and a Roseate Tern was at Titchfield Haven. The low pressure then transferred to the east of Britain, triggering an inland passage of Arctic Terns across the country on 4th: 69 at Fleet

Pond was the largest flock ever recorded in the county and contributed to the highest ever spring total of 250. Purple Sandpipers which had lingered at Sandy Point departed and two exceptionally late Brambling flew east past Milford on Sea.

The warm weather departed on a migration stifling north-westerly airflow; consequently, there was no major migrant fall or big passage day of seabirds in the following week. Nevertheless, a Honey Buzzard struggled over Pennington Marsh on 6th and a Red Kite was also observed there. At least 20 Red Kites were recorded in the county in May, the best month for this colonising species, but just one (again successful) breeding pair was reported. The last of only four migrant Short-eared Owls in spring was at Pennington Marsh on 7th. The last Waxwing was reported nearby at Lymington on 8th; eight remaining from the huge Southampton flock had departed three days earlier. There could not have been a more spectacular avian sight than a count of 15 Hobbies feeding over Ibsley Water on 9th and 12 there the following day. Air pressure rose on 10th, providing a temporary favourable airflow for continental overshoots. A **Tawny Pipit** was at Sandy Point - the first ever county record in spring - and a **Red-rumped Swallow** at Farlington Marshes: both sightings a good return for regular early morning recording at these sites. Air pressure then fell to the south and a chilly easterly wind picked up, prompting the last Hen Harrier east and Merlin north over Hurst on 14th/15th. Forty Eiders were over-summering off Oxey Point (including three displaying pairs) and an adult Spoonbill arrived to summer at Needs Ore. A Temminck's Stint was briefly in Langstone Harbour on 16th and it, or another, was there on 18th. Low pressure, still dominating the weather, brought heavy rain overnight on 19th/20th which drew two **Whiskered Terns** into the county and up to eight in total into southern England. One tern was at Titchfield Haven early afternoon; another at Tundry Pond before moving in the late evening to Fleet Pond. After roosting in Berkshire, it returned to both locations the following day. Previously only five Whiskered Terns have ever been recorded in Hampshire and these were the first for eight years. The first Quail arrivals were on 23rd: a male was seen at Hambledon and another heard near Nether Wallop. Warm southerly winds on 27th generated a very hot day during which a Red-backed Shrike was discovered near Basingstoke. On 28th another Temminck's Stint and the only Wood Sandpiper of the spring were both found at Ibsley Water. A total of 196 species was seen in May (year-to-date total 233).

June began with unsettled conditions and heavy rain, particularly to the east of the county on 3rd, when two **Bee-eaters** previously in Sussex briefly circled Titchfield Haven before leaving north - the 11/12th records for Hampshire. A Black Redstart at Southampton Airport was a breeding possibility; certainly a pair bred in the north-east of the county and successfully raised two broods. Early return wader passage was suggested by a Green Sandpiper at Lower Test Marshes on 8th. From 10th-12th there was an influx of at least nine Quail with a minimum of 23 males in summer on chalk downland - the highest total since 1964.

The BBS volunteer team detected 118 species in 75 2 km transects: the data summary indicates that hirundine numbers were not depressed, despite the poor visible migration. The survey has now been conducted from 1994-2005 and the review conducted for this article shows four species in significant decline across the period, but really holds no surprises (Corn Bunting, Willow Tit, Lesser Whitethroat and Spotted Flycatcher). Conversely, Bullfinch and Marsh Tit detections are apparently increasing when the first and last four-year averages from 1998-2005 are compared. The HOS Corn Bunting surveyors located 74 territories in 36 tetrads (compared to occupancy of 194 tetrads during the 1986-91 Atlas) and confirmed the bleakest forecasts. Other species, however, with very restricted ranges appear to be holding up: 61 singing Nightingales (with a maximum of 9 in and around Botley Wood) and 71 Wood Warblers in the New Forest (but none confirmed breeding elsewhere). Species subject to RSPB monitoring are doing well: for example, Stone-curlew (25 pairs of which 23 definitely bred) and Mediterranean Gull (108 pairs and

165 fledged young on the Langstone Harbour islands). Despite the efforts of wardens, 143 pairs of Little Terns raised just 23 young and the species remains of major concern, with predation thought to be the principal cause of poor productivity. Larger raptors, by contrast, are doing well and particularly Buzzard which appears to have colonised all regions of the county; BBS detections occurred in 53% of squares compared to 14% from 1994-1997. Goshawk, after the first authenticated breeding in 2002, consolidated to six successful pairs breeding in the New Forest. Five pairs of Honey Buzzards bred, raising ten young. Peregrine sightings continue to increase but confirmed breeding was reported from only three sites. Ravens have quickly established themselves; breeding was confirmed at four and pairs present at another seven sites. Little Egrets were breeding at four sites with a minimum of 50 pairs present.

From 17th-20th high pressure returned and south-easterly winds coming off a very warm continent brought the hottest spell of the summer to date. A first-summer male **Red-footed Falcon** was discovered at Martin Down on 18th but had departed the following day. A male **Black-winged Stilt** was watched feeding on Keyhaven/Pennington Marshes on the evening of 21st. It had almost certainly moved up Channel from the Isles of Scilly and made this brief stopover before spending a few days at Dungeness, Kent. A first-summer Spoonbill, however, previously at Needs Ore for three days, transferred to Keyhaven/Pennington Marshes and stayed into July. A total of 174 species was seen in June (year-to-date total 235).

July As in previous years there was a build up in numbers of Yellow-legged Gulls at Lower Test Marshes, with a peak count early in the month of 68. A ridge of high pressure started to push in from the west on 7th and the weather became warm and bright with settled high pressure until the last few days of the month. Another **Serin** was watched feeding in the reserve at Sandy Point. The following day an early returning pair of Whinchat was observed at Farlington Marshes. Crossbills had been seen in small numbers in June but evidence of a larger influx began with a party of 40 at Bramshill on 8th, followed by two parties of 30 and 35 flying west at Itchen Valley CP:17 at Havant; 15 at Acres Down, NF on 9th and around 80 noted later in the New Forest. On 17th the colour-ringed **Great White Egret** returned to Blashford Lakes, it or another was recorded on many dates into December; sightings of a second unringed bird being present from July were confirmed when two were seen together later in the year. At least two **Balearic Shearwaters** were seen from Hurst in Milford Bay on several dates between 17th and 24th and an impressive 568 Gannets were also noted there, mainly east, on the last date. Early returning Arctic Terns were noted, with singles at Ibsley Water on 23rd and at Sandy Point the next day. Larger waders, seemingly, return to the county coast very early in summer as demonstrated by counts made between 22nd and 25th. Curlew numbers build up rapidly in late June with large counts by mid July this year as follows: 426 on Verner Common, 1326 in Langstone Harbour [6.7%], 193 at Needs Ore and 298 on Keyhaven/Pennington Marshes. Icelandic Black-tailed Godwits totalled around 750 [4.9%] (60 Hamble and 94 Beaulieu Estuaries, 278 Keyhaven/Pennington and 325 Farlington Marshes). There were 96 Whimbrel in Langstone Harbour and 39 at Fawley. Inland there were 500 Lapwings at The Vyne, Basingstoke, and the first returning flock of 20 Golden Plovers on Old Winchester Hill. The huge post-breeding flocks of Starling that congregate on Farlington Marshes were estimated to number 10,000 on the evening of 23rd; numbers declined to around a quarter of this peak in subsequent months.

There was heavy rain on 27th and the weather became very unsettled once more until the end of the month; with low pressure dominating, it was much cooler with a south-easterly air flow. A summer plumaged **White-rumped Sandpiper** was located in the evening high tide roost at Farlington Marshes but then departed; it was the 17th county record and the first since 2002. Other early returning wader passage from 27th-30th included: a Wood Sandpiper at Ibsley Water, the first of just 15 in autumn; three Ruff at

Farlington Marshes, two at Titchfield Haven and up to three Curlew Sandpipers at Normandy. By late month there were notable counts of terns returning late in the day to rest in Langstone Harbour: maxima were 220 Little, two Black, 78 Sandwich, 500 Common and one Roseate Tern. On 30th a Merlin at Pennington Marsh was the earliest ever return by three days. On 31st there were five Roseate Terns at Titchfield Haven. Swift departures, evident all month, peaked with 240 south-east at Itchen Valley CP on 31st. A total of 172 species was seen in July (year-to-date total 238).

August The first few days were showery, followed by a short fine spell before two weeks of unsettled weather culminated in some heavy thundery rain overnight into 19th as a cold front moved east. On 2nd a **Pectoral Sandpiper** was at Park Shore, Needs Ore; now annual, this was the 71st record for the county. The following day another **Bee-eater** was watched over Sandy Point in the evening for six minutes before it drifted out of sight. A passage Long-eared Owl was discovered resting at Farlington Marshes on 4th/5th. The Farlington RG trapped an adult and a juvenile **Aquatic Warbler** on 7th and another juvenile on 11th. A Pied Flycatcher was found at Petersfield on 17th; it was the first of 21 in autumn, including three together at Northney later in the month. At Titchfield Haven there was a count of 11 Black Terns - the maximum of the year. The first juvenile Osprey arrived on 18th in Langstone Harbour and stayed for several days. On 20th there were the first sizable returns of Grey Plovers with 400 at Gutner Point, Chichester Harbour, then 425 at Farlington Marshes two days later - surely one of the more memorable images of late summer when these stunning summer-plumaged holarctic breeders arrive suddenly, in such numbers, at this high water roost.

At last - one of Farlington Marshes' notable run of rarities posed for the masses: a juvenile **Wilson's Phalarope**. Previously at Bowling Green Marsh, Devon, from 18th-20th; it followed the widely anticipated convention for south-west peninsular waders and made a stopover in Hampshire. It was present from 21st-25th, the sixth county record. A Spotted Crake was also located close by at the same time and was seen on five dates into September; at least one other was also present in the Farlington Marshes reedbed during autumn. Over the shoulder of the rarity hunters, terns were maintaining their nightly usage of Langstone Harbour, with ever more spectacular counts of Common Terns: over 2500 noted on several dates, including a peak count of 2706. The accompanying cast included up to 40 Little Terns, down to single figures from 18th, five Black, 188 Sandwich, and six Roseate Terns. At Titchfield Haven the ringing programme also had a notable day: 29 Grasshopper Warblers were trapped, a count exceeded by two on 28th; these contributed to an incredible autumn total of 324, together with 543 Blackcap, 45 Garden and 122 Willow Warblers.

There was an unseasonable south-westerly gale on 24th which presaged the arrival of Hampshire's earliest ever **Grey Phalarope**, still partly in juvenile plumage, at Pennington Marsh on 25th/26th. There were also the first returns of Black-necked Grebe, with two at Ibsley Water where one remained until 29th. The first of 14 Wrynecks in autumn was discovered at Farlington Marshes; like others, it was well-photographed during an eight-day stay. A second-winter **Ring-billed Gull** was at Gilkicker Lake on 27th - the 35th county record. A total of 187 species was seen in August (year-to-date total 242).

September A build of air pressure following a weak cold front brought fine and very warm weather throughout the first week. On 2nd another **Pectoral Sandpiper**, a juvenile, was at Titchfield Haven and stayed for six days. At Farlington Marshes there were 13 juvenile Curlew Sandpipers, the peak count of the year, equalled at Keyhaven/Pennington Marshes ten days later. Hirundine departures gathered pace in the first week with the heaviest passage being on 3rd - 1342 Swallows south at Gilkicker Point. There was heavy rain on 9th but then fine weather to the end of the second week. Unusually, there was a female/immature Marsh Harrier on the South Downs, perhaps the same as seen there in the spring; it was one of a total of seven in month and 20 in autumn. At Barton on Sea 1181 House Martins flew east early morning on 10th. Up to four males and a female Blue-headed Yellow Wagtail were seen at Abbottswood, Romsey, from 10th-15th. On 12th the third **Pectoral Sandpiper** of the year flew east over Hurst Beach and was briefly re-located at Oxey Marsh. Concurrently at Hurst Beach/Milford on Sea, the largest Meadow Pipit movement of the autumn was being recorded when 1741 moved south-east; this contrasted with a large north-easterly movement taking place in the east of the county - 1630 over Farlington Marshes and 702 over Sandy Point. The next day sea-watching off Hurst produced a Pomarine and five Arctic Skuas. At Farlington Marshes there were two Tree Sparrows and the following day a juvenile Red-backed Shrike arrived for a three day stay. The shrike departed as a cool north to north-east flow developed on 16th, engendered by a build of pressure to the west of Britain, but conditions remained bright and visibility was good over the next week.

The change in winds on 17th prompted a very early **Richard's Pipit** to make a brief visit to Pennington Marsh before departing eastwards - the earliest autumn record for Hampshire. A **Melodious Warbler** was watched at close range at Cosham for 20 minutes. There was an exceptional movement of small finches throughout the autumn which included an impressive early movement of 486 Siskins over Sandy Point and Farlington Marshes on 12th. Single Lesser Redpolls, too, were at both Titchfield Haven and nearby at Hook-with-Warsash - the first of 331 widely-scattered detections in autumn. A **Corncrake** was glimpsed briefly crossing a footpath at the Kench, Langstone Harbour, and was lost to view; in the harbour 2040 Oystercatchers were logged - the peak count of the autumn. On 18th a first-winter **Barred Warbler** was seen briefly at Solent Court, this just the ninth for Hampshire. Nearby a Tree Sparrow was at Hook-with-Warsash and a Long-tailed Duck at Titchfield Haven; another early autumn first arrival was a Scaup at Ibsley Water. A massive passage of hirundine was witnessed at Barton on Sea when 2170 House Martins and 5730 Swallows flew east; the latter was a large element of the 20,000 logged in autumn. A count of eight Little Stints at Keyhaven Marsh on 20th was the highest of the autumn. An adult female Montagu's Harrier was at Cheesefoot Head on 22nd and the last of the year, a ringtail, was at Titchfield Haven a week later. The last Spotted Crake, of at least four recorded in year, was at Hook-with-Warsash on 24th. A year maximum count of 1350 Lesser Black-backed Gulls was made at Blackbushe Airfield

The week to the end of the month was typically autumnal, with a fresh to strong westerly air flow developing. A first-winter **Caspian Gull** was at Fleet Pond on 25th and an early Black-throated Diver off Hurst two days later. The pre-migratory assembly of Stone-curlew on downland numbered a healthy 54. The Atlantic weather systems brought the prospect of further Nearctic waders reaching Britain, but just two were found: both juvenile **Baird's Sandpipers**, one was in Kent and the other at Keyhaven Marsh on 29th, staying until mid October - the sixth for Hampshire. A total of 197 species was seen in September (year-to-date total 247).

October A south/south-west airflow brought rather cloudy conditions by the middle of the first week. On 4th a juvenile **Red-necked Phalarope** was feeding close to the slipway off Needs Ore. A Yellow-browed Warbler was found at Titchfield Haven, then two more over the next 12 days; one, on a housing estate near Sandy Point, was well-watched during a four day stay and the third was briefly at Cosham. At Gosport the **Ring-billed Gull** returned for its third consecutive winter and another juvenile **Red-necked Phalarope** was found nearby two days later. A second Hoopoe for the year was reported at Broughton on 6th/7th. The first autumn arrivals of Water Pipit were at Pennington Marsh, then another at Titchfield Haven: at least 26 were reported at seven sites in the late year.

It remained unusually mild and largely dry. On 8th the first Slavonian Grebe returns were two in Langstone Harbour and the last Honey Buzzard departed over Gosport. The first autumn returning Great Grey Shrike was at Hampton Ridge, NF and there were two more in the New Forest; later a fourth, around the Broadlands Estate, Romsey, and a fifth at Mottisfont. From 9th a small influx of the distinctive form of Continental Coal Tit was found, first three at Taddiford Gap, then singles at another four sites. On 10th a second **Red-rumped Swallow** was seen very briefly at Keyhaven - one of just three in autumn in Britain. Two in Hampshire in one year doubled the county total – all sightings have been made by single observers so this is a very exclusive club. A build of pressure maintained settled, if rather cloudy, weather. The Little Egret roost at Wade Court/Langstone Mill Pond peaked at 249 - the highest ever count in Hampshire. On 14th another large movement of Meadow Pipits was witnessed when 1220 moved north-east over Farlington Marshes and 737 headed east over the Beaulieu Estuary, then 800 flew east over Sandy Point two days later. A female/immature Snow Bunting was at Needs Ore, the first of the year but 19 were to follow and added to the exceptional numbers of passerines that coastal visible migration watchers were detecting. Goldfinches in particular were now moving in huge numbers; there were 400 east over Hurst on 18th but this was eclipsed by the month's total of 12,203 logged, mainly moving east/north, over Sandy Point. A first-winter Red-necked Grebe arrived to winter in the west Solent. Short-eared Owls were at both Woolmer Pond and Sandy Point. Amongst finch flocks on the move were 400 Goldfinches east over Hurst on 18th and a huge month's total of 12,203 logged, mainly moving east/north, over Sandy Point.

Low pressure became dominant, bringing unsettled weather from 19th. On 21st a **Sooty Shearwater** flew east past Stokes Bay, an excellent record so well into The Solent. The heaviest ever autumn passage of Ring Ouzels was underway with a total of at least 95, including a maximum count of 20 at Old Winchester Hill on 22nd - the highest ever site count. Woodpigeon movements were already underway and massive on 23rd: 15,800 were observed moving south across the west Solent; over Milford on Sea a morning movement of 3650 south-east was observed and over Southampton 1470 south-west. It would thus appear that the movement converged over the Isle of Wight and headed …where? The nature of these movements is the subject of fierce debate and seemingly at variance with 20th century ringing returns; these suggest that British Wood Pigeons, unlike their Scandinavian cousins, rarely stir more than a few kilometres from their natal sites. Clearly, movements are now occurring on a massive scale, but no area appears as yet to be the recipient of such large influxes of Wood Pigeons in late autumn. It is a significant challenge for 21st century ornithology to unravel this seemingly recent phenomenon, but even here there is uncertainty, as an account of an early 18th century observation suggests such movements may have occurred some 300 years ago. Another Nearctic vagrant arrived on 23rd when a juvenile/first-winter **Lesser Yellowlegs** was found at Titchfield Haven – the 12th for Hampshire. The WeBS count at Blashford Lakes logged 1858 Coot [1.0%]. A year maximum count of 5098 Lesser Black-backed Gulls was also made here a week later. A male **Black Brant** arrived in Chalkdock, Langstone Harbour, accompanied by a female Black-bellied Brent and two hybrid young. In the late year there were two additional **Black Brants** in the eastern harbours and two more at Keyhaven/Pennington Marshes. The attentive reader may have noted a change in format here as from mid 2005 assessment of this *branta* form was passed to county record panels by *BBRC*.

The unsettled weather resolved into a severe autumn gale on 24th; a **Long-tailed Skua** flew past Milford on Sea the next day and a first-winter Grey Phalarope joined the **Lesser Yellowlegs** still at Titchfield Haven on 26th. A further Atlantic depression approached Britain on 26th, introducing a vigorous southerly air flow and gales on the coast which triggered further sightings. A record of a Little Ringed Plover at Lower Test Marshes was the latest ever in the county by 18 days. A Lapland Bunting was found at Brownwich near Titchfield Haven and the following day an **Alpine Swift** flew past there – the tenth for Hampshire. Sea-watching at Hordle Cliff and Milford on Sea produced a **Leach's Storm-petrel**. A total of 207 species was seen in October (year-to-date total 258).

November Severe gales persisted and displaced several storm-petrels and Grey Phalaropes into The Solent. The strongest winds were experienced on 3rd when a **Storm-Petrel** was off Hill Head early morning and two in the afternoon at Stokes Bay. Up to six **Leach's Storm-petrels** were seen off Hill Head, five off Hurst Castle and three west of Milford on Sea; at least five **Grey Phalaropes** were located between Southsea and Hurst over the next few days. Inland a Black-throated Diver was over Casbrook Common, Test Valley, doubtless once more storm-driven. A male Snow Bunting at Normandy was seen to be taken by a Peregrine. An early morning watch on high ground just south of Fleet produced a count of 23,544 Wood Pigeons on 5th. This astonishing movement was the highest of the autumn and most birds were noted to fly south/south-west up the Blackwater valley. Major movements were also noted further south in the county, the largest being 4200 south/east over Bedhampton, 1325 south-east over Old Winchester Hill and 803 south over Hurst Castle. Only on Hayling Island was a different general direction of travel noted with 880 west. Thus many Wood Pigeons were once again, apparently, intent on leaving the Hampshire coast. Also on 5th the regular WeBS count at Gosport was rounded off, as normal, late morning at the Boating Lake; the log then read: *Mute Swan-85, Canada Goose-44 and then…Laughing Gull-1* - a new species had been added to the Hampshire list. The first-winter **Laughing Gull** was not entirely unexpected as the sequence of fast-moving Atlantic depressions had transported around 20, an unprecedented number, of this small Nearctic gull, into the south-west of England - at least one had already penetrated to neighbouring Dorset. The Gosport individual rested for a few days and then became more mobile; it was seen at Normandy lagoon 17 days later and was sighted, for the last time, back at Gosport on Dec 11th. Over 50 different Laughing Gulls were thought to be involved in the influx across Britain. The first three Bewick's Swans arrived back in the Avon valley; 12 were present by the end of the year - ominously without any young in the group.

Another period of stormy weather on 11th produced a **Sabine's Gull** from early morning in Hayling Bay and another later off Milford on Sea with a **Red-necked Phalarope,** also watched feeding in the sea there. Another **Grey Phalarope** was subsequently found well inland at Burley, NF. In Langstone Harbour a flock of seven female/first-winter Scaup was located - the largest of the year. The following day four Velvet Scoters were off Hill Head and one remained for the rest of the month. The Azores high pressure built across the UK on 13th and gave a brief spell of more settled weather. At Sandy Point a **Desert Wheatear** was discovered on the beach and later relocated further west at Beachlands (see cover plate) – it was just the second county record since the first 45 years earlier. Early the next morning, unsuccessful searchers for the wheatear received some consolation when a Hoopoe flew west along the beach and was later relocated at Gosport. South-east Hayling excitement had hardly subsided when a Little Auk provided not one but two behavioural excesses. A commotion in the wader roost at Black Point revealed the Little Auk attempting a clumsy landing; after some weird and largely unsuccessful attempts to emulate a roosting wader, it spied a Skylark moving west and flew up to tag along!

Low pressure running into Scandinavia swept weather fronts south across England, producing a cold and showery northerly airflow, before pressure built to give a needed period of settled autumn weather over the next week. The only Twite of the year was seen at Farlington Marshes between 14th and 21st. A Bittern returned to overwinter at Titchfield Haven on 19th when the juvenile Arctic Tern, resting there since the earlier storms, finally departed - the latest ever record for the county. The next day another **Richard's Pipit** was located at Pennington Marsh. The 24th saw a dramatic change with weather of Arctic origin forcing two cold fronts south on a brisk northerly air stream. At least 20 Merlins were present in the county during November, of which 17 appeared to have established winter ranges and, similarly, there were 11 Hen Harriers with perhaps eight overwintering. Another **White Stork** was observed near Beaulieu on 27th. There were records of single **Common Redpolls** from farmland in both the Itchen and Test valleys. Another Lapland

Bunting was at Titchfield Haven on 30th. A total of 187 species was seen in November (year-to-date total 264).

December

Once more a stormy start to the month, with strong southerly winds fuelled by a depression in the south-west of the country dominating the first week: Odiham had 37mm of rain in 24 hours. A first-winter **Glaucous Gull** flew past Milford on Sea on 2nd. There was the highest number of Goldeneye in the county of the year but with 21 in Portsmouth Harbour and other maxima later in month included 45 in Langstone Harbour, 13 in the Lymington/Hurst area and 14 Blashford Lakes[2]. The WeBS total of 400 Red-breasted Mergansers was the highest of the year [4.1%]: 187 in Langstone Harbour, 168 in Portsmouth Harbour and 88 in the Lymington/Hurst area where there were also 1218 Teal [0.6%] - the largest count from any site all year. Around midday on 7th a **Radde's Warbler** was discovered near Sherborne St John; later its identity was confirmed and it was seen in the area until dusk. It was one of the latest dates ever for Britain and extremely rare inland - yet another first for Hampshire. Every effort was made to contact the landowner in order to get approval to release site details but to no avail; the bird in any case had gone the next morning. A garden bird feeding programme in Fareham had an unexpected dividend with a Lesser Whitethroat present from 7th-31st, just the sixth to overwinter in the county.

Several days of drier weather resulted from high pressure centred over the country. A massive fuel explosion occurred at Buncefield, Hertfordshire, just after 0600hrs on 11th; shortly afterwards there was prolonged alarm calling from Pheasants in central Hampshire over 100km away. The UK seismograph network operated by the British Geological Survey recorded significant air (sound) and ground wave disturbances across the country. It is probable that the pheasants sensed the ground disturbance waves which travel about 20x faster than sound. The ageing of Brent Geese flocks renewed in the late year was in complete contrast to the early year - 28% of the flock of around 4000 in Langstone Harbour and 35% of 1700 in the Lymington/Hurst area were first-winters. The last good year for Brent Goose breeding productivity was 24% in 1999 and it is hoped that this one good year will not be an exception as lower productivity in recent years is the primary cause of their decline. At Arlebury Lake there was a brief visit from a first-winter/female Smew, the only late year record. A Lesser Canada Goose was reported from Needs Ore. One was also at Farlington Marshes in the early year on two dates. If these three records *truly* reflect the only contacts in the county all year, then observations on such dates merit further attention!

On 14th, as in the early year, Golden Plovers gathered in large numbers on the mudflats at Northney and a count of 2800 was logged. Nearby, at Havant the following day there were six Waxwings - the only sighting in the late year. On the Beaulieu estuary there were 2044 Wigeon and in Langstone Harbour 20 Avocets – both counts were county maxima for the year. Of three Barn Owls seen near Cheriton on 17th, two were unfortunately road kills. A first-winter Marsh Harrier arriving at Titchfield Haven on 22nd was the first to overwinter in the county since 2002/03. The now familiar visiting **Iceland Gull** returned to Hill Head for its third winter in The Solent on 23rd. Temperatures plummeted in the last few days of the year with heavy snowfalls to the east of the county. On 29th there were at least 4 **Common Redpolls** with a flock of Lesser Redpolls near Romsey. On 31st a peak year site count of 2000 Fieldfare, accompanied by 500 Redwing, was made at the Swanmore Apple Farm, Meon valley. At dusk a male Hen Harrier arrived at Alresford Pond to roost in the reedbed, just squeezing into the Hampshire year. A total of 173 species was seen in December (year total 265).

---

[2] Goldeneye numbers reflect the low levels of recent winters.

SYSTEMATIC LIST OF SPECIES

Species sequence, scientific names and taxonomy comply with the BOU maintained British List published at *www.bou.org.uk/recbrlst.html*. We treat the two taxons Yellow-legged Gull *L. michahellis* and Caspian Gull *L. (a.) cachinnans* as separate species for the purposes of county record keeping. We reflect current usage with respect to the English names for species.

Each species account begins with a brief statement of its status in Hampshire. These use certain terms that have an approximate numerical range attached to them, as shown below:

	Breeding pairs	Winter/Passage
Very rare	Fewer than 5 records	Fewer than 10 records
Rare	Less than annual	Less than annual
Very scarce	1-10 per year	1-20 per year
Scarce	11-100	21-200
Moderately common	101-1000	201-2000
Common	1001-5000	2001-10,000
Numerous	5001-30,000	10,001-60,000
Abundant	30,000+	60,000+

In addition, based on European, national and county classifications, the conservation status is given for resident and regular visiting species in brackets as follows:

ET		European threatened-included in Annex 1 of the EC Directive on the Conservation of Wild Birds, the 'Birds Directive' (79/409/EEC).
SPEC		Species with unfavourable conservation status (*Birds in Europe 2004*, Birdlife International).
	1	Global concern, i.e. classified as globally threatened, near threatened or data deficient.
	2	Population concentrated in Europe and critically endangered, endangered, or vulnerable.
	3	Population not concentrated in Europe but declining, rare, depleted, or localised, in Europe.
WCA1		Included in Schedule 1 of the UK Wildlife and Countryside Act
Red		Red-listed as Birds of Conservation Concern (BoCC) in *The population status of birds in the UK* (*British Birds* **95**, Sept 2002).
Amber		Amber-listed as Birds of Conservation Concern (BoCC) *Ibid*.
NBAP		National Biodiversity Action Plan priority species
HBAP		Hampshire Biodiversity Action Plan priority species

For rare species, three numbers are given in brackets after the status statement referring to the total of individuals recorded: (i) prior to 1951, (ii) between 1951 and 2004 and (iii) in 2005. Observers are reminded of the importance of producing notes of supporting evidence for all sightings of scarce species, as detailed in Guidelines for Submission of Records in this report. A number of claimed sightings of scarce species, appearing on internet and pager services, have either not been submitted or adequately documented and are not included in the systematic list.

The species accounts for many waterfowl and waders include tabulated monthly maxima for various localities - sometimes not the count from the official WeBS count[1] but a larger count on a different day. A locality is listed if it is counted regularly (at least monthly) and if the particular species number exceeds the threshold stated above the table in at least one month. The final row in each table is the county total accumulated from all the official WeBS counts for each month and will include counts from many more sites than those listed. It represents the best estimate of the county population during the counting period (not necessarily but desirably the same day). In the tables, any locality that is indented is a sub-site of the site above. Complete counts for Chichester Harbour

---

[1] Kindly provided with the agreement of the WeBS Partnership (WWT, RSPB, BTO, JNCC) by Keith Wills (inland sites), John Shillitoe (coastal sites), John Clark (Avon valley) and Anne de Potier (Chichester Harbour).

(Hampshire and Sussex) are tabulated but are not included in the county total nor form part of the species account; only the sub-totals for East Hayling, Emsworth and Warblington (Hampshire sites) are included in the county total. Sites in tables are usually arranged coastally from east to west and then inland from west to east. Footnotes to the tables indicate threshold numbers for concentrations of international and/or national significance. An internationally important concentration is defined as a number exceeding 1% of the north-western European winter or passage population of that species. Likewise, a nationally important concentration is defined as greater than 1% of the British winter or passage population of that species. A number followed by an asterisk, e.g. 473* = a record count of that species at the locality concerned.

Reference is made to the terms 'bird-days' and 'bird-months' in the systematic list. 'Bird-days' is normally used to give the sum of the numbers of birds recorded at a well-watched site over a given period e.g. totals of 3, 2, 1, 2, 4, 0, 2 recorded over a week would indicate 14 bird-days, although the number of different individuals involved could actually be anywhere between 4 and 14. 'Bird-months' is normally used to indicate the sum of numbers of birds recorded in a given month over a period of several months or years e.g. totals of 6, 3, 2 recorded in June, July and August would indicate 11 bird-months. In the site lists N, E, S and W = the points of the compass; these are combined with numbers to show the direction in which birds were moving, e.g. 17 W.

The BBS project has been operating annually since 1994; in 2005 there were 75 1-km squares surveyed in the county using the transect counting technique. A transect count is the number of birds of a given species detected on a two km walk completed in 2-3 hours within the 1-km square. It does not purport to detect all the birds within the 1-km square, particularly true of the more secretive species. It provides an estimate for statistical analysis and year-on-year comparisons. Data are provided here for 2005 and earlier years in terms of:

(a)   square occupancy = % of squares where a species is detected.

(b)   mean [number of birds detected] per square = the combined total of transect counts of a species for all squares divided by the number of squares surveyed.

Mute Swan

Cygnus olor

A moderately common resident [Amber].

Monthly maxima at sites where counts exceeded 40 are as follows:

	J	F	M	A	M	J	J	A	S	O	N	D
Chichester Harbour	111	112	97	113	99	96	135	105	166	128	207	184
Emsworth Mill Pond	94	107	97	113		2		74	115	75	76	59
Portsmouth Harbour	127	107	89	65					71	92	104	98
Walpole Park, Gosport	81	84	71	56	49	41	18	18	70	69	85	88
Lymington/Hurst	50	68	43	34	29	14	12	9	30	73	33	74
Sopley/Ringwood	40	53	83						82	106	110	61
Ringwood/Fordingbridge	171	170	71						19	47	141	150
Blashford Lakes	34	21	15	18	41	52	128	97	89	95	66	45
Above Fordingbridge	104	114	137			41			188	108	132	91
Lower Test/Eling/Bury Marshes	8	10	6	14	51	47	43	40	31	39	16	11
Broadlands Estate	82	82	64						43	51	82	98
Stockbridge/Fullerton	31	21	16							40	44	42
Weston/Northam	14	21	16								35	46
Riverside Park/Woodmill	50	57	42			81			66	74	61	41
Alresford Pond	23	2	3		52	73	60	36	10	6	4	
Tundry Pond	4	2	2	8	4	24	40	41	58	70	79	78
Fleet Pond	12	35	40	39	41			17	16	8	11	5
Yateley GP	29	13	10						27	33	42	51
WeBS count totals	988	906	887						848	920	1017	967

Records during the breeding season referred to a total of 77 pairs at 55 sites. Of these at least 20 pairs were unsuccessful, including all three pairs which held territory at Farlington Marshes. One of these pairs hatched four young but all the cygnets died when a Fox killed one of the adults on July 8th. The other adult appeared to abandon the young, although remaining on site - one juvenile survived until July 21st. In the Lymington/Hurst area there were seven territories of which three failed and four raised 11 young. The largest number of territories was 21 at Somerley Lake, Avon valley, where eight broods were noted. There was a total of 135 juveniles reported from successful broods. The largest brood reported was at Tundry Pond with eight juveniles present from May 28th to July 23rd at least and, presumably, among the 16 first-winters there on Nov 21st. Three broods containing white phase cygnets, so called 'Polish Swans', were recorded at Budds Farm SF (1 of 3, July 31st), Elmers Court, Lymington (1 of 5, Aug 1st) and Great Salterns, Portsmouth (2 of 5, Aug 9th).

The highest counts were made at many sites in the second half of the year with maxima of 115 at both Ibsley Water, Avon valley on July 10th and in the Emsworth Mill Pond area on Sep 9th.

Bewick's Swan

Cygnus columbianus

A scarce and declining winter visitor and passage migrant, most frequent in the Avon valley [ET, SPEC 3, WCA 1, Amber, HBAP].

The numbers in the Avon valley flock continue to fall. The peak count in the early winter period was just 12, including two juveniles. The same number was present in the late winter. The table below records the ten-year decline:

Avon valley	95/96	96/97	97/98	98/99	99/00	00/01	01/02	02/03	03/04	04/05
Bewick's Swan	139	88	93	35	45	37	30	24	23	12

In early January there were just eight in the Avon valley group, including the two first-winters, until 15th when two further adults joined the group and another two by 31st. The four late-arriving adults were not seen after Feb 7th but the eight remained in the area until Mar 11th. First returns were three on Nov 5th, increasing to 12 on Dec 30th. All were adults indicating another unsuccessful breeding season. There were no records outside of the Avon valley.

Bean Goose
Anser fabalis

A rare winter visitor (?,203,2).

The flock of 15 of the race *rossicus* at Needs Ore on Dec 12th 2004 was relocated there on Jan 16th (DJU), where it remained until Mar 1st at least. Subsequently, there were at least five there on 11th, two on 13th, three from 15th-25th and two on Apr 2nd (m.o.). Elsewhere, two adults of the race *rossicus* were discovered with ten Bewick's Swans at Ibsley at 1300 hrs on Jan 29th (GH-D). The following morning, the two Bean Geese flew in to Ibsley from the south-east at 0855 hrs, circled low over the meadows and left to the south-west, apparently landing out of sight (M&ZW).

Pink-footed Goose
Anser brachyrhynchus

A rare winter visitor, although presumed feral birds have occurred in most months (24+,118,1).

One was regularly seen with Brent Geese or Greylag Geese at the Beaulieu Estuary between Jan 4th and Apr 24th (NFOC, DJU *et al*). Further records of one flying east over Hurst Beach at 0820 hrs on May 12th (RBW); an unringed and wary bird with Canada Geese at Pennington Marsh on May 15th (RBW) and one flying east over Milford on Sea on Aug 4th (ALo), suggest that a single individual may have been resident in the south-west of the county during the year.

White-fronted Goose
Anser albifrons

A scarce winter visitor and passage migrant which has declined in recent years [Amber, HBAP].

One was reported from Hayling Island on flooded fields Jan 1st. The flock of 12 seen in the late winter at Ibsley Water, Avon valley was relocated nearby in the Bisterne area between Jan 9th and 21st with one addition; it comprised a family group with four first-winters, a second-winter and six further adults. Only two were present on Jan 26th and none was seen thereafter.

An individual of unknown origin was present at Tundry Pond from Jan 16th up to April 26th.

Greylag Goose
Anser anser

A moderately common and increasing resident, passage migrant and winter visitor [Amber].

All records were thought to relate to birds of feral origin. The largest regularly recorded numbers were in the Avon and Test valleys and the Beaulieu estuary. Monthly maxima at sites where counts regularly exceeded 50 are shown below:

	J	F	M	A	M	J	J	A	S	O	N	D
Needs Ore/Beaulieu Estuary	160	130	81	69	113	80	70	150	80	180	203	207
Sopley/Ringwood	229	176	179						350	258	368	215
Ibsley Water	63	11	14	18	9	194	318	51	92			6
Testbourne Lake	92	88			5			74	43			17
Tundry Pond	13	15	22	3	6		64	75	77	72	47	56
WeBS count totals	**453**	**324**	**344**						**468**	**500**	**609**	**585**

Elsewhere there were 81 on Ewhurst Lake on Dec 11th. All other notable counts appear to be associated with movements to sites neighbouring those listed above.

Breeding was under reported. The largest brood count was recorded at Somerley Lake, Avon valley, with eight broods comprising 42 young. The only other reports of multiple broods were from Hawley Lake (7, 7 young), and the Test valley - Houghton (3) and

Longparish (3, 16 young). There were single broods reported from a further seven sites. A breeding attempt was made for the first time at Fleet Pond but was unsuccessful.

Snow Goose   *Anser caerulescens*

A scarce feral resident.

The small feral county population which roosts on Eversley GP outside the breeding season is close to extinction, as the table demonstrates:

January Count	1996	1997	1998	1999	2000	2001	2002	2003	2004	2005
Eversley GP	29	25	19	16	20	19	17	13	11	6

The six in January reduced to five in February and March and again from July to October, then only four in November and December. Elsewhere there was a report of five north-west at Langdown, Southampton Water on Mar 19th, three days after five were last reported from Eversley GP and seven days before six were recorded at Stratfield Saye Park, where they summered. There was just one further report of a single blue phase associating with Greylags at Ibsley Water, Avon valley on Feb 20th.

After no attempts last year, breeding was resumed at Stratfield Saye Park. A pair was seen with four small young and one adult with a single small young. The five seen from July nearby at Eversley GP were all adults.

Ross's Goose   *Anser rossii*

Very rare vagrant or escape.

2001 addition: A juvenile was discovered at Farlington Marshes on October 29th and remained until 31st (RAC, JCr, KC, photo). It associated with both Canada and Brent Geese and, on the last date, became very restless and flighty. The identification has been accepted by *BBRC* and the record has been accepted onto Category D. One discovered with Pink-footed Geese in Norfolk on Nov 2nd was presumably the same individual and remained along the east coast throughout that winter.

This is the first such record for Hampshire but its acceptance onto Category D indicates that the species has not yet been officially accepted as a species occurring in a wild state in Britain. However, the credentials for genuine vagrancy of this individual appear strong. Western Palearctic vagrancy of this Nearctic species is [already] accepted by other members of the Association of European Records Committees (AERC).

(Greater) Canada Goose   *Branta canadensis*

A common resident and partial migrant.

The county population has remained stable since the mid-nineties at around 2000 with counts peaking in early autumn. Monthly maxima at sites where counts regularly exceeded 200 are tabulated below:

	J	F	M	A	M	J	J	A	S	O	N	D
Titchfield Haven area	271	165	39		91	28	25	250	210	525	466	251
Needs Ore/Beaulieu Estuary	200	230	108	142	116		164	127	150	360	350	84
Lymington/Hurst	98	98	7	2	3	79	87	97	13	356	361	235
Avon valley: Sopley/Ringwood	179	160	216						277	178	313	170
Blashford Lakes	250	150	133	88	62	329	495	190	24	25	14	62
Lower Test/Eling Marshes	142	114	10	22	30	400	462	430	252	61	119	121
Tundry Pond	245	121	80	42	45		94	465	410	558	412	300
Fleet Pond	81	145	12						276	410	34	15
Eversley/Yateley GPs	332	273	440						147	139	235	413
WeBS count totals	**2089**	**1802**	**1428**				**1086**	**1114**	**1975**	**2333**	**2503**	**2135**

Maxima exceeding 200 from other sites were as follows: 288 Dogmersfield Lake Oct 8th (Tundry Pond birds); 359 Wellington CP Sep 18th; 360 Frater, Portsmouth Harbour Dec 20th and 280 Testbourne Lake Nov 9th. Reports received of breeding totalled 46 pairs at 23 sites, raising 41 broods containing 172 young - remarkably consistent with 2003 and 2004 reports. The highest brood counts were IBM Lake: (Portsmouth Harbour (7, 24 young); Lymington/Hurst area (7, 21 young); Langstone Harbour (6, 21 young) and Blashford Lakes (6, 23 young). The only count of first-winters reported was 19 in a flock of 37 at Yateley GP on Dec 28th.

Lesser Canada Goose   *Branta hutchinsii*

A very scarce feral resident/escapee.

Single birds were at Farlington Marshes on Jan 9th and 19th (JCr) and Needs Ore on Dec 13th (NFOC).

Barnacle Goose   *Branta leucopsis*

A scarce resident (feral population) and rare winter visitor (wild populations) [ET, Amber].

The feral population in the north-east continues to decline with an early year maximum of 101 reducing to 62 in the late year. There was just one breeding report at Stratfield Saye Park with five broods (14 young) and one nest still being incubated. Monthly counts from the main roost site are as follows:

	J	F	M	A	M	J	J	A	S	O	N	D
Eversley GP	98	101	-	6	11	12	2	30	54	61	56	62

Over the last decade both of the recognised feral populations have declined as follows:

Late year maximum	1996	1997	1998	1999	2000	2001	2002	2003	2004	2005
Eversley GP	135†	176	206	190	183	245	200	96	90	62
Baffins Pond	29	38	42	13	11	10	11	9	9	7

† recorded at Stratfield Saye Park

The remnant Solent feral flock seemingly commutes between Titchfield Haven and Baffins Pond, with excursions to the Beaulieu Estuary. There were eight at Titchfield Haven between the early year and April 6th when they dispersed; five were then seen at Baffins Pond and two on the Beaulieu Estuary later in April. Occasional sightings were then reported from Baffins Pond until Oct 5th (seven). Then six were at Titchfield Haven from Oct 13th rising to seven in December. A single on Lower Test Marshes on three dates between Oct 8th and Nov 6th was possibly part of this group and it, or another, was at Hook-with-Warsash on Nov 1st. Accounting for this flock is an essential precursor before assessing sightings of individuals of possible wild origin. Two first seen at Emsworth on Mar 2nd, Tournerbury on 6th and Farlington Marshes between 8th and 14th were possibly of wild origin. Two were also present at Farlington Marshes from Dec 28th to 31st.

2003 correction: the record of 11 at Alresford Pond from Jan 21st-Mar 24th should be deleted; it referred to Baffins Pond.

Brent Goose   *Branta bernicla*

Dark-bellied Brent Goose (*B.b.bernicla*)

A numerous winter visitor; small numbers summer [SPEC 3, Amber, HBAP].

Monthly maxima at sites where counts exceeded 200 are tabulated below:

	J	F	M	A	M	J	J	A	S	O	N	D
Chichester Harbour	7195	7278	5109	19	8	6	8		13	1000	3031	4792
East Hayling	2880	2391	1674						29	1000	1170	2200
Langstone Harbour	5069	4159	3996	548	6	5	6	6	35	1024	4025	4008
Farlington Marshes	3000		2300	44	1		1		6	14	1800	1400
Portsmouth Harbour	1292	1726	1703	2					1	72	1583	2304
Titchfield Haven area	250	250	490	110	1					368	300	400
Hamble Estuary	704	793	283	40						139	840	548
Calshot	200		516	25							250	136
Needs Ore/Beaulieu Estuary	1498	1319	885	455	3				4	100	1189	1200
Sowley/Lymington	250	301	500	42						119	500	800
Lymington/Hurst	1900	1501	1134	150	2	1				510	1362	1305
WeBS count totals	**13736**	**12371**	**11307**	**599**	**7**	**11**	**6**	**1**	**15**	**2487**	**9168**	**10840**

Sites of international importance: 2,200+; national importance: 981+.

The county total of 13,736 Jan 15th/16th was the highest since December 2002. This represents 6.3% of the world population and 14% of average wintering numbers in Britain (WeBS sources). Since 1985-90 there has been a significant decline both in the world population (approximately 14%) and in the Hampshire numbers compared to the national total, which then averaged 24.7%. Thus the Goose Monitoring Programme conducted by WeBS (Worden, 2005[2]) aims to establish annual breeding success by determining the percentage of first-winters amongst wintering flocks. In Langstone Harbour a total of 2625 was aged in January and February of which 15.2% were first-winters (JCr). One flock of 115 was aged at Tipner Lake, Portsmouth Harbour on Feb 25th, of which 13% were first-winters. In the Lymington/Hurst area east to Pitts Deep a total of 5421 was aged from Jan 1st-20th of which 13.3% were first-winters (EJW). These data contributed to the 2004/05 winter national estuaries count of 68,429 of which just 11.9% were aged as first-winters (Worden, 2005).

Recent wintering county totals are tabulated below:

	96/97	97/98	98/99	99/00	00/01	01/02	02/03	03/04	04/05
December	13500	15300	13300	13400	12800	12500	12200	13400	11400
January	12300	15500	14200	16000	15000	14500	12800	11000	13700

Easterly coastal passage was reported between Mar 2nd and May 4th and was best recorded from Sandy Point and Hayling Bay where a spring total of 1568 included peaks of 180 on Mar 9th, 360 on 17th (including some departing birds from Chichester Harbour), 309 on Apr 3rd and 180 on 4th. Numbers reported at other seawatching sites were small, including 332 past Hill Head/Stokes Bay and just 156 east past Hurst. The first evidence of spring departures from Langstone Harbour was on Mar 17th when 510 left Farlington Marshes high to the south/SSE in the evening. A further 897 were noted departing in directions between south and east on seven dates up to April 3rd. The decrease in numbers was reflected in counts of grazing birds at Farlington where there were 2300 on March 8th, 1300 on 22nd, 300 on 25th (the last three-figure count) and 44 on Apr 2nd. Five summered on Baker's Island and a single on Little Binness Island, Langstone Harbour.

---

[2] Worden, J. 2005, *The breeding success of Dark-bellied Brent Geese in 2005, as assessed in the UK*. Wildfowl & Wetlands Trust Report, Slimbridge.

David Thelwell.

The first return was a single joining the Langstone Harbour summering group on Sep 16th with 20 the next day over Portsmouth Harbour, although the race was not identified and the observer could not rule out Pale-bellied Brents. The first three-figure count was 215 in Langstone Harbour on Oct 4th and in the west 204 in the Lymington/Hurst area on 16th. Numbers built up rapidly thereafter. The only inland records were of one over Fleet Pond on Oct 17th and 40 over Ibsley Water, Avon valley on Oct 29th.

At Gosport on Nov 1st a flock of 53 contained 29 first-winters and at Hook-with-Warsash on 28th a flock of 28 contained five first-winters. These early second winter counts suggested a better breeding season in 2005 than in 2004 and were confirmed in the two main site aging surveys conducted during November and December: in Langstone Harbour three observers (RAC, CC, JCr) made independent counts to a combined total of 18,633, of which 27.7% were first-winters and in the Lymington/Hurst area a total of 1733 was aged of which 35% were first-winters (EJW).

In the late winter birds were feeding on winter cereal on Hayling Island and 400 were on maize stubble at Tanners Lane, Lymington on Dec 12th. The high percentage of juveniles in the late year prompted large numbers of geese to feed inland of Langstone Harbour. The habit was first noted on Nov 21st when 275 arrived into the harbour from inland at dusk, and peaked on Dec 29th when 850 did the same.

Pale-bellied Brent Goose (*B. b. hrota*)

A very scarce winter visitor and passage migrant.

Of ten individuals seen in the early winter period three were first seen on Dec 16th 2004 (adult, 2 first-winters). These birds, a family group, habituated Hayling Oysterbeds where they were joined by an adult Dark-bellied between Feb 27th to Mar 18th; thereafter they were mostly separated from the residual Dark-bellied flock until last seen on Apr 18th. Seven other birds were seen, including another family party, before reports of three singles on passage in April. There were 16 reports in the second winter period from Nov 16th involving at least five individuals, three adults (one colour-ringed) and two first-winters. A summary of all records (adults unless otherwise stated) is as follows:

Black Point, Hayling Island: 1 (female paired with Dark-bellied), Feb 7th and 9th-13th.
Sandy Point, Hayling Island: 1 E, Apr 3rd; 1E, Apr 28th.
Langstone Harbour area: 3 (2 first-winters), Jan 1st--Apr 18th; 1 on eleven dates between Jan 5th and Mar 13th; 1 on six dates Nov 14th - Dec 6th; 1 (colour-ringed), Dec 8th and 22nd; first-winter on four dates, Dec 11th-29th.
Needs Ore/Beaulieu Estuary: 1, Feb 26th; 1, Nov 6th and Dec 4th.
Lymington/Hurst area: 4 (2 first-winters), Feb 5th; 1, Apr 20th; first-winter, Nov 18th.

A presumed returning apparent intergrade between Pale-bellied and Dark-bellied Brent Goose was present at Farlington Marshes on Feb 5th.

Black Brant *(B. b. nigricans)*
A rare winter visitor (0,16,3).

Black Brant with intergrade young, Farlington Playing Fields   © *Jason Crook*

The three adults present in December 2004 remained into 2005. The Gosport bird was seen on the shore of Portsmouth Harbour on Jan 1st, 2nd, 29th and 31st and regularly on HMS Sultan playing field between Jan 2nd and Feb 14th (m.o.). The Northney/ Farlington Marshes bird was at Northney on Jan 2nd (GSAS, SJW), then off Broadmarsh, Langstone Harbour on Jan 16th and between Langstone and Northney, Chichester Harbour, on Mar 7th (JCr). During the interim it had been seen at Thorney Island (West Sussex) where it had originally been found (per CBC). The Black Point individual was regularly recorded between Feb 2nd and Mar 17th (JCr, ACJ), presumably having spent January at West Wittering (West Sussex).

In the late year, five were recorded. A male accompanied by a female *bernicla* and two hybrid young was found in Chalkdock, Langstone Harbour on Oct 23rd. The family remained in the Langstone Harbour/Farlington Marshes area until the end of the year but had lost the female *bernicla* from Nov 9th. The male and two hybrids were also seen at Farlington playing fields on Nov 30th, Dec 7th and 9th and at St John's College playing field on Dec 18th and 29th (JCr *et al*). A new adult was discovered at Farlington Marshes on Nov 19th, and seen subsequently at St John's College playing field on Nov 20th, Farlington playing field on Nov 28th, Brockhampton shore on Dec 8th and back at Farlington Marshes on Dec 25th (JCr). On the last two dates the bird arrived into the area from inland. This was presumably the bird found feeding inland with *bernicla* at Denmead on Dec 3rd (PJS) and Widley on Dec 7th (RAC). The Wittering individual returned and was seen near Black Point on Nov 15th (ACJ). In the west Solent, an apparent female was identified at Keyhaven Marsh on Nov 17th which had been present since Nov 12th. It was last seen on Dec 23rd (RBW *et al*). A second individual, an apparent male, was discovered at Pennington Marsh on Dec 4th and remained in the Pennington/Normandy Marsh area until Dec 18th (RBW, M&ZW *et al*). It is not possible to be sure of the number of new individuals involved in the late year sightings. However, the maximum would appear to be three.

It should be noted that it is of great help to the *HOS Records Panel* if observers continue to submit photographs of this *branta* form. Descriptions may fail to exclude possible intergrades whereas good quality photographs, especially those showing the bird in company with *bernicla*, greatly assist in the assessment process. In this regard, all but the late year Black Point individual had photographs of them available to the Panel.

The apparent *nigricans* x *bernicla* intergrade which has wintered in the Langstone area since 1998/99 was recorded in the north-east of the harbour on seven dates between Jan 4th and Mar 18th. It returned to the same area on Oct 25th when it was accompanied by a *bernicla* and four hybrid young. However, the only subsequent sighting of the family was at Warblington shore, Chichester Harbour on Dec 14th (JCr).

Egyptian Goose   *Alopochen aegyptiaca*

A scarce feral resident.

Following last year's unprecedented numbers, record counts were again made in successive months of the late summer and autumn roost at Eversley GP, reflecting recently established breeding in the north-east of the county:

	J	F	M	A	M	J	J	A	S	O	N	D
Eversley GP	19	14	26	10	14	19	38	41	50	66*	36	43

In the north-east, single pairs bred at three sites, including Stratfield Saye Park for the first time. A summary of breeding/summering birds follows:

Stratfield Saye Park: Pair Mar 26th with five 1/3rd grown young July 20th not reported thereafter.
Hartley Wintney GC: Pair raised nine young, first seen May 3rd -all survived, fully-fledged June 28th.
Tundry Pond: Pair with seven young Apr 7th which did not survive, subsequently raised a second brood of 11 which fledged in October.

A summary of all other records (away from the north-east) follows:

Sandy Point, Hayling Island: 1E, Mar 26th.
Curbridge, Hamble Estuary: 1, Apr 20th.
Titchfield Haven: 1, on many dates Jan 7th to June 12th and then from Aug 7th-13th (1 SE Brownwich, Apr 29th is assumed to be the same individual).
Blashford Lakes: 1, June 9th and July 3rd and 3 from Aug 21st-Sep 1st.

Ruddy Shelduck
Tadorna ferruginea

A very scarce (possibly feral) visitor or escape.

The only record was an unringed female in Langstone Harbour on Nov 1st and at Budds Farm SF the following day (JCr).

In addition, a Ruddy Shelduck x Shelduck hybrid was recorded in Langstone Harbour on January 5th and 9th and Feb 12th and 20th, and at Farlington Marshes on Mar 14th (DB, JCr, SJW). Possibly, the same bird was reported at Normandy Marsh from Feb 21st-Feb 22nd and Feb 26th (M&ZW).

Shelduck
Tadorna tadorna

A scarce breeder and common but declining winter visitor [Amber].

The underlying trend for Shelduck in Hampshire is one of gradual decline. The maximum count for the year was again in Langstone Harbour, with 524 on Jan 15th. Numbers in the Avon Valley were also lower than in recent years. Notable coastal counts came from Calshot (50 on Feb 4th) and Fawley (60 on Mar 16th). Monthly maxima at coastal sites where counts regularly exceeded 50 and inland sites where counts exceeded 10 are as follows:

	J	F	M	A	M	J	J	A	S	O	N	D
Chichester Harbour	729	825	595	208	99	124	25	18	21	35	83	359
East Hayling	199	168	166	4	2	18			10	33	36	49
Langstone Harbour	524	450	377	199	106	55	7	17	24	77	295	344
Portsmouth Harbour	242	86	144	5							5	84
Needs Ore/Beaulieu Estuary	68	63	89	85	96		8	4	1	1	1	20
Lymington/Hurst	94	91	80	73	96	21	1	14	4		42	42
Blashford Lakes	1	7	21	16	14	4	4		1			
Alresford Pond	14	16	13	14	7	13	4				14	10
WeBS count totals	**1263**	**1022**	**949**	**459**	**359**	**114**	**36**	**28**	**27**	**167**	**194**	**247**

In spring, a total of 16 was recorded moving east at Sandy Point and Stokes Bay between Mar 21st and Apr 30th.

Broods were reported from 15 sites as follows (number of broods-young where available): East Hayling, 1-10; Budds Farm, 1-6; Farlington Marshes, 1-6; Langstone Harbour islands, 3 pairs; RNAD Frater, 1-7; Titchfield Haven, 2 broods; Hook-with-Warsash, 1-1; Lower Test Marshes, 1 pair; Needs Ore, 2 broods; Normandy Marsh, 2-11; Hurst Castle, 1-15; Milford on Sea 1-5; Ibsley North Lake, 1-11; Testwood Lakes, 1 pair; Leckford Reservoir, 1-5. In addition, a pair was present at Itchen Valley CP Apr 1st-May 31st but no young were seen.

Other inland records included: Avon Causeway (5, Jan 9th; 2, Feb 6th; 1, Mar 6th); Frankenbury Walk (2, June 4th); Testwood Lakes (present throughout much of the year with maximum counts being 14 on Apr 10th and 24th); Broadlands Lake (4, Jan 24th; 2, Feb 18th; 4, Mar 15th; 1, Dec 4th); Fishlake Meadows (6, Mar 31st); Kings Somborne (4, June 4th); Marsh Court (maximum of 17 on Mar 6th and 14 on May 21st); 2 at Bishopstoke on Apr 17th and Highwood Reservoir on May 17th (presumably the Itchen Valley CP pair); Lakeside CP (2, Apr 20th; 1, Nov 10th); Arlebury Lakes (2, Jan 18th, Feb 3rd and 14th); Fleet Pond (1, Apr 30th); Stratfield Saye Park (3, Feb 15th, 2 Mar 5th, 2, June 1st); Woolmer Pond (3 south Apr 16th, 2 May 17th, 1 juvenile, Aug 4th); The Vyne lake (maximum of ten on May 5th); The Vyne Watermeadow (maximum up to 6 during April and May).

Mandarin
Aix galericulata

A scarce or moderately common and increasing resident.

In total, 190 records were received from 63 sites (compared with 189 records from 51 sites in 2004). This species is probably under-recorded in areas such as the New Forest and the north-east of the county, from where there were many scattered records. The highest count of the year came from Liss Forest with 73 on Oct 24th.

Monthly maxima at sites where counts regularly exceeded 10:

	J	F	M	A	M	J	J	A	S	O	N	D
Eyeworth Pond	10	17	10		6		14					14
Clamp Kiln Farm	13	6	6	1	2	5	4	9	20	13	13	6
Liss Forest	22	8	5	2	4	1		25	59	73		32
Headley Mill Pond	23	1	4						5	12	66	29
WeBS count totals	**29**	**2**	**18**						**24**	**20**	**68**	**35**

Outside the breeding season, other notable counts included the following: Roydon Woods NF (22, Jan 24th); Sowley Pond (11, Sept 14th); Broadlands Estate (12, Nov 6th); Fleet Pond (14, Jan 22nd); Deadwater Valley, Bordon (15, Feb 16th); Passfield Pond (14, Sept 18th); Yateley GP (23, Feb 14th).

Wigeon   *Anas penelope*

A common winter visitor and passage migrant; a few summer each year [Amber].

The Beaulieu estuary again produced the highest count of the year with 2044 on Dec 15th. Numbers at some coastal sites were slightly higher than the corresponding periods during 2004 (for example, Langstone Harbour and Lower Test/Eling/Bury Marshes). However, numbers in the Avon valley were lower. This is presumably a reflection of the level of flooding in the valley. As usual, there was a rapid reduction in numbers by late March.

Monthly maxima at sites where counts exceeded 100:

	J	F	M	A	M	J	J	A	S	O	N	D
Chichester Harbour	1801	2235	1124	6	2	2		10	480	1353	1100	2548
East Hayling	99	307	23						126	316	118	230
Langstone Harbour	793	1205	1150	65	2	4	4	19	500	680	693	1430
Portsmouth Harbour	448	351	295						257	269	190	460
Titchfield Haven	300	605	100	24	3	3	2	1	29	120	231	471
Hamble Estuary	349	420	379	27				2	65	73	225	411
Lwr Test/Eling/Bury Marshes	1149	1550	975	12					74	335	742	1310
Dibden Bay/Calshot	336	146	97	5					6	140	101	277
Needs Ore/Beaulieu Estuary	1482	939	995	8	8		3		144	700	1053	2044
Sowley/Lymington	260	220	200						35	417	433	250
Lymington/Hurst	900	1094	617	41	1		1	5	136	450	450	905
Sopley/Ringwood	565	899	825						31	343	410	457
Blashford Lakes	1182	1097	920	23	7			6	253	416	730	955
Broadlands Estate	30	50	52						21	79	150	130
Dogmersfield Lake	79	10	2						1	205*	122	50
Tundry Pond	93	91	61						8	76	173	104
Eversley GP	398	247	18	12	1	0	1	15	88	284	398	432
Yateley GP	101	16	8						38	70	116	160
WeBS count totals	**8417**	**8438**	**6342**	**66**	**14**	**4**	**4**	**17**	**1040**	**3904**	**5257**	**8591**

Spring passage was recorded at Sandy Point, Gilkicker Point and Hurst Beach, and totalled 65 east during Mar 6th–Apr 27th. In addition, 60 left high SSE from Langstone Harbour on Mar 17th and marked the period of peak departure from this area.

The first returns were noted at several sites from Aug 17th onwards, with significant arrivals being recorded at some localities during October and November. Peak counts (50+) at selected localities not included in the table above were as follows: Fishlake Meadows (50, Jan 1st and 30th); Allington GP (56, Jan 16th); Bramshill Plantation North Pool (73, Feb 20th); Stratfield Saye (140, Jan 16th); The Vyne Watermeadow (59, Dec 4th).

2004 correction: the November maximum at Tundry Pond was 74, not 59.

Gadwall
Anas strepera

A moderately common winter visitor and scarce breeder [SPEC 3, Amber, HBAP].

Numbers in the county remained at a high level during 2005 although the numbers recorded at Blashford Lakes in the late year were rather lower than the corresponding period during 2004. Numbers were also slightly lower at Eversley and Yateley GPs compared with recent years; it will be interesting to see if this is a temporary reduction or a halt to the longer term trend of increasing numbers. Monthly maxima at sites where counts exceeded 50 are as follows:

	J	F	M	A	M	J	J	A	S	O	N	D
Titchfield Haven	16	13		15	17	31	8	11	3	16	37	53
Lymington/Hurst	59	52	28	6	8				6	26	76
Blashford Lakes	708	720	372	49	49	19	40	92	109	267	433	658
Testbourne Lake	6	16						7	35	45	24	60
Allington GP	68	16							43		53	38
Winchester SF	72	103	118	95	55	40	21	41	58	43	24	36
Alresford Pond/Arlebury Lakes	56	61	20	15	27	46	4		58	8	80	12
Dogmersfield Lake	111	43	34		4				0	80	145	112
Tundry Pond	77	90	75	10	2		1		26	17	76	31
Stratfield Saye	110	25	8			34	6			42		2
Wellington CP	9	121	22							2	7	2
Eversley GP	139	215	26	16	10	12	11	20	44	76	174	84
Yateley GP	117	58	35						51	87	196	149
WeBS count totals	**1519**	**1394**	**816**						**427**	**661**	**923**	**1333**

Sites of international importance: 600+; national importance: 171+.

Notable counts from elsewhere included: Hook-with-Warsash (39, Oct 2nd); Fawley Refinery (30, Feb 16th); Needs Ore (30, Jan 25th), Fishlake Meadows (60, Oct 31st); Overton Lagoons (57, Jan 15th); Highwood Reservoir (45, Aug 20th); Ovington (46, Oct 13th); Bramshill Plantation North Pool (43, Jan 16th).

Evidence of spring passage came from Sandy Point and Hurst Beach where a total of 19 flew east between Apr 23rd and May 13th.

A total of 25 broods was reported as follows: Titchfield Haven, 1; Blashford Lakes, 5; Timsbury Lake, 1; Gavelacre, 1; Bransbury Village, 2; Longparish, 1; Longstock, 2; Southington Lane, 2; Winchester SF, 2; Alresford Pond, 1; Woolmer Pond, 1; Stratfield Saye, 6. It is likely that this species is under-recorded in the breeding season along the main river valleys.

Teal
Anas crecca

A scarce resident and common winter visitor [Amber].

Overall numbers were lower during the year compared with 2004. The highest counts were again recorded at Lymington/Hurst where, conversely, the peak counts were actually higher than the previous year with 812 on Feb 13th, 995 on Nov 6th and 1218 on Dec 4th. Small numbers remained during the summer and breeding was again confirmed at Woolmer Pond where a brood of seven young was reported. Monthly maxima at sites where counts regularly exceeded 100 are tabulated below:

	J	F	M	A	M	J	J	A	S	O	N	D
Chichester Harbour	1084	819	811	51		1		40	280	711	827	1366
East Hayling	199	270	287						235	138	174	210
Langstone Harbour	240	365	391	182	8	15	12	84	402	430	384	399
Portsmouth Harbour	186	89	216	4					9	5	23	167
Titchfield Haven	308	262	140	55	8	19	7	10	96	200	226	320
Hamble Estuary	236	113	127	8				48	122	180	255	180
Lower Test/Eling/Bury Marshes	183	246	84	11	3				66	80	136	168
Dibden Bay/Calshot	242	149	112					1	130	40	99	404
Needs Ore/Beaulieu Estuary	516	254	199	20			17	75	175	600	786	723
Lymington/Hurst	787	812	599	228	22		10	33	362	511	995	1218
Sopley/Ringwood	160	134	107						184	405	189
Blashford Lakes	239	53	41	5		4	32	170	117	80	253

(continued)	J	F	M	A	M	J	J	A	S	O	N	D
Broadlands Estate	30	15	5						3	4	40	152
Fishlake Meadows	100	46	15	15			2	18	14	100	20	30
The Vyne Watermeadow	162	101	54	34	2		3	63	117	110	102	106
Alresford Pond	44	66	16	5	1		2		150	67	100	39
Avington Lake	81	7	5						21	10	103	189
Stratfield Saye Park	100+	125	4+							117		145
WeBS count totals	**3888**	**2901**	**2481**	**420**	**9**		**29**	**258**	**1689**	**2555**	**3790**	**4708**

Peak counts (100+) at selected other localities were as follows: 120, Dark Water (Lepe), Nov 13th; 600, Sowley Pond, Dec 28th; 152, Broadlands Estate, Dec 5th.

Green-winged Teal   *Anas carolinensis*

A rare vagrant (0,17,1).

A male was found with a group of 30 Teal at a partially frozen Bramshill Plantation North Pools at 1000 hrs on Feb 20th (JMCk). This was a very high number of Teal for the site; they evidently moved on as none was found later in the day or on subsequent visits during the following week.

Mallard   *Anas platyrhynchos*

A common resident and winter visitor.

Monthly maxima at sites where counts exceeded 200 or more are tabulated below:

	J	F	M	A	M	J	J	A	S	O	N	D
Chichester Harbour	332	274	223	21	48	85	61	73	95	350	421	475
East Hayling	120	86	78			2			59	156	118	208
Titchfield Haven	348	266	287				120		145	228	267	310
Lymington/Hurst	201	124	126	55	66	107	86	64	56	159	184	204
Sopley/Ringwood	80	56	39						190	325	178	154
Blashford Lakes	179	131	78	45	75	70	77	125	113	195	206	248
Broadlands Estate	210	250	70	40					250	360	310	340
St Mary Bourne	103	45	37						80	203		109
Stratfield Saye Park	50	77				20	19			227		156
WeBS count totals	**2907**	**2479**	**1841**						**2326**	**3521**	**3129**	**3712**

Counts of 100 or more from other localities included: Emsworth Pond (190, Nov 13th); Langstone Harbour (100, Sept 17th); Baffins Pond (130, Feb 8th); Portsmouth Harbour (154, Jan 15th); Lower Test Marshes (130, July 23rd); Beaulieu Estuary (155, July 26th); Sadlers Mill (133, Nov 6th); Stockbridge-Fullerton (117, Nov 11th); Andover Rivers (192, Sept 4th); Anton Lakes (100, Oct 16th); Testbourne Lake (140, Oct 14th); Avington Lake (148, Nov 10th); New Alresford (112, Feb 3rd); Alresford Pond (100, Nov 5th); Petersfield Heath Pond (164, Nov 30th); The Vyne Watermeadow (103, Dec 4th); Bramshill Police College (113, Oct 16th); Foley Manor, Liphook (200, Aug 29th – released for shooting); Yateley GP (125, Dec 21st).

Pintail   *Anas acuta*

A moderately common winter visitor and passage migrant; occasional in summer [SPEC 3, Amber].

Numbers were considerably down on 2004 with peak counts in the early and late year both coming from the Lymington area, with 401 on Feb 13th and 351 on Dec 4th. The population in the Avon valley was much lower than in some recent winters (such as 2001 and 2003) and this is, presumably, a reflection of lower water levels in the valley. Numbers declined rapidly during March, with the last at Farlington Marshes on Apr 22nd. The first returns (3) were also reported from Langstone Harbour on Aug 25th, with the main arrival commencing in late September. Monthly maxima at sites where counts exceeded 10 are as follows:

	J	F	M	A	M	J	J	A	S	O	N	D
Langstone Harbour	78	268	249	7				3	41	95	119	148
Titchfield Haven	24	29	26						1	1	5	14
Hamble Estuary	14	14	6						14	14	27	25
Fawley	40	30	56							14		32
Needs Ore/Beaulieu Estuary	70	50	60	1					40	38		37
Sowley/Lymington	46	77	54						2		14	41
Lymington/Hurst	250	401	275	3					5	106	306	351
Ibsley Water	65	34	2						2	4	2	10
Fishlake Meadows, Romsey	22	9									3	8
WeBS count totals	**448**	**544**	**431**						**56**	**215**	**366**	**493**

Sites of national importance: 279+.

Away from the Avon valley and Fishlake Meadows, inland records were reported as follows: Testwood Lakes (2, Feb 13th-25th); Broadlands Estate (5, Dec 5th); Lakeside CP (1, Sept 12th); Allington GP (1, Feb 13th); Itchen Valley CP (1, Feb 26th); Alresford Pond (1, Dec 6th); Kings Pond, Alton (1, Aug 1st); Tundry Pond (1, Oct 16th, Nov 30th, Dec 4th and Dec 16th); Bramshill plantation (1, Oct 16th); Eversley GP (2, Dec 28th); Woolmer Pond (2, Feb 18th-Mar 14th).

In addition, up to two were reported from Portsmouth Harbour/IBM Lake between Mar 10th and Apr 9th.

Garganey

Anas querquedula

A scarce passage migrant and summer visitor. Occasionally breeds [SPEC 3, WCA 1, Amber].

A better year than 2004 (although there were no March records), with a minimum of 21 during the spring and 20 during autumn. The first of the year were seven at Keyhaven/Pennington on Apr 1st, with the last at Farlington Marshes on Oct 25th. There were no records indicative of breeding.

A summary of all records is as follows:

Coastal Sites:
Farlington Marshes: male, Apr 21st; male, Apr 28th-29th; male, May 1st-2nd, 5th and 16th-28th (also seen at Thorney Island (West Sussex) during the month); up to 5 juveniles July 25th-Aug 25th with an adult female, July 27th-31st; juvenile female, Oct 25th.
Titchfield Haven: male, Apr 26th; male, June 11th-12th.
Hook-with-Warsash: Male and juvenile, Sept 10th-13th; juvenile, Sept 20th-26th.
Lower Test Marshes: female, July 29th-Aug 13th.
Lymington/Hurst area: 7, Apr 1st (5 males, 2 females, with a pair remaining the following day); male, Apr 7th-16th; pair, Apr 23rd-May 1st; male, east with 8 Common Scoter, Apr 29th; male, May 3rd; up to 4 (female and 3 juveniles), Aug 19th-29th.

Inland Sites:
Ibsley Water/Mockbeggar Lake: pair, May 1st; juvenile, Aug 14th-26th; female Sept 2nd-5th; up to 3, Sept 21st-Oct 8th.
Testwood Lakes: female, Aug 29th and Sept 4th.
The Millfield, Basing: male, Apr 20th.
Fleet Pond: male, May 14th.

Shoveler

Anas clypeata

A moderately common winter visitor and passage migrant; formerly a few pairs bred annually [SPEC 3, Amber].

Numbers were average in the early year but were rather low in the late year period. The highest individual counts were 149 at Blashford Lakes on Jan 9th and 134 in Langstone Harbour on Feb 18th. Monthly maxima at sites where counts exceeded 50 are as follows:

	J	F	M	A	M	J	J	A	S	O	N	D
Langstone Harbour	45	134	89	41	5	5	9	44	86	62	67	54
Budds Farm SF	40	80	61	4							17	23
Baffins Pond, Portsmouth	12	64	6							29	12	18
Titchfield Haven	39	39	20	22	20	14	1	10	12	23	26	36
Needs Ore/Beaulieu Estuary	43	100	68	57	4		2	5	50	3	49	67
Lymington/Hurst	61	88	53	20	1			10	18	18	73	70
Blashford Lakes	149	130	123	39				6	99	98	57	109
Fishlake Meadows, Romsey	32	35		16	7	4	1	12	30	50	40	18
Eversley GP	29	50	31	7				7	9	20	22	36
Yateley GP	48	55	63							9	34	16
WeBS count totals	429	502	386	136	5	2		56	253	237	307	383

Sites of national importance: 148+.

Maximum counts at some other sites were as follows: Fawley Refinery (24, Feb 28th); Hook-with-Warsash (21, Sept 10th); Testwood Lakes (14, Mar 6th); Broadlands Estate (15, Feb 13th); Allington GP (18, Jan 16th); Winchester SF (29, Dec 31st) and Alresford Pond (18, Jan 13th and Sept 24th).

Spring passage was poorer than in 2004 but was recorded from both ends of The Solent. At Hurst Beach a total of 25 flew east between Apr 2nd and May 10th (peak 20 E on Apr 2nd) while at Sandy Point/Hayling Bay a minimum of just 8 moved east on Apr 2nd and 4th. In addition, 19 left east from Langstone Harbour on Mar 17th.

Definite breeding was reported from Titchfield Haven where a brood of nine young was seen in late May and early June. There were no other reports of breeding but juveniles were noted at Farlington Marshes on July 4th and at Keyhaven and Ibsley Water/Mockbeggar Lake during August, although these may have been migrants.

Red-crested Pochard   *Netta rufina*

A scarce feral resident, rare passage migrant and winter visitor, or escape.

There was a marked increase in the number of records received, although analysis reveals that most of the records were generated by two long-staying and a handful of wandering birds.

A male was on the Hamble Estuary between Mar 5th and Apr 16th (JF) and, presumably the same individual, later transferred to Titchfield Haven where it remained between Apr 24th and May 30th, having been seen in flight between the two estuaries, over Chilling, on Apr 27th. It was not seen during most of June, when it presumably moulted at an unknown location, before settling on the Beaulieu Estuary between June 24th and July 20th. In the north of the county a male was observed dropping into pools in the Loddon valley by Millfield, Old Basing at 0830 hrs on May 13th (JKA).

Later in the year, a well-watched female was at Tundry Pond between Aug 22nd and Oct 23rd, before being relocated on Dogmersfield Lake on Nov 11th (m.o.). A female was in Itchen valley on Nov 11th (A&JDG). Another female, a leucistic individual, was on Yateley GPs at the end of the year (JMC).

Pochard   *Aythya ferina*

A scarce breeder and moderately common winter visitor [SPEC 2, Amber, HBAP].

Numbers continue to decline. Blashford Lakes remains the principal location in the county, where numbers peaked at 199 on Feb 17th, dwindling to a single on May 11th. In the second winter period 70 had returned by Oct 23rd, with 140 there on Dec 18th.

Just five reports of breeding as follows: Alresford Pond, where one pair raised three young; Romsey, where two pairs raised a total of six young; Leckford, where one pair raised seven young; Longstock, where there were two broods and Mottisfont, where one pair raised four young.

Double-figure counts also came from (annual maxima in parenthesis): Budds Farm SF (26, Feb 4th); Sowley Pond (33, Dec 28th); Dogmersfield Lake (40, Jan 16th); Fleet Pond (23, Mar 13th); Testwood Lake (38, Nov 11th).

Monthly maxima at sites where counts exceeded 50 in some months are as follows:

	J	F	M	A	M	J	J	A	S	O	N	D
Titchfield Haven	92	80	87	7					7		36	62
Blashford Lakes	163	199	131	3	1	1	4		6	70	108	140
Eversley GP	57	51	26	2			3	3	4	61	68	43
Yateley GP	27	50	60	1					2	6	41	34
WeBS count totals	**479**	**360**	**312**	**7**	**1**	**1**	**1**		**15**	**154**	**281**	**346**

Tufted Duck   *Aythya fuligula*

A moderately common breeding species whose numbers increase considerably in winter [SPEC 3].

Blashford Lakes and Eversley/Yateley GPs remain the principal locations for Tufted Duck in the county. There were higher numbers during the first winter period than during the second, with numbers on Blashford Lakes peaking at 411 on Feb 17th, 167 on Eversley GP on Feb 15th and 225 on Yateley GP on Jan 19th. In the second winter period peak counts were: 318 on Blashford Lakes on Dec 18th, 138 on Eversley GP on Dec 11th and 144 on Yateley GP on Dec 21st.

A total of 59-61 pairs were reported breeding as follows: Ibsley North Lake (8 pairs, 58 young); Mockbeggar Lake (7-8 pairs, 48 young); North Somerley Lake (1 pair, 5 young); Foley Manor (1 pair, 2 young); Liphook (1 pair, 2 young); Alresford Pond (1 pair, 6 young); Arlebury Lakes (1 pair, 4 young); Keyhaven (1 pair, 4 young); Budds Farm SF (at least 5 pairs, 22 young); Milton Reclamation (1 pair, 7 young); Frith End (1 pair, 2 young); Selborne (4 pairs); Stratfield Saye (2 pairs, 9 young); Eversley GP (7-8 pairs); Woolmer Pond (1 pair, 3 young); Sherfield Links GC (1 pair, 7 young); The Vyne (2 pairs, 4 young); Bransbury (2 pairs, 8 young); Chilbolton (1 pair, 5 young); Romsey (4 pairs) and Testbourne Lake (4 pairs).

Monthly maxima at sites where counts exceeded 50 are as follows:

	J	F	M	A	M	J	J	A	S	O	N	D
Baffins Pond, Portsmouth		98	78	36		22				22	38	48
Titchfield Haven	31	40	53	5			2	1	1	16	33	43
Blashford Lakes	354	411	319	312	152	109	90	87	121	282	304	318
Ibsley/Bickton	26	111							4	2	105	147
Testwood Lakes	14	10	9		23				26	118	30	16
Broadlands Estate	76	45	46						3	18	31	25
Stockbridge/Fullerton	37	51	41							13	25	10
Testbourne Lake		19		10		30		28	78	8	14	15
Alresford Pond	9	78	7	20	16		10			5	4	2
Arlebury Lakes	80	66	55		12			1	1	12	6	38
Wellington CP	5	59	29			5				11	21	44
Eversley GP	135	167	82	70	79	45	64		30	110	101	138
Yateley GP	225	131	162	51					27	105	128	144
Camp Farm GP/SF	41	19	13						62		11	61
WeBS count totals	**1334**	**1284**	**662**						**403**	**592**	**921**	**1210**

Scaup   *Aythya marila*

A very scarce winter visitor and passage migrant [SPEC 3, WCA 1, Amber].

An average year. During the first winter period, a total of nine were observed on the coast, all passage individuals with one inland. In the late year there was a total of 16 including one flock of seven.

In January five, two males and three females, flew west off Hurst Beach on Jan 15th; a male was at Normandy on 21st. A pair was briefly in Chichester Harbour on Feb 12th before departing south and a female was in Langstone channel on Feb 26th. Inland a female/immature was on Yateley GPs on Mar 12/14th.

There was a very early record for the late year period with a female on Sep 18th at Ibsley Water (on the same date as the first returning Long-tailed Duck at Titchfield Haven). On the coast the first arrival and the highest count of the year, was a flock of seven, all female/first-winter types, first seen in Langstone Harbour before relocating into Cams Bay on Nov 11th. The next were two female/first-winters on Dec 14th, one at Titchfield Haven and the other at Normandy, Lymington where it remained to the end of the year. Two females and a first-winter female were in The Solent off Pennington Marsh on Dec 18th, with a first-winter male there on Dec 24th; finally two female/first-winters were in Stokes Bay from Dec 28th to the year end. The annual totals of bird-months from 1996 to 2005 are as follows:

	1996	1997	1998	1999	2000	2001	2002	2003	2004	2005
Scaup	34	78	23	1	19	3	18	20	11	26

Aythya hybrids

The drake Ferruginous Duck x Pochard hybrid was present intermittently at Budds Farm SF and Farlington Marshes from Feb 4th to Mar 17th and at the former site from Oct 20th to Nov 23rd (JCr, MC *et al*). This individual was first recorded on Nov 23rd 1999 and has now over-wintered in the Langstone Harbour area for the last seven years.

A female type Scaup x Tufted Duck hybrid was on Dogmersfield Lake on Feb 13th (KBW). The female Scaup-type hybrid, previously identified as a Scaup *Aythya marila*, was on its usual pool by Southington Lane from Feb 11th to May 2nd (PEH).

Eider   *Somateria mollissima*

A scarce but increasing winter visitor and passage migrant; small numbers usually summer; bred for the first time in 2003 [Amber].

As usual, most birds favoured Hill Head in winter and Lymington/Hurst/Milford on Sea (West Solent) in summer. During the first winter period, numbers off Hill Head peaked at 97 on Feb 6th, before decreasing to 28 on May 26th, with none thereafter. The first significant numbers in the West Solent were on Apr 3rd, when 39 flew east off Hurst, although there was an early high count of 17 flying east there on Jan 16th. The highest counts subsequently from the West Solent area were of 79 at the roost off Normandy on Apr 10th, and 41 on Aug 23rd, but indications were that no more than 40 birds actually over-summered. There was no repeat of the attempted breeding of 2003, but three pairs were noted displaying off Oxey Point on May 15th (RBW). The first returning birds off Hill Head were noted on Aug 25th, with significant numbers appearing from mid September, and peaking at 107 on Dec 11th. The corresponding decline in numbers in the West Solent commenced after Oct 16th, on which date 20 were counted, with monthly maxima thereafter of 5 on Nov 4th and 4 on Dec 17th.

Monthly maxima at all sites where counts exceeded five are as follows:

	J	F	M	A	M	J	J	A	S	O	N	D
Chichester Harbour	1	2	2						6	3	5	5
Sandy Point/Hayling Bay	2	1	7	6			1				6	5
Gilkicker/Hill Head/Warsash	90	97	90	70	50			1	43	52	90	107
Lepe/Needs Ore			28	9	27	1	3		6		2	
Sowley/Lymington			9				1	4	2	3	1	
Lymington/Hurst	3		33	79	27	31	36	41	28	20	5	4
Hurst/Milford on Sea	17		33	40	28				11	5		
Approx. county totals	**113**	**100**	**203**	**204**	**132**	**32**	**41**	**46**	**96**	**83**	**109**	**121**

Long-tailed Duck   *Clangula hyemalis*
A very scarce winter visitor and passage migrant.

An average year. During the early year period a first-winter male was in Emsworth Channel, Chichester Harbour from Jan 13th-15th, while in the west of the county another first winter male resided in the Keyhaven/Hurst Castle area until Mar 13th, having first been seen on Nov 3rd 2004. The only spring passage records came from Hurst Beach/Castle where a female or immature flew east on March 22nd and a group of four moved east on Apr 11th.

The earliest record from the second winter period was of a male at Titchfield Haven on Sep 18th. Subsequently a female was in Emsworth Channel and Sweare Deep, Chichester Harbour, between Oct 23rd and Nov 20th; an adult male was in Langstone Harbour between Oct 26th and Oct 31st; and one (unsexed) was at Needs Ore on Dec 20th. The annual totals of bird-months from 1996-2005 are as follows:

	1996	1997	1998	1999	2000	2001	2002	2003	2004	2005
Long-tailed Duck	4	8	15	6	23	12	6	14	15	14

Common Scoter   *Melanitta nigra*
A moderately common passage migrant; small numbers usually summer [WCA 1, Red, NBAP, HBAP].

It was a below average year with a light eastward passage in the spring. There were records in every month off Hill Head with small numbers of long-staying birds and a maximum of 22 on Dec 3rd. The largest resting flock of the year was an arrival from the west of 139 off Hurst Beach on Apr 22nd.

The only double-figure counts during the first winter period were at Sandy Point with 12 east on Jan 26th and ten on Feb 18th.

Eastward passage commenced with 140 bird-days noted in the second half of March including 16 off Southsea on 20th; 65 passed Hurst on 24th and 34 passed Sandy Point on 30th. In April there were 425 bird-days of eastward movements with the highest counts on 22nd with 48 east past Sandy Point and 139 off Hurst Beach. In May resting flocks off Hurst Beach included a total of 123 on 1st and 100 on 10th (with 93 reported there as late as June 10th); eastward movements off Sandy Point included 71 on 15th and the last double-figure count of 56 on 25th. The spring passage included movements through The Solent, with for example 80 bird-days recorded from Stokes Bay, but perhaps the majority south of the Isle of Wight, the 370 bird-days recorded at Sandy Point being composed of sightings from both movements.

The highest counts in the second half of the year were from Hurst Beach: 43 on July 17th and 48 counted east into The Solent on Oct 19th, 50 off Needs Ore on Sep 21st and 22 off Hill Head on Dec 3rd.

Monthly maxima at sites where counts exceeded ten in some months are as follows:

	J	F	M	A	M	J	J	A	S	O	N	D
Sandy Point/Hayling Bay	13	10	52	118	218	7	49	5	5		11	12
Stokes Bay				54	26					5		
Hill Head/Brownwich	4	4	7	13	5	6	15	20	2	10	6	22
Needs Ore/Beaulieu Estuary		5	2				9		50	1	7	
Hurst/Milford on Sea	4		139	239	123	93	57	19	4	48	5	4
Barton on Sea			8	21	20				1	4		
Approx. county totals	22	31	200	445	392	106	133	55	62	74	39	48

In Southampton Water there were two off Cracknore Hard on Feb 5th, a pair at Langdown on Apr 10th, one off Town Quay on Nov 5th, up to five females/immatures off Western Shore on four dates between Nov 20th and Dec 29th and two off Calshot on Dec 10th. Other sightings included three males in Langstone Harbour from July 27th/31st. and just possibly the same Sweare Deep, Chichester Harbour on Oct 26th; two females at Port Solent, Portsmouth Harbour on Nov 19th, three females at Hook-with-Warsash spit on Dec 3rd and finally there were nine bird-days in the early and late year from Sowley shore to Tanners Lane, Lymington with a maximum of four on Feb 22nd.

Velvet Scoter   *Melanitta fusca*

A scarce passage migrant and winter visitor.

An average year with 51 records submitted. During the first winter period two were on the sea off Sandy Point on Feb 18th before leaving east; a third moved west with Common Scoter. Spring migrants were noted at both ends of The Solent with a total of 25 bird-days recorded and a minimum of 21 individuals. At Sandy Point a group of six flew east on Mar 21st and at least two males were there on 25th. At Hurst Castle a pair first noted on Mar 21st was probably the same in that area until 26th and, possibly this pair sighted with an immature, or second female, on Apr 2nd. Other April records from Hurst beach or Milford on Sea were of one on 11th, three on 16th (adult male, first-summer male and female) and four on 21st.

The first return, a long-staying first-winter, was seen intermittently between Hill Head and Brownwich from Nov 12th to Dec 19th; in addition, there was a count of four there on Nov 12th, two on Dec 2nd and a male on 3rd. The additional three at Hill Head on Nov 12th could have been the same as those seen off Needs Ore on 15th, and then off Hurst Castle on 20th. Apart from the records at Hill Head/Brownwich, there were four further records in December: four in Chichester Harbour on 14th; one at Weston Shore on 17th, which flew off towards Town Quay early morning; two at Hurst Castle on 18th and two at Sandy Point on 30th. The annual totals of bird-months from 1996-2005 are as follows:

	1996	1997	1998	1999	2000	2001	2002	2003	2004	2005
Velvet Scoter	22	10	22	11	33	11	44	47	47	39

Goldeneye   *Bucephala clangula*

A scarce and declining winter visitor [Amber].

Another poor year, with numbers continuing to decline. During the first winter period, the best counts were of 44 in Langstone Harbour on Feb 18th and 29 at Ibsley Water on Feb 17th. The last was recorded on Apr 2nd and the first to return was noted on Oct 16th. In the second winter period 45 were in Langstone Harbour on Dec 13th, but 14 on Dec 18th was the peak count at Ibsley Water.

Notable inland records were of two at Heath Pond, Petersfield on Jan 15th, one at Waggoners Wells, Grayshott on Nov 16th and one at Winchester SF on Dec 15th.

Monthly maxima at sites where counts exceeded ten are as follows:

	J	F	M	A	M	J	J	A	S	O	N	D
Chichester Harbour	12	13	5								4	16
East Hayling	7	11									4	2
Langstone Harbour	30	44	22							1	20	45
Portsmouth Harbour	23	16	18									21
Lymington/Hurst	10	10	5								7	13
Blashford Lakes	17	29	29	21						7	8	14
WeBS count totals	**94**	**66**	**64**	**9**						**8**	**21**	**47**

2004 addition: A female/first-winter was at Tundry Pond on Nov 25th.

Smew   *Mergus albellus*

A very scarce winter visitor [ET, SPEC 3].

Another poor showing with only two redheads recorded - one on Ibsley Water on Feb 6th

(TMJD) and one on Arlebury Lake on Dec 13th (MS). The annual totals of bird-months from 1996-2005 are as follows:

	1996	1997	1998	1999	2000	2001	2002	2003	2004	2005
Smew	22	79	24	13	4	1	8	7	1	2

Red-breasted Merganser   *Mergus serrator*
A moderately common winter visitor and passage migrant; rare inland [HBAP].

Red-breasted Merganser   © *George Spraggs*

An average year. During the first winter period numbers in the eastern three harbours were low with maxima of 95 in Chichester Harbour on Mar 12th, 107 in Langstone Harbour on Feb 17th, and 65 in Portsmouth Harbour on Mar 12th. Most had left by mid April, and two remained throughout the summer in Langstone Harbour. In the Lymington/ Hurst area, the highest count was 51 on Mar 22nd at the Oxey Lake roost. There were few counts of spring passage birds, but these included ten at Sandy Point (including nine east) on Apr 22nd and the last at Hurst Beach on May 13th.

Birds started to return to the eastern harbours and Lymington/Hurst in mid October but the main arrival began in November. Numbers then returned to their expected levels, with the 3rd December WeBS count in the eastern harbours revealing 50 in Chichester Harbour, 187 in Langstone Harbour and 88 in Portsmouth Harbour. Next day 57 were also counted at Lymington/Hurst.

Monthly maxima at sites where counts exceeded 50 are as follows:

	J	F	M	A	M	J	J	A	S	O	N	D
Chichester Harbour	147	124	194	13						15	68	168
East Hayling	30	48	95	13						6	30	50
Langstone Harbour	91	107	69	46	3	2	2	2	2	20	99	187
Portsmouth Harbour	75	76	65	19						14	27	88
Lymington/Hurst	41		51	40						5	47	57
WeBS count totals	**261**	**226**	**312**	**70**	**2**	**2**	**2**	**2**	**2**	**23**	**131**	**400**

Other double-figure counts were of 15 at Dibden Bay on Feb 13th, 12 at Brownwich and 14 at Needs Ore each on Mar 13th.

Goosander
Mergus merganser

A scarce winter visitor and very scarce breeder [HBAP].

A good year, but with no evidence of a repeat of last year's successful breeding. During the first winter period numbers at the Ibsley Water roost peaked on Mar 6th when 60, including 18 adult and four first-winter males, were recorded - the highest count ever for the site. Few were reported from the coast, suggesting overland dispersal, although three were off East Hayling on Jan 16th/17th, three were off Weston Shore on Feb 17th and one passed Hurst Castle on Apr 15th. During the second winter period, numbers at Ibsley Water peaked at 48 on Dec 31st, while at Eversley GP there were 37 on Dec 28th (including 19 males).

The Ibsley Water roost had mostly dispersed by Apr 2nd, when numbers had dropped to just five, with very few recorded in June and none at all in July. The first significant return was on Nov 19th when 15 were counted. The Eversley GP roost dispersed earlier with no birds after Mar 14th, but the first returns, on Nov 19th, were on the same date as those at Blashford Lakes.

Away from the main concentrations, the highest counts were of six on Broadlands Lake on Feb 13th and five between Stockbridge and Fullerton on several dates - suggesting there is a small number (up to 10) wintering in the Test valley. Other notable local records were on Petersfield Heath Pond and Testwood Lakes, but there was no repeat of last year's influx of birds onto Eyeworth Pond, where just one was noted on Feb 2nd.

Monthly maxima at sites where counts exceeded ten are tabulated below, including the two known roosts:

	J	F	M	A	M	J	J	A	S	O	N	D
Ibsley Water (roost)	37	46	60*	13	4	3		1		1	30	48
Eversley GP (roost)	31	27	22								7	37
Yateley GP	12	8	5									15
Tundry Pond	13	6									2	2
Bramshill Plantation	7	12	10								3	
Approx. County Totals	**66**	**85**	**92**	**17**	**4**	**3**	**0**	**1**	**0**	**3**	**44**	**95**

2004 addition: eight or nine were at Tundry Pond on Nov 25th; a female/first-winter was at Alresford Pond on Dec 27th.

Ruddy Duck
Oxyura jamaicensis

A scarce resident and winter visitor.

At the two main locations the monthly totals are holding up, despite the implementation of plans to eradicate this species from Britain. The peak counts were 26 at Ibsley Water on three dates in February and nine at Alresford Pond on Apr 9th. Breeding was recorded at Ibsley Water where an unfledged juvenile was seen on Aug 19th, and at Alresford Pond, where three pairs bred, including one pair which raised a brood of five. Elsewhere, there were scattered records from coastal sites, together with small numbers on Fishlake Meadows, Romsey and at The Vyne.

The monthly maxima at the two principal sites are shown below:

	J	F	M	A	M	J	J	A	S	O	N	D
Blashford Lakes	26	26	24	14	3	2	1	1	7	9	12	12
Alresford Pond				9	6	6	5	4	2			

Elsewhere maxima were as follows: three males at Fishlake Meadows between May 13th and 20th; three males at The Vyne on July 9th; two males at Zions Hill Pond, Chandlers Ford, Mar 25th; two males in the Lymington/Hurst area, Apr 25th; two juveniles at Normandy lagoon, Sep 13th; two at Testbourne Lake, Nov 9th and two on the sea at Hook-with-Warsash, Nov 19th. Another five sites had single bird-days. The annual total of 207 bird-months comprised 167 bird-months from the two principal sites and 40 bird-months elsewhere. The ten-year annual total of bird-months is as follows:

	1996	1997	1998	1999	2000	2001	2002	2003	2004	2005
Ruddy Duck	276	139	87	77	108	172	102	115	139	207

Red-legged Partridge   *Alectoris rufa*

A common resident; numbers are supplemented by releases [SPEC 2].

The largest coveys in the early year were reported along the South Downs at Cheesefoot Head (33, Jan 3rd) and Beacon Hill (40, Mar 25th). Most reports were from coastal localities where the species is not as numerous as inland. There were seven in the Lymington/Hurst area on Jan 30th and eight at Hayling Oysterbeds on Feb 1st. Singles at Leaden Hall NF on Mar 12th, Brownwich on Mar 21st and Hook-with-Warsash on Apr 16th were notable at these localities.

The BBS survey data reveal that 32% of surveyed squares were occupied compared to the previous ten-year average of 23%. The last atlas survey 1987-91 registered 64% of the county tetrads occupied, but there are significant differences in survey methodology. There was a report of coastal breeding with a pair in fields behind Hurst Spit.

As usual, doubtless as a result of releases of many thousands of birds, large coveys were observed in autumn. On the coast there were counts of 70 at Lepe on Aug 18th and 100 there on Sep 15th. Inland there were counts as follows: 126 on two farms at Nether Wallop on Oct 5th; further east at Drayton Camp 54 on Aug 31st and 64 on Dec 29th; nearby at Firgo Farm 55 on Oct 14th and 52 on Dec 6th; 98 on Watership Down on Sep 21st; 75 in two coveys at Medstead on Sep 18th; the largest covey noted on the South Downs was 50 at Old Winchester Hill on Oct 29th and in the Greywell/Upton Grey area there were counts of 250 on Oct 9th, including 160 in one field, 70 on Nov 5th and 62 on Dec 12th.

Grey Partridge   *Perdix perdix*

A moderately common, but declining, resident with numbers supplemented by releases [SPEC 3, Red, NBAP, HBAP].

The GCT is attempting to augment wild stock by the release of captive-bred birds. Details have not been released of the GCT scheme to date but it appears to be having some impact on numbers, mainly in the late year.

In the early year, the few counts of note were clustered in the Bransbury Common (6), Newton (6) and Barton Stacey (8) area in the north-west of the county. Along the Test Way on the Broadlands Estate, where 97 had been released at the end of 2004, nine were detected on Feb 23rd. The only other reports, from seven widely scattered sites, all in the north of the county, totalled just 22 birds.

In the breeding season there were six territories on Martin Down and a further three nearby at Whitsbury Down; nineteen pairs/territories were noted at 13 sites in the Porton Down/Stockbridge area, including three at both Danebury and Nether Wallop; just to the east there were two pairs at Farley Mount and another nearby at King's Somborne; other records were widely scattered, with two pairs at Ashley Warren, Old Winchester Hill, and single pairs at Bramley Frith and Roke Farm. There were reports of possible breeding from another eleven sites. The BBS indicated 9% of survey squares occupied (ten-year average 7%). The last atlas survey 1987-91 registered 58% of the county tetrads occupied. Even allowing for differences in survey method (see Red-legged Partridge), it is clear that there has been a significant range contraction, despite conservation measures being put in place.

In the late year 118 were reported from 17 sites. In the north-west there were reports from Chilbolton (4, Aug 21st), Jacks Bush (18, Sep 1st), Hare Warren Farm (6, Sep 5th), Wildhern (a covey of 16, Sep 26th), Cholderton (12, Dec 3rd) and Linkenholt (4, Dec 18th). There were 17 reported from Moorcourt Farm, Broadlands Estate on Oct 22nd but just nine there on Dec 10th. At Cheesefoot Head there were five at Fawley Down on Nov 21st and 18 at Gander Down on Dec 26th. In the north-east there were 12 at Bentley on Sep 24th and eight at Hillside on Dec 25th. There were reports of two from a further three sites and singles at another two.

Quail   *Coturnix coturnix*

A normally very scarce but erratic summer visitor; very rarely recorded in winter [SPEC 3, Schedule 1, Red, HBAP].

It was the third highest year for registration of calling males since county recording began in 1951 (notable years being 1989, 63; 1964, 47 and 1998, 22). It can be difficult to differentiate between newly-arrived and long-standing, unpaired males; calling is believed not to occur after a pair bond is established. There were three main occupied areas, all comprising chalk downland: at Martin Down (3-5); further north in the Porton/Nether Wallop area (9-12) and on the South Downs around Cheesefoot Head (7-8). With four males noted elsewhere, a conservative estimate of the county total is 23 calling males - almost certainly an under-estimate - as is the year total of 25. A total of 41 records was received.

The first noted arrivals were a male seen at Hambledon and another heard near Nether Wallop on May 23rd, also possibly the same individual on June 3rd. The next was calling at Martin Down from May 28th/31st, but there were also other arrivals with two at Kimpton Down and one at Longwood Warren between these dates. The next was at Fawley Down on June 5th, followed by a small influx on June10th/12th with three at Martin Down; two at Over Wallop; singles at Ashley Warren and Gander Down, with two at Longwood Warren on 14th and one calling on two dates thereafter to 26th. In the Porton/Nether Wallop area a minimum of five were calling on 26th/27th, with three at least at Jacks Bush. There was a gap of two weeks before the next report; this was from Kimpton Down and nearly a month before the next reports from July 25th/27th, with three at Fawley Down (two still calling on 29th), one at Martin Down and another at Frankenbury Walk above Fordingbridge. These are late but typically expected arrival dates for young males fledged in southern Europe in the early year. There were three records in August, one in the Normandy, Lymington area on 9th; one still calling at Martin Down on 13th and another at Hambledon on 16th – surely not that first arrival in May, if so it was a very lonely summer for this male!. The last was flushed underfoot at Brownwich on Oct 10th. Annual county totals of singing males are as follows:

Quail	1996	1997	1998	1999	2000	2001	2002	2003	2004	2005
Singing males	8	15	22	12	6/7	4	5	11	15	23

Pheasant
Phasianus colchicus

An abundant resident, the wild population being supplemented by releases.

Many thousands of birds are captive-reared and released every year. The BBS survey data reveal 84% of squares occupied at a mean of 4.4 per square: these are close to the previous ten-year average (77%, 4.3). Other breeding data came from five sites as follows: Longmoor Inclosure, 17 territories; Titchfield Haven, 16; Lower Test Marshes, 20; Farlington Marshes, 4 and Langstone Harbour islands 1. This reflects the second year of breeding in Langstone Harbour.

Unusual, even bizarre sightings, included three March observations from well-frequented sea watch points: Hurst Castle on 4th, Black Point on 15th at the Lifeboat Station and Gilkicker Point on 17th.

Lady Amherst's Pheasant
Chrysolophus amherstiae

Once a very scarce feral resident; now presumed extinct.

A male, presumably an escape, was in a Langstone garden on Apr 28th and was later found dead on a nearby road (ACG).

Red-throated Diver
Gavia stellata

A scarce winter visitor and passage migrant [ET, SPEC 3, WCA 1, Amber].

A total of 137 records was received, referring to about 116 individuals, an increase on the 76 recorded in 2004. The majority of winter sightings came from open coasts at the eastern and western extremities of the county. Eastward spring passage was moderately good and took place between early March and mid May.

Peak counts in the early year included nine moving east off Hurst on Jan 16th and three off Sandy Point on Jan 26th. An oiled bird was seen in the Sandy Point and Langstone Harbour area on Feb 2nd and 6th; what may have been the same, was seen flying up Southampton Water from Hill Head on Feb 26th. Spring passage saw a good total of 43 moving east along the coast between Mar 6th and May 15th, peaking at five east off Hurst on Apr 26th.

The first return was an adult, possibly oiled, seen off Hill Head on Sep 28th. Numbers in the late year subsequently built up to an approximate total of 23 in December, with peaks of eight moving east off Sandy Point on Dec 29th and up to three in the Lymington-Hurst area in November and December.

The approximate monthly totals are shown below:

J	F	M	A	M	J	J	A	S	O	N	D
14	6	12	29	10	0	0	0	1	5	16	23

Black-throated Diver
Gavia arctica

A very scarce winter visitor and passage migrant [ET, SPEC 3, WCA 1, Amber].

A total of 34 records referring to about 24 individuals was received, making it the best year for two decades. Most were seen on eastwards spring passage but an individual inland in the late year was notable.

The year opened with a single east at Hurst Castle on Jan 1st and a further four moving east there on Jan 16th. Spring passage began on Apr 3rd and comprised a maximum of 14 moving east up to May 2nd, all from the coast between Hurst and Barton on Sea. The peak passage period was Apr 17th-27th, with 11 recorded.

The first return was a moulting adult at Hurst Castle on Sep 28th, followed by singles off Hurst on Oct 17th and 21st. The next record was the most surprising, being a single flying high south-east over Casbrook Common in the Test valley on Nov 4th. In December, one frequented the Lymington-Hurst area from 4th to16th, one was at Calshot on 10th and one was in Chichester Harbour the following day.

2004 correction: the individual recorded off Sandy and Black Points in December 2004 was first recorded on Nov 28th. All records were thought to relate to the same long-staying bird.

Great Northern Diver   *Gavia immer*

A scarce winter visitor and passage migrant [ET, WCA 1, Amber].

An average year with 45 records received, referring to about 32 individuals. Small numbers were recorded in both winter periods from the open coast, the eastern harbours and inside The Solent, and a small spring passage was noted.

In the first winter period, three records of singles off Hurst from Jan 2nd-16th may have referred to the same individual, while other singles were off Sandy Point on Feb 18th and in mid-Solent on Mar 2nd. Spring passage saw a total of 11 moving along the coast between Apr 3rd and May 15th, with two east off Hurst and one east off Gilkicker Point on Apr 3rd, up to five east off Sandy Point on Apr 25th/26th, one west off Taddiford Gap on May 2nd and singles east off Hurst on May 11th and 15th.

The first return was off Hurst on Nov 5th/6th, followed by singles in Chichester Harbour on Nov 7th and Hill Head on Nov 9th. Numbers built up to a total of 12 in December, with records of single birds coming from several widely scattered coastal sites, and two each at Needs Ore, Chichester Harbour and Portsmouth Harbour.

The approximate monthly totals are tabulated below:

J	F	M	A	M	J	J	A	S	O	N	D
3	1	1	8	3	0	0	0	0	0	4	12

The annual totals 1996-2005 of bird-months for diver species are as follows:

Annual bird-months	1996	1997	1998	1999	2000	2001	2002	2003	2004	2005
Red-throated Diver	108	93	101	67	79	116	98	128	76	116
Black-throated Diver	17	12	14	16	14	9	15	23	13	24
Great Northern Diver	22	19	73	53	52	52	41	35	19	32

Little Grebe   *Tachybaptus ruficollis*

A moderately common resident, passage migrant and winter visitor.

Numbers at the main coastal and river valley wintering sites showed no overall pattern of change compared to 2004, and the total of 94 breeding pairs/territories was also comparable. It should be noted that the true total of breeding pairs of this species is likely to be significantly higher than recorded, as good numbers are present throughout the major river valleys, and these are only partly surveyed.

Monthly maxima at sites where counts exceeded 20 are tabulated below:

	J	F	M	A	M	J	J	A	S	O	N	D
Langstone Harbour	42	35	41	14	12	18	15	15	32	29	30	32
Portsmouth Harbour	36	18	36	12					13	27	31	57
Lymington/Hurst	30	30	30	8	8	8	6	8	19	30	30	30
Blashford Lakes	13	28	23	22	20	19	29	61	75	46	14	27
Ibsley/Bickton	22	7							33	5	28	20
Broadlands Estate	7	10	13	8					16	24	22	25
Stockbridge/Fullerton	39	49	45							76	46	52
Testbourne Lake								7	22	12	12	9
Eversley GP	2	3	2	6			22	27	21	8	4	3
WeBS count totals	295	247	283						258	339	262	299

The only other sites with counts above 15 were either at the coast, e.g. Chichester

Harbour (17, Nov 5th) and Lakeside Holiday Village, South Hayling (15, Jan 16th and Nov 14th), or in the Itchen and Test Valleys, e.g. Itchen Woodmill (19, Dec 5th), Chilbolton (18, Jan 1st), Leckford (31, Oct 16th), Longstock (28, Oct 16th) and Marsh Court (17, Sep 28th).

A total of 91 breeding pairs was recorded as follows: Blashford Lakes, 14; Longstock (R. Test), Itchen Valley CP, Eversley GP, 7 each; Romsey Water Meadows, 5; Lymington-Hurst, Saddlers Mill Romsey, 4 each; IBM Lake Cosham, Warsash, Broadlands Estate, Fishlake Meadows (Romsey), Timsbury (R. Test), Houghton (R. Test), Bourley Reservoir, 3 each; Needs Ore, Longparish (R. Test), Basingstoke Canal (Greywell), 2 each; Milton Reclamation, Brown Loaf NF, Casbrook Common, Timsbury Lake, Huntage Copse (Soberton), Holywell House (Soberton), St Clair's Farm Pond (Corhampton), Frith End SP, Springhead (Greywell), Sherfield Links GC, The Vyne Lake, Stratfield Saye, 1 each.

Great Crested Grebe   *Podiceps cristatus*

A moderately common resident, passage migrant and winter visitor.

The total of 42 breeding pairs was slightly down on the 51 recorded in 2004, but numbers at the main wintering sites were similar. As usual, records were also received of up to three birds passing the main coastal watchpoints during the spring.

Monthly maxima at sites where counts exceeded 20 are tabulated below:

	J	F	M	A	M	J	J	A	S	O	N	D
Chichester Harbour	22	15	37	28	11	4		1	2	1	32	34
East Hayling	16	15							2	1	8	22
Sandy Point/Hayling Bay			22	24	3	1	1	1		1	6	10
Langstone Harbour	21	46	29	23	26	8	31	66	60	110	68	99
Southampton Water	30	14	26	5	1				7	21	31	30
Pitts Deep/Hurst	12	12	6	4	7		4	6	7	12	25	16	26
Blashford Lakes	52	59	76	54	60	64	70	52	60	67	61	66
Fleet Pond	13	13	25	23	18	16	15	19	27	34	34	14
Yateley GP	15	22	26						18	20	22	26
WeBS count totals	**211**	**206**	**316**						**201**	**250**	**281**	**305**

A total of 42 breeding pairs was recorded as follows: Blashford Lakes, 13; Somerley Park, 10; Wellington CP, 4; Ewhurst Lake, 3; Frankenbury (R. Avon), Breamore Mill (R. Avon), Hale Park Bridge (R. Avon), Houghton (R. Test), Mottisfont (R. Test), Rooksbury Mill (R. Test), Lakeside CP, Fleet Pond, Camp Farm GP, Eversley GP, Yateley GP, Stratfield Saye, 1 each.

Red-necked Grebe   *Podiceps grisegena*

A very scarce but regular winter visitor and passage migrant [Amber].

A series of long-staying individuals meant that a minimum of just eight birds generated a total of 53 records, most from the south-east or south-west coastal areas.

In the early year one or two were seen, intermittently, in the Sandy Point/Black Point area between Jan 13th and Mar 25th, with another off Tanners Lane on Mar 6th and nearby off Lepe on Mar 8th.

In the late year, a first-winter was regularly seen in the Lymington/Hurst area from Oct 15th-Dec 11th and, unusually, was heard calling at dusk on the latter date. A moulting adult bird was off Hurst Castle on Oct 31st, and one was again in the Chichester Harbour/Sandy Point/Black Point area from Nov 5th-Dec 22nd. Further singles were seen off Hill Head/Brownwich on Dec 4th-6th and Needs Ore on Dec 15th.

Slavonian Grebe   *Podiceps auritus*

A scarce winter visitor and passage migrant [ET, SPEC 3, WCA 1, Amber, HBAP].

Another poor year with a total of 124 records received, most coming from three areas: Chichester Harbour/Hayling Bay, Needs Ore and Lymington/Hurst. The annual bird-month total of 81 was similar to that for 2004, which was the lowest for over a decade.

The peak count in the early year was of 11 in Hayling Bay on Feb 28th, with no more than five recorded elsewhere. The last spring records were two off Normandy Marsh on Apr 2nd and one off Hill Head on Apr 8th.

The first returns were two in Langstone Harbour on Oct 8th. The peak counts in the late year were of five in the Lymington/Hurst area on Dec 7th and four at Needs Ore on Dec 11th.

Monthly maxima from the main sites and approximate monthly totals are shown below:

	J	F	M	A	M	J	J	A	S	O	N	D
Black Point/Hayling Bay	7	11	9									2
Langstone Harbour	5	2	2						2		1	2
Lepe/Needs Ore	2	5	4								3	4
Lymington/Hurst	2	4	5	2								5
Approximate monthly totals	**17**	**22**	**20**	**3**	**0**	**0**	**0**	**0**	**0**	**2**	**3**	**13**

The only other records were of singles at Dibden Bay on Jan 11th and Hill Head on Apr 8th.

Black-necked Grebe   *Podiceps nigricollis*

A scarce winter visitor and passage migrant; rare in summer but has bred [WCA 1, Amber, HBAP].

A total of 75 records was received, most referring to the regular wintering flock in Langstone Harbour.

In the early year, numbers in Langstone Harbour peaked at 18 on Mar 9th. Elsewhere, up to six were at Needs Ore/Lepe, and singles were seen, intermittently, in the Lymington/Hurst area between Jan 2nd-Mar 22nd and off Sandy Point on Apr 1st. The last spring record at the coast was one in Langstone Harbour on Apr 9th, while one was seen at an inland site on Apr 28th. There were no records in the late spring or early summer period indicative of breeding.

The first returns were two at Ibsley Water on Aug 25th, with one remaining there until 29th. The first returning bird in Langstone Harbour arrived on Sep 21st, and numbers there built up to a peak of 17 from Dec 14th onwards. The small group at Needs Ore/Lepe peaked at six on Dec 20th. Elsewhere, singles were seen at Eversley GP from Oct 7th-9th, Ibsley Water on Oct 22nd, Chichester Harbour on Nov 5th and off Hurst on Dec 13th.

Monthly maxima at the main sites and approximate monthly totals are tabulated below:

	J	F	M	A	M	J	J	A	S	O	N	D
Langstone Harbour	14	16	18	1					1	7	14	17
Lepe/Needs Ore	1	6	5	1							4	6
Approximate monthly totals	**16**	**23**	**24**	**4**	**0**	**0**	**0**	**2**	**1**	**9**	**19**	**24**

The annual totals 1996-2005 of bird-months for the scarcer grebes are shown below. Numbers of Red-necked and Slavonian Grebes are showing a steady decline, possibly due to the recent run of mild winters, while Black-necked Grebe totals have remained fairly constant:

Bird-months	1996	1997	1998	1999	2000	2001	2002	2003	2004	2005
Red-necked Grebe	37	24	15	12	8	11	9	13	7	8
Slavonian Grebe	168	156	113	87	89	120	116	84	79	81
Black-necked Grebe	195	91	106	112	106	137	136	119	124	122

Fulmar
Fulmarus glacialis
A scarce passage migrant, most frequent in spring and early autumn.

A good year with a total of 73 records received, mostly referring to birds moving along the coast during spring passage.

The only sightings in the early year were a dead and oiled individual at Hurst Castle on Jan 8th and one moving east off Sandy Point on Mar 7th. Spring passage began on Mar 21st and peaked in late April, with counts of 12 moving east off Hurst on 23rd and 13 there and 12 off Sandy Point on 26th.

Summer records included six off Hurst on July 24th (*cf* Gannet). Numbers then declined with just three in September and the last of the year off Hurst on Nov 11th.

J	F	M	A	M	J	J	A	S	O	N	D
1	0	8	56	39	4	11	6	3	0	1	0

Sooty Shearwater
Puffinus griseus
A rare vagrant (0,9,1).

One flew east past Stokes Bay at 0826 hrs on Oct 21st (TFC).

Manx Shearwater
Puffinus puffinus
A scarce, but in some years locally moderately common, passage migrant.

Only 13 records were received, and the annual total of 65 is the lowest since 1998. All sightings were made from Hurst Beach and related to birds moving through Bournemouth Bay between late April and late July.

The first spring records were six moving east off Hurst on Apr 24th, followed by a further two east and one west on 26th and one west on 28th. In May, a total of 14 moved west on 5th, followed by eight west on 13th and six west on 20th.

Summer records included five moving south out of Bournemouth Bay on June 4th, with three west on 10th and seven moving north into the bay on 12th. A total of 12 on July 20th consisted of seven moving north in the morning and five moving south in the evening, possibly indicating movement of the same individuals in and out of the bay.

The approximate monthly totals are shown below:

J	F	M	A	M	J	J	A	S	O	N	D
0	0	0	10	28	15	12	0	0	0	0	0

The annual totals 1996-2005 of bird-months are as follows:

Bird-months	1996	1997	1998	1999	2000	2001	2002	2003	2004	2005
Manx Shearwater	50	34	13	68	125	686	308	669	192	65

Balearic Shearwater
Puffinus mauretanicus
A very scarce but regular passage migrant, mostly late summer/autumn (0,60,2).

Two, possibly three, individuals were recorded in the Hurst/Milford on Sea area in the latter half of July. The first, a rather dark individual, was feeding off Hurst Beach between 0820 and 1020 hrs on July 17th (TP, MPM, M&ZW). The following day, presumably the same bird was off Hurst from 0749 hrs (MPM, RM). A second, paler individual was seen at 0847 hrs and then both were observed in Milford Bay at 0909 hrs (MPM). What was presumably the original rather dark individual was seen close in as it moved south-east along Hurst Beach at 1853 hrs on July 23rd (MPM). The next morning this bird was seen again at close range off Milford on Sea on three occasions between 0924 and 0933 hrs; it or another was then seen heading south-east towards the shingles at 0952 hrs (MPM).

Shearwater sp. *Puffinus sp.*

Records of unidentified small shearwaters (either Manx *P. puffinus* or Balearic Shearwaters *P. mauretanicus*) were received as follows: one was off Hurst Beach on July 26th (MR) and a flock of seven flew west off Pennington Marsh towards Hurst Narrows at 1830 hrs on Aug 16th (DR).

Storm-petrel *Hydrobates pelagicus*

A very scarce visitor, usually appearing after autumn gales (1,147,4).

Gales in early November produced a total of four sightings in the eastern Solent. At Hill Head, one moved east at around 1400 hrs on Nov 2nd (RAK, BSD) and another moved west there at 0840 hrs the next day (RM, BSD). Later that day, at nearby Stokes Bay, two singles flew east between 1350 and 1525 hrs (GC).

Leach's Storm-petrels © *Dan Powell*

Petrel sp. *Hydrobates/Oceanodroma sp.*

One was seen at long range off Pennington Marsh from Hurst Castle at 1002 hrs on Nov 3rd (MPM) and on the same day another bird, considered to be a Storm-petrel by one observer, was off Milford on Sea at 1025 hrs (MDD, MC).

Leach's Storm-petrel *Oceanodroma leucorhoa*

A very scarce autumn and winter visitor, usually appearing after gales (11,140,18).

Late autumn gales produced a small influx into the county of around 18 individuals. Presumably the same individual was observed off Hordle Cliff and Milford on Sea on Oct 30th (EJW, StB). Further birds were recorded on Nov 3rd, with three singles west off Milford on Sea between 0925 and 1300 hrs; four east into The Solent at Hurst Castle between 0735 and 0855 hrs; then singles west there at 1225 and 1552 hrs; six west and one east off Hill Head between 0935 and 1240 hrs (RM, AG, JDG, MJP *et al*); singles moving east in Stokes Bay at 1130 and 1157 hrs, the former being blown over a nearby playing field (JAN, PNR) and two west there between 1350 and 1525 hrs (GC). Finally, on Nov 6th, one was blown east past Hurst Castle at 1035 hrs (MPM, GCS) and another was there at 1227 hrs (MPM).

Gannet *Morus bassanus*

A moderately common passage migrant and non-breeding summer visitor; scarce but increasing in winter.

This species is well-recorded, with 195 records submitted for 2005. Typically, the largest numbers were seen off Hurst in summer, although increasing numbers of birds are penetrating into The Solent.

The year opened with small numbers seen off Hurst and Sandy Point through January and February, peaking at 18 off the latter site on Feb 20th. Numbers then built up steadily through March and April, with 116 off Hurst on Apr 28th being the first three-figure count.

The regular feeding flock off Hurst consistently reached monthly peaks of between 100-200 birds from May through to early September. However, the largest movement there occurred on July 25th, with 568 recorded, of which 454 moved east.

Unusually high numbers were seen moving in late October, including 159 (139 west and 20 east) off Hurst on 21st, 96 west off Sandy Point on 24th, 186 west off Hurst on 30th and 35 east inside The Solent at Stokes Bay on the same date. These birds evidently moved on rapidly, as just two were seen in November and December.

Monthly maxima at the main localities are tabulated below, followed by the 1996-2005 annual county totals, again based on monthly maxima at the main sites:

	J	F	M	A	M	J	J	A	S	O	N	D
Sandy Point/Hayling Bay	5	18	30	20	43	10	28	15	12	96		2
Gilkicker/Hill Head				7	18	23	8	10	6	35		
Hurst/Milford on Sea	5	0	8	116	152	117	568	211	167	186	2	

Annual total	1996	1997	1998	1999	2000	2001	2002	2003	2004	2005
Gannet	583	534	501	835	997	3336	875	1227	1608	1918

Cormorant *Phalacrocorax carbo*

A moderately common non-breeding resident, passage migrant and winter visitor [Amber].

The highest counts of the year were all recorded at the Langdown/Hythe roost on MOD land with 98 birds on Jan 16th, 106 on Aug 30th, an unusual date for the highest count of the year, and 100 on Dec 28th. No records were received in 2005 from Horse Sand Fort in the Solent that held the largest roost in 2004. Inland, early in the year Yateley GP held 82 birds on Feb 12th and in the last quarter 90 birds on Dec 28th. Two birds of the form *sinensis* were present at Ibsley Water/Mockbeggar on Mar 19th but it is likely that many more go undetected or unreported throughout the county. The monthly maxima at sites where counts exceeded 30 are tabulated below:

	J	F	M	A	M	J	J	A	S	O	N	D
Black Point area	26	61	56	3	1			1	5	5	7	19
Langstone Harbour	34	50†	49†	26†	12	20	44	56	80	42	43†	60†
Portsmouth Harbour	50	25	42	7					48	34	49	62
Titchfield Haven	25†	31†	22†								32†	32†
Bury/Eling/Lower Test	94†	62	38	9	3	8	17	27	19	17	38	48
Langdown/Hythe	98†	86†	32†	35†	3†	12†	76†	106†	35†			100†
Blashford Lakes	59	68	81	31	23	26	40	40	71	78	58	57
Petersfield Heath Pond	33	21	2		2		1	1	1	2	8	28
Testw'd Lakes/Broadlands	32	16	9	3					18	32	57	28
Alresford Pond	19	20	9	4	4	2	11	8	17	16	31†	14
Fleet Pond	59†	55†	27†	23†	5†	9†	12†	22†	28†	52†	55†	54†
Yateley GP	52†	82†	73†						7	32	86†	90†
WeBS count totals	**513**	**362**	**389**						**342**	**415**	**463**	**525**

† = night roost

Traditionally the largest roosts in Hampshire have been on man-made structures or islands. Although this species has been found roosting in trees in the Avon valley as far back as 1959, the count of 52 birds roosting in poplars *Populus sp* at Bickton on Dec 18th is of note.

2003 correction: the monthly maximum for Alresford Pond should read 28, not 70.

Shag   *Phalacrocorax aristotelis*

A scarce winter visitor and passage migrant [Amber].

The 90 records submitted more than doubled the number in 2004 and included two double-figure counts, both of birds moving east past Sandy Point; 18 on Aug 24th and 21 on Oct 23rd. Horse Sand Fort was visited in February and held no birds; it is possible and likely that a double figure roost still exists in the Solent.

Monthly maxima for the main sites are tabulated below:

	J	F	M	A	M	J	J	A	S	O	N	D
Sandy Point	0	0	1	1			1	18	9	21		1
Langstone Harbour/Eastney	2	2	2	1						1	1	3
Southsea Castle	1										1	2
Hurst/Barton on Sea	1		1	3	6				3	6	1	0

The only records distinctly away from the above sites are listed below.

Chessel Bay, Itchen Valley: 1, Dec 12th.
Hill Head: 1, Nov 11th.

Approximate monthly totals:

J	F	M	A	M	J	J	A	S	O	N	D
6	4	3	4	10		3	18	12	28	6	6

Bittern   *Botaurus stellaris*

A very scarce but regular winter visitor [ET, SPEC 3, WCA 1, Red, NBAP, HBAP].

The number of sightings of this elusive winter visitor declined marginally for the second year running. Titchfield Haven recorded the most sightings and was the only site to report two individuals present. Birds were seen up to Apr 17th and again from Nov 19th. Other individuals were found at the now annual sites of Farlington Marshes, Fawley Reservoir and Alresford Pond.

All records are summarised below.

Farlington Marsh: 1 on 4 dates, Jan 15th-Feb 26th.
Titchfield Haven: 1-2, Jan 1st-Apr 17th; 1-2, Nov 19th-Dec 31st.
Hook-with-Warsash: 1, Jan 13th.
Fawley Reservoir: 1, Feb 28th and Mar 10th.
Pylewell Lake: 1, Jan 24th.
Testwood Lakes: 1, Dec 4th.
Alresford Pond: 1, Jan 29th and Mar 5th.
Bransbury Common: 1, Dec 3rd.
Testbourne Lake: 1, Dec 6th.

Approximate monthly totals:

J	F	M	A	M	J	J	A	S	O	N	D
4-6	4-5	3-4	1-2							1-2	4-5

Little Egret   *Egretta garzetta*

A widely observed winter visitor and passage migrant; some non-breeding individuals over-summer. The small breeding population continues to increase [ET, Amber, HBAP].

Little Egret   © Russell Wynn

Throughout the year the nocturnal roost site at Wade Court/Langstone Mill Pond held the highest numbers. As usual, the autumn roost counts were the largest and peaked at 249 on Oct 13th - a new county record (JCr). Fifty percent of these birds arrived from Langstone Harbour at dusk as shown by a count of 125 leaving towards this roost on Oct 4th. Numbers utilising the major river valleys and elsewhere inland throughout the winter months appear to be stabilising, with the exception of the Alresford area from where fewer records were received, particularly data regarding the Arlebury roost.

Monthly maxima at sites where counts regularly exceeded 10 are as follows:

	J	F	M	A	M	J	J	A	S	O	N	D
Chichester Harbour	6	21	10	15	9	12	12	11	47	75	55	7
Wade Court/Langstone Mill Pond†	92	85	28	47	30	22	67	188	224	249	82	56
Langstone Harbour	3	2	15	10	13	22	57	70	63	125	71	8
Portsmouth Harbour	29	17	17	16	1			51	69	50	45	15
Titchfield Haven	1	2	2	7	15	6	10	5	7	14	7	1
Hamble Estuary	2	1	6	3	2	1	11	23	18	16	10	3
Lower Test Marshes	4	5	10	8	6	13	14	12	21	13	11	6
Fawley†/Ashlett Creek†											19	15
Needs Ore/Beaulieu Estuary	15	3	7	14	16	17	44	49	35	19	10	23
Sowley Pond†	16		7						22	8		25
Lymington/Hurst	33	26	23	11	13	11	22	29	35	44	18	30
Bickton†	40	30	46								28	45
Allington GP†/Highwood Reservoir	17†	20†				2	6	23	23	9†	10†	21†
Alresford area	6	7	7					1	1	7	1	2
Mislingford†	32	5+									21	
Basing†	23†	16	10	1								1

As a breeding species within the county, the recent expansion continues with a minimum of 50 pairs recorded from four sites, an increase from the three known locations in 2004. The Fort Elson colony continues to hold the largest breeding population with 25-30 pairs in 2005.

2004 additions: A roost near the Hen and Chicken fishing pond in the Wey valley held six on Jan 24th, seven on Feb 11th and 13th, five or six on Mar 23rd, five on Apr 1st, and the last one at nearby Alton on Apr 13th. In the late year, roost counts were not made and the only records were of single birds in the area on Oct 23rd and Nov 19th.

Great White Egret   *Ardea alba*

A rare vagrant (0,13,0)

The colour-ringed individual found at Mockbeggar Lake on Nov 5th 2004 remained in the area until Jan 23rd. It was most frequently observed on Ibsley Water but was also seen adjacent to the River Avon at Ibsley on Jan 2nd, 9th and 23rd. It returned on July 17th and was recorded almost daily until Dec 7th. The presence of a second unringed bird was suspected; there was an unconfirmed report of two on July 29th and two were seen together at Ibsley Water on Oct 11th (LC).

Grey Heron   *Ardea cinerea*

A moderately common resident, passage migrant and winter visitor. It is widely observed but a local, colonial breeder.

The highest count from the first quarter was 53 birds at RNAD Frater by Portsmouth Harbour on Feb 7th, probably relating to nesting pairs establishing themselves at the Fort Elson colony. Post breeding concentrations were noticeable in the Avon valley with Ibsley Water/Mockbeggar holding 53 birds on July 10th and 66 birds on Aug 10th; Blashford Lakes having 59 birds on July 16th, 79 on Aug 21st and 53 on Sep 25th. A percentage of these birds at both sites will be from the two known heronries in the Avon Valley at Midgham and Somerley Park. Fishlake Meadows, Romsey, produced the highest count in the last quarter with 31 birds on Dec 24th.

Monthly maxima at sites where counts regularly exceeded 30 are as follows:

	J	F	M	A	M	J	J	A	S	O	N	D
Blashford Lakes	16	8	6	5	5	41	59	79	53	28	17	12
Fishlake Meadows	36						12	35		24		31
WeBS count totals	146	100	58						166	134	139	89

Coverage of the known breeding colonies for the BTO Heronry Census was good in 2005, with the exception of the Fawley Refinery area, that could not be accessed to establish whether the colony in that area had moved elsewhere or abandoned the area totally. All of the colonies appear stable given the fickle breeding habits of this species and difficulties of accurate nest census. The small Fleet Pond colony first noted in 2002 continues to grow annually.

Heronry Census	2005	2004	2003
Tournerbury, Hayling Island	9-12	18	11-13
Fort Elson, Gosport	40-60	85	83-87
Midgham Wood	28-47	23-42	42
Fawley Refinery	Not assessed	0	15
Sowley Pond	15	12	8
Holt Pound Inclosure	9	14	12
Fleet Pond	6	3	2
Kettlesbrook, Steep	3	2	2
Heronry Census (continued)	**2005**	**2004**	**2003**
Arlebury Park	6-8	14	8-12
Elvetham	15-25	25	23
Somerley Park	15-18	3-5	11
Efford	5	5	7

2003 correction: There were 12-13, not 8, nests at Arlebury Park. The record of 1 nest at Highbridge should be deleted.

White Stork   *Ciconia ciconia*

A rare vagrant (5,29,4).

It was a very good year for sightings, with four individuals recorded. One circled over Titchfield Haven and left east at 1550 hrs on Apr 27th (VL, BC). One was seen moving north-west over the west end of Portsdown Hill from 0935-0940 hrs on June 25th (KWM) and almost certainly the same bird was seen later that day in a flooded field at Upper Titchfield Haven (MDR). One flew north over the Rose Bowl stadium, Hedge End at 1130 hrs on Aug 20th,(ASJ), and finally one was over Lower Exbury at 1440 hrs on Nov 26th (MRa).

Spoonbill   *Platalea leucorodia*

A scarce visitor, most frequent in spring and autumn but recorded in every month (20+,126, 12) [ET, SPEC 2, WCA 1, Amber].

There was an increase in the number of birds in Hampshire compared with 2004, though when viewing numbers from the previous decade, it was an average year with between eight and 11 recorded.

In the early year an adult was at Titchfield Haven from Mar 13th-Apr 2nd. Another adult was at Needs Ore from May 15th; this was joined by a first-summer on June 18th and both remained until 26th. The first-summer relocated to the Lymington/Hurst area, where it was present from June 29th-July 5th, before reappearing at Needs Ore from July 16th-29th and Aug 12th-16th. It was possibly this bird that was seen flying east over Farlington Marshes on Aug 17th. This or another individual was back at Needs Ore on Aug 30th. Two presumed passage migrants were present at Titchfield Haven on Aug 31st/Sep 1st, departing south after being flushed by an Osprey. Titchfield Haven also held a further individual from Sep 11th-23rd. Possibly the same individual seen at Hurst Castle on Oct 19th was later recorded as a juvenile over Hurst Beach on 23rd, at Needs Ore on 24th and Nov 1st and at Lower Test Marshes on Oct 27th. The final records of the year came from Farlington Marshes where three juveniles were recorded on Oct 30th, Nov 3rd and 4th, then rising to four (possibly the wandering late October bird), the last sighting of the year on Nov 9th.

Annual totals are shown below:

Annual total	1996	1997	1998	1999	2000	2001	2002	2003	2004	2005
Spoonbill	21	3	7	4	9	29	18	11-14	4-6	8-11

Honey Buzzard   *Pernis apivorus*

A very scarce summer visitor and passage migrant (ET, WCA 1, Amber, HBAP).

The earliest recorded was one flying north over Pennington Marsh on May 6th. The first was noted in the New Forest the next day. Further migrants were a pale phase west over Fishlake Meadows, Romsey on 8th; then two north at Milford on Sea and later an adult north at Hurstbourne Tarrant on 9th. Late passage was evident with a male west over Emsworth Channel and Farlington Marshes on May 27th and one west over Milford on Sea on June 8th.

Successful breeding was noted in the county with five pairs raising ten young; one or two further pairs held territory throughout the season but did not breed. In addition, one pair occupied a territory early in the season but not subsequently; two or three other individuals were also present for much of the summer. There was a further record of one in a separate area on July 8th.

There were three post-breeding dispersal or return migrant records - all singles: a juvenile north-west over Sholing, Southampton on Aug 7th, south over Old Winchester Hill on Aug 23rd, and over Dibden Bay on Sep 5th. There were five early October records - again all singles: an adult south over Titchfield Haven on 1st; another south-west over the Meon valley on 2nd, with possibly the same at Needs Ore mid afternoon; one east over Sandy Point on 3rd and the last at Bridgemary, Gosport on 8th.

2004 additions: Approximately 17-19 individuals were present in the county during the breeding season; five pairs were successful in rearing a minimum of nine young. Three other pairs held territory, one of which definitely built a summer nest.

Red Kite   *Milvus milvus*

A scarce and increasing visitor that is becoming more frequent due to the presence of released birds in southern England, and now an established but very scarce breeder (ET, SPEC 2, WCA 1, Amber, HBAP).

There were 150 widespread reports from across the county, with most sightings occurring in May, when at least twenty individuals were present; the peak count was again recorded at Faccombe with eight on Dec 13th. Seven untagged adults present at Five Lanes End, Greywell (hunting with Buzzards over new mown hay) on May 30th is presumably evidence of colonisation of the north-east from the Chilterns release scheme (MJP). Breeding occurred at one location for the third successive year and at least two young were fledged. Display flight was observed at another site. Birds were present in other likely breeding areas throughout the spring and summer, but no further reports of breeding success have been received. Analysis shows that around sixteen inland site clusters or "ranges" (up to 10km in radius) were involved during the year, with five additional occurrences at coastal sites. These sightings are summarised below:

	J	F	M	A	M	J	J	A	S	O	N	D
Sites/Range	4	4	9	11	13	7	4	5	3	2	7	4
Individuals	9	8	12	10	20	9	7	7	4	5	7	12
Bird-days	19	14	26	18	40	19	20	7	6	8	12	13

It remains extremely difficult to make an accurate estimate of numbers. The request made in HBR 2004 to submit details of age and wing-tagging with each record is again emphasised. Only two reports stated that there was no evidence of wing-tagging. Such information is extremely helpful - all observers are requested to note in their submissions "no wing-tags", if they are fortunate enough to obtain good views. Only one observer reported wing-tagged birds, with two seen together in the east of the county on May 30th.

The picture emerging is of a small resident population augmented both by spring arrivals, possibly of continental origin, and mid-winter dispersion from neighbouring counties' successful reintroduction programmes. Of note is the first breeding in modern times in Sussex in 2004. All details of movements and ageing of individuals submitted are summarised below; details are withheld if it is thought that the location currently holds a breeding site, or is a possibility for future colonisation.

In the early year there were four or five winter territories, with at least nine individuals involved with a count of five at one site. Additionally, one was north of Fleet Pond on Jan 20th/21st and an adult was also at Stephens Castle Down on the latter date.

Possible new arrivals included one north-east over Rownhams on Mar 21st and a first-summer over Itchen Valley CP on Apr 2nd. Between Apr 24th and May 10th six singles were involved in a northerly movement: Botley Wood on Apr 24th and another at Gosport 35 minutes later; Tweseldown on May 1st; Hurst Castle on 6th; Hook on 7th and Denmead on 10th. A first-summer was at Romsey Water Meadows on May 22nd. A possible return migrant was over Langley on Aug 17th. The only individual to be aged in the late year was a probable juvenile at Allington GP on Nov 26th. There were again four winter territories in the late year.

Marsh Harrier   *Circus aeruginosus*

A scarce visitor, most frequent in spring and autumn but occasional in mid winter; has bred once, in 1957 (ET, WCA 1, Amber).

There were 72 bird-days with 19 individuals estimated in spring, 20 in autumn and one in each winter period. Notes provided on age, time of observation and direction of movement were essential in establishing the estimates above and observers are both congratulated and encouraged to continue the good work. There was naturally some variation in estimates of age, and even gender, and so a note on the time of observation is most helpful in carrying out the assessment of records. Although there were no long-staying individuals, in late March an adult male was at Itchen Valley CP, also a male and female lingered at Titchfield Haven for a six-day period into April; a juvenile was at Farlington Marshes from Oct 29th to Nov 1st and a first-winter appeared at Titchfield Haven in late December.

In January an adult female was seen over Keyhaven Marsh and, presumably, the same on the east side of the Lymington river on 21st. There were no further records until the start of spring passage with the arrival of an adult male at Itchen Valley CP on Mar 20th. Up to 20 individuals were reported in spring, notably a female well inland at Danebury on May 7th and another female, or immature, at Westend Down on Apr 21st. The last of spring was at Titchfield Haven on May 22nd.

The first juvenile was noted at Farlington Marshes on July 31st, with the last of autumn passage also there on Nov 1st. A female, or immature, was again at Westend Down on Sep 7th and was perhaps the spring individual returning. The first-winter at Titchfield Haven from Dec 22nd to the end of the year was the first to over-winter in the county since 2002/03.

Monthly totals are shown below followed by a list of all records:

	J	F	M	A	M	J	J	A	S	O	N	D
Birds	1	0	6	11	3	0	1	5	7	6	1	1
Sites	2	0	4	8	2	0	1	4	6	4	1	1
Bird-days	1	0	10	14	3	0	1	6	7	21	1	7

Coastal Sites:
Chichester Harbour, Emsworth: male NW, Apr 1st.

Chichester Harbour, East Hayling: immature male W, Apr 4th; juvenile W Sep 1st.

Langstone Harbour/Farlington Marshes: 2, male W and later female E, Mar 31st; female from E, Apr 21st; juvenile, July 31st; juvenile male W, Aug 4th; adult male SW, Sep 12th; 1W, Sep 23rd; adult male, Oct 10th; juvenile, Oct 29th/31st.

Gosport, Bridgemary: 1W, Apr 24th.

Gilkicker Point: female, Sep 3rd.

Titchfield Haven: female/immature, Mar 23rd; 2, immature/female and male, Mar 30th/31st and 1 to Apr 3rd; immature/female, Apr 16th/17th; immature/female, May 3rd; female N, May 22nd; 1, Aug 3rd; 1, Sep 4th; immature, Oct 9th/21st; first-winter, Dec 22nd/28th.

Calshot: 1W, Sep 10th.

Needs Ore: immature/female, Aug 21st and 23rd.

Lymington/Normandy/Keyhaven Marsh: adult female, Jan 21st; first-summer male N, Apr 29th; adult female W, Aug 7th; juvenile SW, Aug 19th; 1, Oct 14th; female, Oct 27th.

Hurst Beach: adult male N, Mar 22nd; first-year N, Apr 2nd; first-year N, Apr 22nd.

Barton on Sea: female W, Oct 5th.

Inland Sites:

Danebury: female, May 5th.

Beaulieu Road NF: 1E, Apr 24th.

Itchen Valley CP: male on three dates, Mar 20th-31st.

Westend Down: female/immature N, Apr 21st; female/immature S, Sep 7th.

Hen Harrier   *Circus cyaneus*

A scarce winter visitor and passage migrant (ET, SPEC 3, WCA 1, Red, HBAP).

Numbers were lower in the winter of 2004/05 with up to 14 individuals (eight males, six females/immatures) in the county compared with 22 in 2003/2004. In the second winter period numbers were lower still with a peak of five males and seven ringtails.

There were 135 records: 62% originated from the New Forest; 14% from the Wealden heaths;10% from the South Downs; 8% from coastal sites and the remaining 5% from the west and north-west of the county - a total of 192 bird-days. In the early year just two nocturnal roosts were discovered, both were well-watched. A New Forest roost peaked at six as follows: an adult male, another male (possibly adult), a female, two third-year males and a juvenile. In a Wealden heath roost there was a peak of three: one male and two ringtails. Away from these sites there were the following reports: two further males elsewhere in the New Forest on several dates in January and February; a ringtail along the South Downs on four dates; a male at Sowley on Jan 17th and a ringtail at Hannington on

Feb 4th. In March a ringtail was at Martin Down on 1st and one was at Faccombe on 23rd. The last of the early year in the New Forest were two males to roost on Apr 4th: on the eastern heaths a male was seen until 12th. A ringtail lingered on the South Downs until May 2nd and the last of the early year was a male arriving from Dorset and east over Hurst Beach on 14th.

The first return was a ringtail to the South Downs on Sep 29th. A juvenile was hunting over the Langstone Harbour islands before it left high north-west and two returned to a New Forest roost on Oct 4th. A ringtail flew south over Barton on Sea on Oct 16th; another was in the Martin Down area on 17th, followed by a male over Hurst Beach on 27th and a ringtail flew west over Emsworth Channel on Nov 11th. There were further late year coastal records: at Langstone Harbour a female arrived from the east and departed south-west on Nov 30th, one was at Hurst Castle on Dec 8th and a ringtail was at Needs Ore between 11th and 15th. There were three roosts: six at the New Forest roost included three adult males, an adult female and two further ringtails; the Wealden heath roost held just one ringtail; a male into the Alresford Pond reedbed on Dec 31st was the only record of the year from this previously occupied roost. The monthly totals are tabulated below together with the male/ringtail splits:

	J	F	M	A	M	J	J	A	S	O	N	D
Sites	7	5	7	3	2				1	7	7	8
Bird-days	35	27	26	5	2				1	26	28	42
Ringtails	5	6	6	2	1				1	6	5	7
Males	8	6	5	2	1					4	5	5
Approx County Total	**13**	**12**	**11**	**4**	**2**				**1**	**11**	**11**	**13**

Montagu's Harrier   *Circus pygargus*
A very scarce passage migrant and summer visitor (ET, WCA 1, Amber, HBAP).

An above average year with nine individuals reported. The first of spring was an adult pair observed moving high north along Hurst spit on May 2nd (MPM,TP). A first-summer female moved north-west over Emsworth on 14th (CBC) and a male flew north over Bishop's Dyke NF on 28th (TP). There were no reports of breeding although an adult male was seen in likely habitat at two locations on May 30th and a female was in the same area quartering fields on June 14th. A ringtail in the north-west of the county on Aug 23rd may have been on passage although may also have related to breeding outside the county. An adult female was at Cheesefoot Head on Sep 22nd (GHH) and the last, a ringtail, was at Little Posbrook, Titchfield Haven, on 29th (RJC).

2003 additions: A female or first-summer male was near Bolderwood on May 25th (SGK) and an adult female was seen from a view point in the north of the New Forest on May 31st (DR); the latter on the same date as a melanistic female was seen at Keyhaven. These records were submitted punctually and inadvertently omitted from the 2003 report.

Goshawk   *Accipiter gentilis*
A very scarce resident (WCA 1).

There was a total of 59 reports relating to 60 bird-days; just six reports were from outside the New Forest. Six pairs bred successfully in the New Forest, of which five were known to have raised at least nine young; a male was present at a seventh site. Successful breeding pairs have increased in number each year since the first documented breeding successes in 2002. Elsewhere there was one report from a possible breeding site.

From the watch-point reports submitted, the majority of New Forest bird-days was in the first half of the year (47) and the maximum count, made on four dates, was three. Elsewhere, one was at a site in the north-east on Apr 11th and another was over Hill Head on June 27th. One was in the Southampton Water area on Aug 16th and, possibly, the same individual again on Sep 22nd. September and October were the only blank months in terms of New Forest records submitted, but one was reported from a site in the north-east of the county on Oct 29th. There were no reports of juveniles in the late year. There follows a summary of bird-days across the county, excluding reports from breeding as summarised above:

J	F	M	A	M	J	J	A	S	O	N	D
2	9	13	5	9	9	2	4	1	1	3	2

Sparrowhawk   *Accipiter nisus*

A common resident, passage migrant and probable winter visitor.

In the New Forest fieldwork coverage for the breeding monitoring programme was similar to recent years. Twenty five occupied sites were found, with twenty three pairs proceeding to egg laying or beyond. Of these twenty three nests, four failed (three during incubation and one with four dead chicks around two weeks old). Productivity was 3.2 young per successful nest or 2.5 for all nests found. Just 46 young were ringed, 27 males and 19 females (59 in 2004). The breeding pair count was up on the very poor total last year, but productivity was lower and both were below the long term average of the last two decades. At present no single cause can be pinpointed for this decline with confidence. However, remains of three Sparrowhawks where predation by Goshawk was suspected have been found in recent years. Timber felling also meant that some sites were less suitable. If this alone does not limit breeding, it is making location of nesting attempts harder for fieldworkers, as pairs may occupy atypical sites which traditionally would never have been searched (*per* AGP).

Breeding reports from elsewhere totalled 22 territorial pairs, of which 11 were confirmed to have bred, including seven reports where young were seen or heard. BBS data indicated detections in 17% of the squares surveyed – the same as the ten-year average.

No reports of migratory movements were received, but there were four reports from locations where records of bird-days were kept on a monthly basis throughout the year as follows:

	J	F	M	A	M	J	J	A	S	O	N	D
Sandy Point, Hayling Island	1	2	14	14	7	4	6	26	18	30	14	5
Titchfield Haven	12	9	18	15	13	14	23	30	16	17	15	19
Regents Park, Southampton	10	3	5	3	2	5	10	10	9	4	4	-
Normandy, Lymington	2	3	2	1	0	1	4	5	5	3	6	7

Observers with similar data for regularly watched sites are encouraged to provide this information so that both seasonal variation and long-term trends can be monitored.

Buzzard   *Buteo buteo*

A common resident, passage migrant and probable winter visitor.

During the year there were 90 reports from 58 sites of five or more in any count; double-figure counts from 33 of these are reported below. There were reports of 125 breeding territories in the county, the most ever recorded and this, probably, a considerable under-estimate of the total number. Detections were made in the majority of BBS 1-km squares for the first time (53%, double the ten-year average and nearly three times greater than the % occupancy ten years ago). There were 87 occupied territories in the New Forest - astonishing considering the poor showing in 2004 (35) and easily exceeding the previous highest number recorded in 2003 (59).

In the early year large soaring groups included 12 at Cheesefoot Head, with seven further east at Old Winchester Hill on Jan 23rd (17 there on Feb 2nd), ten at Farley Mount

Buzzard, Normandy Marsh, October 2005 © *Marcus Ward*

on Jan 25th and 20 at Faccombe on Feb 8th. The New Forest early winter survey maximum count was 48 from 15 sites on Jan 22nd. In March there were counts of ten over Brandsbury Common on 2nd and Pipers Wait, NF on 12th, 12 at Linkenholt and ten over Butser Hill on 22nd. The last week of March included some possible group movement, with a peak of 16 north of Titchfield Haven on 25th, the same count over the Pennington/Keyhaven Marshes the same day, 16 at Cheesefoot Head on 26th and 14 at Bramley Heath on 28th. There were several movements across the coastal strip in April and May - all singles unless stated otherwise: north over Havant on Apr 3rd; over Langstone Harbour/Farlington Marshes on six dates between Apr 3rd and May 29th, east off Barton on Sea on May 9th; six north from the Isle of Wight across the West Solent later on 9th, north over Hurst Beach on 11th and east over Alverstoke on 15th..

In the New Forest, within the 87 occupied territories, 72 pairs were known to have bred successfully, producing 122 young (brood statistics for 65 pairs were 8x3, 31x2 and 26x1). Four nesting pairs failed to produce young and six more nests were occupied but there was no incubation. This, the best year ever, reflected a good seed crop the previous autumn and a consequent high density of small mammals, combined with a large rabbit population (*per* JMT). There were ten territories in Somerley Park, Avon valley (13, 2004). Other territories were reported as follows: 14 widely distributed across the north-east of the county; four along the South Downs (surely under-recorded); three in the upper Test valley, two at Botley Wood and Hayling Island. First breeding was noted at Hayling Island (two young) and at Thornhill Park (1 young). There were eight other reports of territories with fledged young - totalling 16 (1x3, 6x2, 1x1). A further seven reports of single territories across the county have not been classified above. In addition, through the summer there were day counts to a maximum of 12 at Romsey Water Meadows and six around the Beaulieu estuary.

In the early autumn large groups included eight at Titchfield Haven on Aug 21st and 12 sighted inland from Keyhaven reedbed on 28th, 12 at Testwood Lakes on Sep 4th (14 on 25th) and 12 at Lyeway, Ropley on 24th. Movements noted were five north-west at Bitterne on Aug 28th, one north-east over Farlington Marshes on Sep 9th (also one west, Sep 18th), three south over Keyhaven/Pennington Marshes on Sep 17th and five high south there on Sep 21st.

In the late year ten were at Old Winchester Hill on Oct 22nd and Nov 15th. Movements noted were one west over Farlington Marshes on Oct 9th (also one east there on 15th) and four south-east over Keyhaven/Pennington Marshes on Oct 22nd (also one south on Dec 3rd). The New Forest late winter survey maximum count was 19 from 11 sites on Nov 19th.

Osprey   *Pandion haliaetus*

A scarce passage migrant.

It was an above average year with 110 records received, registering 105 bird-days between Mar 19th and Nov 7th. In spring there was a probable 28 individuals. Up to three were present on several dates through the summer: two, on one date, and a single later appeared to be engaged in building a late summer nest. In autumn there was a probable 18 individuals, at least four of which were juveniles.

The first of the spring was at Tanners Lane, Lymington on Mar 19th, followed by two singles on 24th at Marwell Trout Fishery and Woolmer Pond. There were four further sightings in March, with one at Pennington Marsh and then Oxey Marsh on 25th, at Waterlooville on 28th, Ibsley Water on 30th and Langstone Harbour on 31st. April arrivals, all singles, were as follows: Fleet Pond on 5th which left north-east in the evening; Heath Pond, Petersfield 6th; Lower Test Marshes and then Broadlands Lake, 9th/10th and, probably, the same at Longstock later on 10th; Lakeside, Eastleigh, 11th; Titchfield Haven, 13th/14th and another north early morning 16th; north-east over Keyhaven Marsh, 14th; Langstone Harbour from 14th to 17th; west Longparish, 17th; Mottisfont, 22nd; north over Woolmer Pond, 24th; Lower Test Marshes, 26th and north-east over Ellingham meadows, Avon valley on 30th. In May there were nine arrivals: at Titchfield Haven, one on 1st and another south-west on 8th; north-east over Hayling Oysterbeds on 6th; Chichester Harbour then west over Langstone Harbour on 9th; one over Ibsley Water left north-west on 15th; a first-summer was at Fleet Pond on four dates from 16th to 20th and close by at Potbridge on 22nd; one left east from Needs Ore on 17th; one fishing in Langstone Harbour left east and was seen both at Hayling Oysterbeds and in Chichester Harbour on 17th and the last of spring was one north at Itchen Valley CP on 21st. The first-summer at Fleet Pond on June 6th/7th was presumably the individual from mid May; this or another was at Romsey Water Meadows on June 13th, Alresford Pond on 22nd and Old Winchester Hill on 24th.

On July 22nd one was at Redbridge and was, probably, the individual faithful to Lower Test Marshes from July 26th to Aug 7th, also returning to Redbridge on three dates and at Calshot on July 27th.

A very early juvenile was in Langstone Harbour and, occasionally, Chichester Harbour from Aug 18th-26th and one unaged also from 28th-31st. One roosted overnight at Itchen Valley CP Aug 18th/19th and was possibly the individual responsible for sightings at Needs Ore, Inchmery, Exbury between 21st and 24th and at Normandy on 26th. One was also fishing at Keyhaven Marsh on 22nd but was seen heading west over Hurst narrows on 23rd. The only autumn bird at Titchfield Haven was fishing there on Sep 1st. A second juvenile was at Langstone/Chichester Harbour on Sep 4th and 6th, seen off East Hayling, Mill Rythe, Northney and Farlington Marshes. One at Needs Ore on Sep 5th was joined by a second on 6th. An adult in moult was out to sea off Hurst Castle on 9th. Reports from Langstone Harbour and Hook-with-Warsash on 13th could have been the same individual. A juvenile roosted overnight on 15th/16th at Itchen Valley CP and was, possibly, the same reported from the Beaulieu estuary and the Sowley area on 17th, with additional sightings on 20th, 21st and 26th. One was over Botley Wood on Sep 18th. Two were off East Hayling on Sep 23rd; a juvenile there on Oct 6th was, probably, also responsible for the sighting on 13th; the last there was between 26th and 29th. At another site one was present from Oct 18th to Nov 7th.

Monthly totals for probable number of individuals, number of sites visited and bird-days are as follows. Details of one site have been withheld:

	J	F	M	A	M	J	J	A	S	O	N	D
Probable Number			7	13	9	2	2	6	11	3	1	
Sites			7	13	11	4	6	9	12	4	1	
Bird-days			7	20	16	5	10	15	18	17	7	

2004 addition: one was at Tundry Pond on Sep 24th.

Kestrel   *Falco tinnunculus*
A common resident, passage migrant and winter visitor (SPEC 3, Amber).

The BBS data indicate 28% of squares occupied at a mean of 0.4 per square, both figures close to the ten-year average. Breeding was reported from 25 sites with 32 territories noted, of these at least 14 pairs bred successfully, raising at least 28 juveniles. Coastal movements were apparently very scarce: two east at Taddiford Gap on Apr 23rd; one north-east off the sea at Sandy Point on 24th and in autumn one high south over Langstone Harbour on Sep 16th. At well-watched sites there was a marked increase in bird-days at both Sandy Point and Normandy between August and October; no obvious changes at Farlington Marshes (mostly three to four present) and monthly bird days at Titchfield Haven, ranging between a low of seven in January and a high of 27 in July.

Red-footed Falcon   *Falco vespertinus*
A rare non-breeding summer visitor and passage migrant (2,48,1).

A first-summer male was discovered at Martin Down at 1500 hrs on June 18th (ASMS, RB, ISE). It was found independently at 1845 hrs by A&JDS and was reported at 2100 hrs in a tree at nearby Bockerly Dyke. It was not seen there again. The record has been accepted by *BBRC* and is the 51st for Hampshire. The most recent was a female at Acres Down on May 31st 2003.

Merlin   *Falco columbarius*
A scarce winter visitor and passage migrant (ET, WCA 1, Amber, HBAP).

As last year there were 330 records - a significant increase compared to the recent past. There were probably rather fewer individuals than 2004 in the early year, with 30 estimated throughout the period. There were rather more, 43 individuals present, or arriving, in the late year; most of these were moving through with around 17 wintering.

In the early year there were records from the Avon valley (a male on Jan 23rd and a female on two dates): up to five across the New Forest with a roost of two males and a female on Feb 19th. One on the South Downs on eight dates to Mar 12th, probably, involved sightings of the same male. Single date records came from another eight inland localities. Along the coast a male was at Black Point on Feb 15th, probably the same on 28th but another on Apr 14th. There were 46 bird-days recorded in Langstone Harbour until Apr 20th - at least two individuals were involved, an immature male and a first-winter female. Two at least were also recorded at Titchfield Haven with singles on eight dates between Jan 23rd and Apr 28th, including a male on Feb 16th and a female on Apr 6th. There were three records in February from Southampton Water, involving both a male and female. Just three records were received from Needs Ore - one on Feb 15th and a female on Feb 27th and Mar 8th. Between Sowley Pond and Hurst Castle there were 26 bird-days, probably, involving two individuals, an adult female and a female/immature until Apr 19th. In addition, one moved north over Hurst Castle on Mar 11th; another off Hurst Beach on Apr 28th and the last of the early year in off the sea, then north on May 15th.

The first return was at Pennington Marsh on the earliest county return date ever - July 30th (S&SL) - followed by two in August; an immature male at Langstone Harbour from 12th into September and a juvenile at Ibsley Water on 22nd. On the coast in September, one moved high ENE over Farlington Marshes on 2nd and a juvenile male was there on 14th; both were different to the resident bird remaining from August. A female/immature was at Hurst Castle on Sep 8th and the same, probably, at Barton on Sea on 10th with another at Needs Ore between 8th and 13th and the same, probably, at Park Shore on 21st. Inland, in autumn a male was at Marwell Zoo on Sep 1st; one was over Fleet Pond moving NNE on Sep 17th and another was at Over Wallop on Oct 3rd. The first return in the New Forest was a male on Oct 8th, then a female on 20th. Arrivals in mid October, all singles unless stated otherwise, were widespread: Lower Test Marshes on 13th, with two south there on 16th; Titchfield Haven area from 14th; east at Sandy Point on 15th; Kings Somborne and another at Eling on 16th; Old Basing on 17th; Westend Down and another at Tanners Lane on 19th and three were around Hurst Castle on 25th. New arrivals in November included a juvenile at Sinah Common on 3rd, one east at Sandy Point and another at Riverside Park on 5th, one over Sweare Deep, Chichester Harbour, on 8th, and two females north/north-west over Langstone Harbour on 15th. Inland, one was at Itchen Valley CP on 3rd, possibly the same at Chandlers Ford on 14th and another on Brandsbury Common on 31st. In December, new arrivals included one at Chilbolton Airfield on 5th; Eastleigh SF on 10th (although this could be the individual seen at Itchen Valley CP and Chandlers Ford in November), Beaulieu Road NF on 17th/18th and Ashmansworth Downs on 20th and South Warnborough on 22nd. A female or immature was at Northney on Dec 17th and 31st and was perhaps that seen at Sweare Deep in November or possibly the same at Langstone Harbour islands on Dec 29th. The approximate monthly totals together with monthly bird-days are summarised below:

	J	F	M	A	M	J	J	A	S	O	N	D
Sites	14	16	10	6	1		1	2	6	14	16	15
Estimated Number	14	18	10	6	1		1	3	5	18	20	17
Bird-days	46	44	46	16	1		1	14	22	53	46	37

Hobby   *Falco subbuteo*

A moderately common summer visitor and passage migrant (WCA 1, HBAP).

The first arrivals of spring were on two early dates: Bishops Waltham Moors on Apr 2nd and Bolderwood, NF on 3rd. One flew north over Fareham and another was at Testwood Lakes on Apr 10th; the next was near Basingstoke on 13th. There were four on Apr 21st: singles at Basing and Woolmer Pond and two north over Hayling Bay - the first noted off the sea. From Apr 25th onwards arrivals were virtually on a daily basis as the weekly bird-day counts show below:

	April				May				June			
Weekly Period	1-7	8-14	15-21	22-28	29-5	6-12	13-19	20-26	27-2	3-9	10-16	17-23	24-30
Sites	2	2	3	8	15	20	20	13	16	6	8	13	8
Bird-days	2	2	4	8	33	58	26	19	20	6	10	16	8

Groups of three or more were as follows, all in May unless stated otherwise: three, Woolmer Pond on Apr 29th and five there on 1st; four, Ibsley Water on 6th with 15 feeding there on 9th, 7 on 10th and 12 on 11th (some, or all, moved north on 9th thus the counts on the 11th, at least, involved new arrivals); four, at one New Forest site on 15th and three at a second on 25th; finally three were seen at Romsey Water Meadows on 27th. Ibsley Water appears to be a migratory stopover point for this species offering spectacular views.

At least 16 territories in the county were occupied and at least twelve pairs bred successfully. In the New Forest seven breeding territories were reported with successful breeding strongly suspected at five sites.

Along the coast, the first juvenile was in the Lymington/Hurst area on Aug 23rd; another at Farlington Marshes on Sep 16th, then four in October: Testwood Lakes on 2nd; Gilkicker Point on 3rd and Titchfield Haven on 5th and 9th. In September, there were 31 bird-days at coastal sites from a county total of 61, with three at Fawley Refinery on 9th. There were 21 bird-days (16 on the coast) in October - the last sightings were as follows: in the north-east at Tundry Pond on Oct 2nd; on the South Downs at Old Winchester Hill, with two on Oct 13th, and, along the coast, the last two for the county at Taddiford Gap on Oct 16th.

Peregrine   *Falco peregrinus*

A scarce but increasing resident; numbers are augmented throughout the year by visitors from neighbouring counties. Probably also a passage migrant and winter visitor from further afield (ET, SPEC 3, WCA 1, Amber).

Just over 500 records were submitted, with reports coming from all areas of the county. A total of 675 bird-days, excluding breeding sites, was logged and the only evidence of seasonal variation was a low in the summer months. This may suggest that several breeding in adjacent counties establish non-breeding territories within Hampshire.

Breeding was reported from three sites with one pair raising two young and two failing, one at the egg stage and one when the young died in the nest. Breeding behaviour was noted at three other sites but no evidence of nesting has been presented.

Estimating how many individuals were present in the county remains extremely difficult: the following account is based on notes supplied with records. On East Hayling, both an adult male and female were seen in the early year and an immature on one date; in the late year, an adult female roosted in Emsworth channel. Of the 675 bird-days recorded, 217 were at Langstone Harbour and, with the detailed field notes submitted (JCr, CC), it is possible to establish that at least ten individuals were involved. In the early year an adult pair, at least one first-winter male and first-winter female were present. In addition, a pale adult female first noted in 2004 frequented the Langstone Harbour islands. Two adult males were together on Apr 25th, and then two second-calendar year females were seen together on May 9th. A juvenile female was first seen on the early date of May 29th; thereafter a juvenile male was first seen on July 9th, then two were hunting together on Sep 7th when they took a Bar-tailed Godwit.

An adult female and juvenile were at Chilling on Aug 6th and an adult male was nearby at Titchfield Haven on Nov 28th. A juvenile was observed hunting in Southampton from July 17th. In the early and late year, an adult pair and second-calendar year female were noted in the Lymington area. A juvenile male was off Barton on Sea on Oct 5th.

Inland in the Avon valley, first-winter females were noted in the early and late year, with the latter, presumably, first seen on Sep 18th and noted as a juvenile. A pair and first-winter were in the New Forest during the early year, an adult male in summer and an adult pair in the late year. On the South Downs, there were at least four individuals during the year: three adults in the early year, two males and a female; an adult female in August; an adult male and a first-winter in the late year. An adult and juvenile were to the east of Fishlake Meadows, Romsey, on June 17th and an adult female was nearby at Testwood Lakes on Sep 18th. In the Itchen valley, below Winchester, an adult male was present in the early year and in mid summer. In the north-east, at Crondall, an adult pair was present in September, with a male seen later on three dates. A male was also at Drayton on Sep 9th. The table below summarises monthly bird-day totals for localities where sightings averaged at least two per month: breeding data have been excluded:

	J	F	M	A	M	J	J	A	S	O	N	D
Chichester Harbour (Hants)	6	8	7	4	3	1	1	1	1	2	4	1
Sandy Point	3	4	6	4	2			1		1	1	1
Langstone Harbour	12	16	19	19	20	22	14	12	33	20	17	16
Portsmouth Harbour	1	3		7	1	1		4	2	1	6	2
Southampton Water	8	6	8	3	1	1	9	1	2		4	1
Beaulieu Estuary/Needs Ore	5	5	1	1			1	2	2	4	1	5
Lymington/Hurst	7	4	3	1	2	1		5	7	7	8	13
Normandy	2	4	3	1	2	1		5	7	7	6	13
Avon valley	4	5	1	1		4	2		2	3	4	2
New Forest	13	8	6	1	9	4		1	1	5	11	3
South Downs	8	5	2	1	1		1		3	1	11	6
COUNTY TOTAL (Bird-days)	70	71	62	50	51	38	31	34	59	67	74	68

In addition to the sites listed above, annual bird-day totals were recorded as follows: Titchfield Haven, 19; Hamble estuary, 8; lower Itchen valley, 19; lower Test valley, 20; Test valley above Romsey, 8; north-east, 22 and a total of 33 bird-days from another 16 localities.

The table below summarises annual totals of bird-months for selected raptors in the ten-year period 1996-2005, excluding any breeding birds:

	1996	1997	1998	1999	2000	2001	2002	2003	2004	2005
Honey Buzzard	6	3	14	8	124	11	16	16	17	24
Red Kite	50	26	37	83	48	59	73	50	91	99
Marsh Harrier	32	34	27	34	55	62	30	51	32	41
Hen Harrier	97	102	113	112	89	54	75	96	90	78
Montagu's Harrier	9	5	5	6	7	3	16	5	4	9
Osprey	35	36	36	74	50	55	56	35	70	53
Merlin	83	64	70	69	83	94	86	108	141	113

Water Rail   *Rallus aquaticus*

A scarce resident, moderately common passage migrant and winter visitor (Amber).

There were reports from 58 sites during the year, with a peak monthly count of 115 in November. The highest site count in the early year was 12 at Fleet Pond in January/February; then in the late year, at least 40 at Titchfield Haven. The number of sites with reported birds and monthly maxima were:

	J	F	M	A	M	J	J	A	S	O	N	D
Sites	21	17	14	6	7	8	8	7	6	26	24	22
Monthly maxima total	44	42	27	7	15†	36†	14†	16	11	73	115	67

† Total includes territorial pairs at breeding sites but not juveniles

First breeding was noted at Fishlake Meadows, Romsey (two pairs, one brood of two juveniles) on May 13th, with the next at Stephill Bottom NF (pair, four juveniles) on 18th. There was a further 17 territories reported: two territories, juveniles noted, Farlington Marshes; seven territories, young seen, Titchfield Haven; four territories, Lower Test Marshes; pair with young, Needs Ore; two pairs and a juvenile, Pennington Marsh; one territory, Longstock and summer records from another five sites.

In November there were at least 40 calling at Titchfield Haven: 23 at Fishlake Meadows, Romsey; eight, both at Fleet Pond and in the upper Test valley between Houghton and Fullerton; five at Normandy and four, both at Hook-with-Warsash and Calshot.

Spotted Crake   *Porzana porzana*

A very scarce passage migrant (most frequent in autumn) and winter visitor; has bred (?,128,3).

At least four were recorded. One was heard calling at night at one site from Apr 30th-May 5th. Another, or the same, was heard nearby on May 8th. In autumn one was at Farlington Marshes from Aug 21st-23rd (HV, JCr et al, photo): discussion of the photographic evidence suggests that this bird was a possible adult although aging is particularly difficult in this species. There were further sightings on Aug 27th, Sep 4th and 9th - presumably of this individual. An almost certainly different individual was seen during ringing activities in the reedbed at dawn on Sep 10th (PJL). One from Sep 17th/21st (JCr) was aged as a first-winter on 19th (GC & DLB) and last seen on 23rd (MG). Another was present at Hook-with-Warsash on Sep 24th (RKL).

2002 addition: one was singing at an undisclosed site around May 1st.

2003 correction: the bird at Hook Lake from Aug 12th-15th was not the first at the site. An adult was there on July 20th 1992 but the location was not disclosed in the report for that year (DAC).

2004 addition: one was singing at an undisclosed site on May 1st.

Corncrake   *Crex crex*

A rare passage migrant; formerly bred (?,61,1).

One crossed a path at the Kench at 0930 hrs on Sep 17th (TAL); it was not relocated despite a thorough search. This observation was the first since 2001.

Moorhen   *Gallinula chloropus*

A numerous resident and winter visitor.

The highest count of the year was 105 at Farlington Marshes on Dec 29th. Monthly maxima at sites where counts exceeded 50 are as follows:

	J	F	M	A	M	J	J	A	S	O	N	D
Langstone Harbour	34	7	31	30	24	23	47	91	75	44	35	105
Portsmouth Harbour	48	61	65	27					14	49	56	72
Sopley/Avon Causeway	54	17	10						8	6	38	
Avon above Fordingbridge	60	61	50						159	93	94	99
Stockbridge/Fullerton	30	33	48							52	42	31
Yateley GP	32	19	24						21	27	42	55
WeBS count totals	**530**	**454**	**443**						**618**	**608**	**497**	**632**

Other maxima greater than 25 were recorded at: Titchfield Haven (45, Dec 23rd); Broadlands Estate (42, Oct 15th and Dec 5th); East Hayling (28, Jan 15th); Testbourne Lake (28, Dec 6th); Lower Test Marshes (28, Jan 15th); Tundry Pond (27, Sep 18th) and Hamble Estuary (26, Oct 15th).

Counts of breeding territories in double-figures were 33 at Titchfield Haven and 15 in the Lymington/Hurst area. BBS data indicated 25% of squares occupied at a mean detection of 0.6 per square – which is the same as the previous ten-year average.

Coot

Fulica atra

A common resident and winter visitor.

The late year counts at Blashford Lakes were close to last year's totals and sustain the site as one of national importance for this species. Monthly maxima at sites where counts exceeded 100:

	J	F	M	A	M	J	J	A	S	O	N	D
Portsmouth Harbour	104	54	50	24					18	11	34	49
Blashford Lakes	1572	775	343	202	116	184	472	699	1421	1858	1827	1807
Stockbridge/Fullerton	123	113	108							229	221	182
Wellington CP	54	107	78		10				30	31	42	25
Dogmersfield Lake	110	55	21					65	103	136	145	102
Eversley GP	223	128	88						173	183	196	208
Yateley GP	129	106	99	30					171	187	216	245
WeBS count totals	**3015**	**2030**	**1446**						**2610**	**3229**	**3383**	**3410**

Nationally important number 1730

Other peak counts greater than 60 were recorded at: Alresford Pond (98, July 16th); East Hayling (93, Jan 15th); Langstone Harbour (85, Aug 20th); Broadlands Estate (75, Nov 6th); Testbourne Lake (67, Dec 6th); Anton Lakes (64, Oct 16th); Allington GP (62, Dec 5th) and Avon above Fordingbridge (62, Dec 18th). Highest ever counts were recorded at Cams Hall, Portsmouth Harbour (55, Nov 20th) and Petersfield Heath Pond (46, Nov 30th).

Double-figure counts in the breeding season were recorded at: at Longstock (34 broods); Titchfield Haven (18 breeding territories); Stratfield Saye Park (12 pairs) and Wellington CP (10 pairs). BBS data indicated 17% of squares occupied at a mean detection of 0.9 per square – the same as the previous ten-year average.

Common Crane

Grus grus

A rare vagrant (0, 89+, 0).

2004 additions: one flew north over Needs Ore Point at 1610 hrs on Apr 13th (BG *et al*); what was presumably the same bird was seen to arrive from the south at Bransbury Common at 2020 hrs, circle and apparently drop into roost (KP, DAT). It was not seen the next morning, due to thick fog, but was reported by the local keeper as coming into roost on Apr 14th. A sub-adult arrived over Needs Ore from the south-west at 1410 hrs on May 11th; it circled for five minutes before continuing to the north-east (AH, AM, PO, MAO, PW). These are the first records since 2000, when one was seen over the Avon Causeway on Dec 26th.

Black-winged Stilt

Himantopus himantopus

A rare vagrant (5,15,1).

A male was watched feeding at the Fishtail Lagoon, Pennington Marsh, on the evening of June 21st. (AFB) The record has been accepted by *BBRC* and is the 21st for the county. This may well have been the bird present at Denge Marsh, Kent on June 22nd/23rd.

Collared Pratincole
Glareola pratincola

A very rare vagrant (0,2,1).

One was seen in flight over Farlington Marshes at 1315 hrs on May 1st before landing on the Deeps. It remained there until 1355 hrs when it was flushed, along with all the other birds, by a Peregrine. It was chased by the falcon but took effective evasive actions. It was, however, lost to view over the harbour islands to the east, and was not relocated (RAC, JCr photo). The record has been accepted by *BBRC* and is the third for Hampshire. The only previous occurrences were at Fleet Pond on May 28th 1974 and Ibsley Water/Mockbeggar Lake on May 6th 1987.

Collared Pratincole   © Jason Crook

Oystercatcher
Haematopus ostralegus

A moderately common breeding resident, common passage migrant and winter visitor [Amber].

The wintering population at the end of 2004 stayed fairly constant into January, when the count was similar to the previous year. The February total was noticeably down on 2004, but thereafter spring counts were higher. Numbers around the coast in September were generally higher than 2004 and the population in December was also up.

Approximately 126 possible passage migrants were reported, mainly from the east of the county, between Mar 7th and May 15th. The peak passage was of 35 east at Sandy Point on Mar 10th and a notable 21 north inland over Testwood Lakes on Apr 17th.

A total of 132 breeding pairs was reported from the following sites: Sandy Point and Sinah Common, 1 each; Hayling Oysterbeds, 5; Langstone Harbour islands, 41; Farlington Marshes and surrounding minor islands, 5; Titchfield Haven, 6; Warsash area, 5; Beaulieu estuary, 17; Pitts Deep to Hurst, 49 and Ibsley Water and Testwood Lakes, 1 each. This total is up from 107 in 2004 – mainly due to an increase in the Hurst area and reporting from the Beaulieu estuary sites.

Only six were reported on autumn passage, including inland records of three north-west over Fleet Pond on July 23rd and one heard in fog at Petersfield on Aug 2nd.

Monthly maxima at the main high tide roosts are tabulated below:

	J	F	M	A	M	J	J	A	S	O	N	D
Chichester Harbour	965	864	664	291	345	282	765	1221	1465	698	1013	1574
East Hayling	150	127	70	107	19	41	9	198	397	238	160	151
Langstone Harbour	1416	1174	488	502	381	336	780	1437	2040	1645	1376	1277
Portsmouth Harbour	499	189	236	43					277	257	545	606
Titchfield Haven	143	130	120	109	103	95	92	165	188	170	160	159
Hamble Estuary†	36	131	214	130	17		233	265	324	305	252	316
Southampton Water	895	466	657	240	59	40	112	40	330	509	470	1013
Needs Ore/Beaulieu Estuary	276	247	189	214	207	71	266	123	139	138	350	124
Lymington/Hurst	178	184	132	52	84	84	134	100	194	67	73	88
WeBS count totals	**3486**	**2438**	**2089**	**1178**	**741**	**469**	**1315**	**1670**	**3027**	**2866**	**2962**	**3364**

† = low tide counts

Avocet
Recurvirostra avosetta

A scarce passage migrant and winter visitor; bred for the first time in 2002 [ET, WCA 1, Amber].

Records of this species continue their dramatic increase of recent years. The approximate number of bird months, based on monthly maxima, was 166. This is an increase of more than 50% over the figures for the last two years.

After the recent successes, breeding was rather disappointing this year. It was only reported from the second of the two recent nesting sites, where three pairs attempted to breed. Of these, one pair raised four young. The nest of another pair was trampled by a Canada Goose. The Avocets were seen removing the broken eggs.

Birds on spring passage were reported from Sandy Point (4 E, Apr 2nd) and Hurst (5 NE, May 15th). In the autumn at Hurst spit, there were 4 E on Aug 31st and 5 W on Sep 30th.

Monthly maxima at sites with regular records:

	J	F	M	A	M	J	J	A	S	O	N	D
Langstone Harbour	14	10	4	2	1	1	1		5	2	18	20
Titchfield Haven/Hook		1	4	6	7	10	8		1			
Needs Ore/Park Shore		1	1	3	1	2	3	4	1	2	6	5
Lymington/Hurst			4		3	1	3	1	1	1	2	8

Away from sites mentioned above, one was reported from Sowley Marsh on Jan 3rd and Feb 13th; the latter was probably also seen at Tanners Lane on the same date, also one there on Mar 20th. One was seen at Fawley on July 27th.

Stone-curlew
Burhinus oedicnemus

A scarce summer visitor [ET, SPEC 3, WCA 1, Red, NBAP, HBAP].

The first arrival recorded by RSPB survey staff was on Mar 14th. A grounded migrant was present for much of the day at Sandy Point, Hayling on Mar 30th (ASJ *et al*, photo).

During the breeding season, 25 pairs were located, of which 23 were proved to breed and made a total of 29 breeding attempts. A total of 12 chicks was ringed which subsequently fledged; three further juveniles of unknown origin were recorded in post-breeding flocks. These were noted by RSPB survey staff at three sites during August-October with a maximum total of 54 on Sep 28th. The latest recorded by them was on Oct

9th but other reports indicate that birds remained at one site until early November. Other reports, no doubt involving the same birds, were of 18 at one site on Sep 1st and 17 nearby on Sep 29th.

Away from the core breeding area, there were reports of one heard at 1315 hrs on Apr 30th at a former breeding site, another heard at 2130 hrs on May 1st in suitable habitat, one which was seen from a moving vehicle as it flew across a road and landed on the heath near Beaulieu Road, NF on June 13th, one heard over the observer's garden in Overton on June 19th, and one heard over Holmsley Ridge, NF on Sep 18th.

Little Ringed Plover   *Charadrius dubius*
A scarce summer visitor and passage migrant [WCA 1].

The first sighting was one at Farlington Marshes on Mar 16th, with singles at Lower Test Marshes and Woolmer Pond two days later. Six were reported from Blashford Lakes on Mar 20th; thereafter there was passage reported until around the middle of May. Away from potential breeding sites, passage was mainly coastal, with records of ones and twos from Hurst, Lepe, Titchfield Haven and Farlington Marshes. Inland passage records came from Winchester SF (two, Mar 23rd) and Fleet Pond (singles on Apr 2nd, May 14th, 15th and 17th).

A total of 17 territorial pairs was located, compared to 22 last year. Some of these did not nest, but a total of 16 young was seen at five sites, the same as 2004.

Return passage and dispersal was apparent from the end of May, although breeding activity was still being recorded into August. Ibsley Water had double-figure counts through most of July, with a peak of 26 (including 12 juveniles) on 23rd (LC), the largest gathering ever recorded in the county. Elsewhere, the only double-figure count was ten at Keyhaven Marsh on July 24th. Passage continued until mid September, including five at Ibsley on 2nd. However, the latest ever record for the county was of a single at Lower Test Marshes on Oct 29th (SSK), 18 days later than the previous latest occurrence of Oct 11th 1980.

Ringed Plover   *Charadrius hiaticula*
A moderately common breeder, common passage migrant and winter visitor [Amber].

Numbers at the beginning of the year were up on 2004, mainly due to higher counts in the Langstone/Hayling area. Numbers were also up during the second winter period, apart from in December when the main roosts were not located at Langstone Harbour during the WeBS count. Apart from wintering birds tabulated below, there were singles inland at Black Gutter Bottom, NF on Feb 7th and Ibsley Water on Feb 6th and Mar 20th, plus flocks of 200, 130 and 100 on the seafront at Sandy Point on Oct 20th, Nov 19th and Dec 16th respectively, presumably disturbed from Black Point.

Coastal spring passage went almost unreported this year, apart from four west at Browndown on Apr 24th and one at Sandy Point on May 11th. However, a total of 28 were seen to move inland from Langstone Harbour on five dates between Apr 28th and June 6th and other groups of migrants were present on several dates between mid May and mid June including two of the northern race *C. h. tundrae* (or "Tundra Ringed Plover") at Hayling Oysterbeds on June 13th. Singles were seen at Ibsley Water on four dates in May and eleven bird-days were recorded at Woolmer Pond between Mar 18th and May 15th. One was at Testwood Lakes on May 16th, two were at Winchester SF on May 24th and a single (presumably a late migrant) was at Woolmer Pond on June 10th.

About 43 breeding pairs/territories were reported as follows: 8, Hayling Oysterbeds; 12, Langstone Harbour RSPB islands; 2, Sinah Common; 1 each at Hook and Marchwood; 2, Pitts Deep; 12, Lymington/Hurst; 6-7, Hurst Castle and 1 at Ibsley Water. This is down from the 51 reports of last year. Breeding success was reported from Hayling Oysterbeds, where one pair fledged two young, and Ibsley Water, where one pair raised three young, while single young were raised at both Marchwood and Normandy.

Most inland autumn records came from Ibsley Water, with monthly maxima of six, 16 and 14 between July 2nd and September 24th; on the last date 13 left south at 0700hrs. Other records were of one south-west at Lakeside on July 29th, one at Winchester SF on Aug 5th and one or two there from Aug 26th-Sep 1st, three at Woolmer Pond from Aug 25th-27th and up to five at The Vyne Watermeadow between Sep 13th and 18th.

Monthly maxima at the main high tide roosts are as follows:

	J	F	M	A	M	J	J	A	S	O	N	D
Chichester Harbour	157	117	52	44	118	13	46	400	266	250	227	122
Black Point	157	104	52	5	10	0	46	148	266	250	227	112
Langstone Harbour	226	170	105	22	22	40	16	85	52	78	230	1
Portsmouth Harbour	23		1		1			12	8		11	41
Titchfield Haven area	63	50	70				5	50	99	55		95
Hamble Estuary	8	22	13	4		2	66	80	18	21	26	13
Langdown/Fawley/Calshot	60	76	21	4		2	8	22	36	52	87	10
Beaulieu Estuary		9	30	2	12		15	35	11	37	22	9
Sowley/Lymington		22	32	20			5	100	50	51	12	32
Lymington/Hurst	122	70	78	30	27	18	32	300	250	286	200	167
WeBS count totals	470	349	253	56	62	24	64	429	206	488	513	311

Golden Plover
Pluvialis apricaria

A common winter visitor and passage migrant; very scarce in summer [ET, HBAP].

The build up of numbers on the coast at the end of 2004 continued into 2005, with very large gatherings at the north of Hayling Island and Chichester Harbour during the first two months. A flock of at least 5000, probably 5200, was on the mudflats at Northney on Feb 4th (JCr). This is the largest ever recorded in the county, the previous high counts being of an estimated 4000 at Thruxton airfield on Jan 28th 1993 and 3500 near Winchester SF on Dec 5th 1996. Numbers in the Lymington/Hurst area also increased from the late 2004 influx, with a peak count of 2000 at Pennington on Jan 21st - easily surpassing the previous highest ever site count a month earlier. Late year numbers were also up on 2004, with large flocks in the Hayling area again, reaching a peak of 2800 at Northney on Dec 14th. Inland the counts at Eastleigh and Odiham were substantially higher than 2004.

Monthly maxima at sites where counts regularly exceeded 100 are as follows:

	J	F	M	A	M	J	J	A	S	O	N	D
Chichester Harbour	4000	5000	1310				1	11	18	1150	1273	3586
East/North Hayling	4000	5000	1300						1	1150	1200	2800
Titchfield Haven area	150	403	330						1	85	143	100
HMS Sultan, Gosport	266	55	157								119	20
Hamble Estuary	14	105	270					1	70	317	325	285
Lymington/Hurst	2000	1550	1000	100	1				13	240	440	868
Michelmersh/King's Somborne			12	50						200	300	230
Wide Lane, Eastleigh	350	340	300						15	7	250	395
Hillside, Odiham	455	350	670							120	160	350
WeBS count totals	236	1656	720					1	2	498	889	737

Sites of national importance: 2500+.

Counts exceeding 200 from other areas in the early year were as follows: 400 over Itchen Valley CP on Jan 27th; 375 at Lyeway on Feb 18th; and 240 at Kingsley Sand Pit on Feb 20th. The last sightings of spring were a summer plumaged bird at Keyhaven on May 1st, three at Needs Ore on May 5th and one there on 8th.

The first autumn sighting was of one east at Hurst on the early date of July 3rd. Other July birds were singles at Langdown on 22nd and 20 at Old Winchester Hill on 25th. There were two singles in August and three in early September, before widespread sightings from 16th. Notable counts in the late year, away from the regularly reported areas, were:

81

250 at Bentworth on Oct 8th; 210 SE over Old Winchester Hill on Oct 22nd; 200 near Basingstoke on Oct 28th; 250 leaving Firgo Farm, Longparish to the east on Nov 9th; 340 at Penton Grafton on Nov 22nd; 210 at Newlands Farm, Fareham on Dec 18th; 200 SE over Itchen Valley CP on Dec 25th; 200 at Plastow Green on Dec 27th and 200 at Battle Green, Hambledon on Dec 29th.

Grey Plover   *Pluvialis squatarola*

A common winter visitor and passage migrant, with a variable (usually small) summering population [Amber, HBAP].

In spite of a winter peak in February, numbers in each of the first three months were lower than in 2004. The numbers for the late year showed a welcome reversal of recent trends, with significant increases compared to 2004. This was also reflected in higher counts from the West Sussex area of Chichester Harbour producing the highest combined Chichester/Langstone harbour count since 2001-02.

Spring passage was again light. All records of movements during the year are as follows:

Hayling Bay: 7 E, Mar 21st; 2 W July 29th.
Lepe: 11 E, Apr 1st.
Langstone Harbour: 7 NE, Apr 19th; 8 E, Apr 23rd; 3 E May 15th.
Hurst: 21 E, May 2nd; 10 E May 12th; 12 E May 13th

There were 42 in Langstone Harbour on May 29th decreasing to 17 by June 26th but increasing to 61 by July 23rd; almost all of these were first-summer birds. The first major arrivals of returning adults were in August with 400 at Gutner Point, Chichester Harbour, on Aug 21st and 425 on Farlington Marshes at the high water roost on 22nd. Monthly maxima at the main high tide roosts are as follows:

	J	F	M	A	M	J	J	A	S	O	N	D
Chichester Harbour	1265	1286	1420	68	21	11	67	400	381	854	1168	2026
East Hayling	488	440	556	3			44	400	53	428	858	400
Langstone Harbour	579	701	326	11	48	17	61	425	500	584	658	798
Portsmouth Harbour	17	15	6							7	3	9
Hamble Estuary	45	56	30	2			1	6	14	36	24	32
Dibden Bay/Langdown	80	55	53	11		5		4	31	8	48	62
Needs Ore/Beaulieu Estuary	297	560	54	6	11		9	61	108	154	200	176
Sowley/Lymington	100	150	220	160	52	3	16	150	50	100	70	100
Lymington/Hurst	193	176	150	10	14	8	7	63	106	153	238	228
WeBS count totals	**1800**	**2099**	**1296**	**118**	**65**	**17**	**137**	**609**	**801**	**1363**	**2027**	**1787**

Sites of national importance: 530+. International Importance 2500.

The annual combined harbour peak count is summarised below:

Winter peak	86-91†	91-96†	96-01†	00-05†	00-01	01-02	02-03	03-04	04-05	LY-05
Chichester Hbr	2046	2944	1945	1790	2180	3180	1700	1515	1286	2026
Langstone Hbr	1455	1713	1550	792	671	819	982	1119	701	798
Combined Hbrs	-			**2507**	**2630**	**3541**	**2637**	**2240**	**1987**	**2824**

† mean of annual peak counts over a five year period

Lapwing   *Vanellus vanellus*

A common but decreasing breeder and numerous winter visitor [SPEC2, Amber, HBAP].

Numbers as usual peaked in January when the county total was very similar to 2004. The February and March totals were noticeably higher than 2004 suggesting that more birds remained later in the county. Counts for the last four months of the year were broadly similar to 2004.

Monthly maxima at sites where counts exceeded 200 are as follows:

	J	F	M	A	M	J	J	A	S	O	N	D
Chichester Harbour	2837	3473	1271	30	31	177	303	172	230	927	738	2513
North/East Hayling	828	650	230	6	7	50	83	8	129	179	126	455
Langstone Harbour	1100	807	600	55	40	90	110	185	173	260	600	625
Titchfield Haven area	370	210	126	55	60	40	50	54	54	100	164	400
Hamble Estuary	375	267	35	8	8	8	30	40	30		234	238
Lower Test/Eling/Bury Marshes	1000	1106	42			45	67	51		43	120	680
Needs Ore/Beaulieu Estuary	3815	1447	429	65	92	15	69	145	101	400	1000	1449
Lymington/Hurst	1600	2473	766	14	12	20	93	44	134	395	301	786
Blashford Lakes/Ibsley	220	392	60	20	19	56	270	212	250	429	15	1
Testwood Lakes/Broadlands	61	330	130			40	35		154	100	102	100
Winchester SF	433	19	4						168	250	141	400
Woolmer Pond	35	100	370	10	22	30	111	84		40	70	80
The Vyne Watermeadow	61	262	26	10	8	223	500	514	450	498	325	23
Tundry Pond	207	175	36			4		5		40	40	135
Eversley GP	700	400	250	18	18	82	180	238	300	184	365	540
WeBS count totals	9497	7672	2508	160	260	198	689	642	1483	2384	3004	5496

Apart from the areas listed above, the largest flocks reported were: 650 at Greatbridge, Romsey on Jan 1st; 435 at Pylewell on Jan 8th; 410 at Elvetham Park on Jan 14th; 400 at Hare Warren Farm on Feb 1st; 1000 at Roke Manor, Romsey on Feb 6th; 470 at Paper Mill Farm CB on Feb 14th; 400 at Exton on Mar 9th; 500 at Old Winchester Hill on Mar 14th; 700 near Longparish on Nov 9th; 400 at Roke Manor, Romsey on Dec 18th and approximately 1000 at Winchester SF on Dec 19th.

Breeding is widespread in the county but reports are received from few areas. Extensive survey work in the north-west of the county, around Chilbolton, located 71 breeding pairs in 13 1-km squares. There were 22 pairs reported from Titchfield Haven and 15 from Beaulieu. Elsewhere, a further 62 territories were reported from 20 sites - all with eight pairs or fewer. This gives a total of 155 pairs; the equivalent number in 2004 was 117 pairs. The increase may reflect an increase in recording effort, as the results from the BBS report a different picture. Detections were made in 31% of 1-km squares, down slightly from 2004 but equal to the ten-year average. However, numbers were down at a mean of 6.0 per occupied square, which was substantially lower than 9.4 last year and less than the ten-year average of 7.1.

Lapwing chick   © Richard Ford

Knot
Calidris canutus

A moderately common winter visitor (to the eastern harbours) and passage migrant [SPEC 3, Amber, HBAP].

After the low numbers of last year, counts were more normal with the largest gatherings again being noted in Langstone Harbour where a maximum of 1144 was recorded on Dec 3rd.

Spring coastal passage movements were light with very little diurnal movement being reported, though grounded birds were seen regularly at Keyhaven Marsh with a maximum of 20 on May 13th. The largest apparent movement observed was of 34 seen circling high over Langstone Harbour on Apr 9th. This site also recorded some noteworthy numbers later in April, with 100 on 10th which included a flock of 70 which arrived from the south, and 140 on 16th which included 40 roosting (unusually) with Black-tailed Godwits at Farlington Marshes. At Hurst the only spring passage record was of 10 east on May 10th.

The first double counts of the autumn were 19 at Farlington Marshes on Aug 19th and 89 at Keyhaven Marsh on Aug 21st.

Elsewhere there were records from just four sites: Tipner Lake (1, Feb 25th; 1, Nov 26th, Dec 3rd and 10th); Hill Head (3, Mar 31st; 1, Aug 25th; 4, Aug 31st; 2, Sep 3rd and 1, Dec 10th); Hook-with-Warsash (up to 3 during January and February; 2, Sep 3rd; 1, Sep 20th and 1, Nov 19th and 25th) and Weston Shore (1, Jan 15th and 1, Nov 12th).

Monthly maxima at the main high tide roosts:

	J	F	M	A	M	J	J	A	S	O	N	D
Chichester Harbour	400	421	833	127	22		5	8	6	47	300	400
East Hayling	400	400	310	1					6	15	300	400
Langstone Harbour	800	1100	150	100	8	4	6	25	41	13	50	1144
Park Shore/Beaulieu Estuary	300	590	24		19	9		1	6	4	14	
Sowley/Lymington	250	100	33	47	19			2	8		3	
Lymington/Hurst	29	25	77	5	20	5	7	89	18	4	9	14
WeBS count totals	**441**	**1627**	**473**	**30**	**19**		**4**	**93**	**68**	**55**	**105**	**1144**

Sanderling
Calidris alba

A moderately common passage migrant and winter visitor [HBAP].

An average year with the majority of the wintering population again confined to the eastern harbours, the peak count of 255 being noted at Sandy Point on Jan 22nd.

Spring movements, mostly passage but including inter-coastal movements between Mar 20th and June 11th, were below average and were observed at four sites as follows: Sandy Point (78 E, May 1st-16th); Stokes Bay (54 E, Apr 19th-May 21st and 109 W, Apr 19th-May 2nd); Hill Head (only 48 birds noted moving east, but grounded birds were regular with a maximum of 90 seen on Apr 21st); Hurst Beach (246 E, Mar 20th-June 10th, max 60, May 12th). Other coastal records included: Hook-with-Warsash, four dates with a maximum of four on Apr 1st and Aug 3rd; ten at Park Shore, Needs Ore on May 10th and 14 on 15th and also at Lepe; three at Barton on Sea on May 7th.

The only inland record was of a single bird seen at Ibsley Water on May 22nd.

Monthly maxima at sites where counts exceeded 20 are as follows:

	J	F	M	A	M	J	J	A	S	O	N	D
Chichester Harbour (WeBS)	28	54	77	32	4	4	29	100		48	65	15
Black Point (HW roost)	63	40	100	61		2	47	20	3	159	140	
Sandy Point/Hayling Bay	255	68	150	100	78	50	190	160	160	140	250	152
Eastney/Southsea	100	45	85									
Gilkicker Pt/Stokes Bay			36	39	39						2	154
Hill Head/Brownwich				90	16	18						
Needs Ore/Beaulieu Estuary		5		1	14			9		50	1	7
Hurst/Milford on Sea	4		4	139	221	93	43	14	6	48	3	4
Approx. county totals	**260**	**150**	**220**	**330**	**350**	**160**	**240**	**170**	**210**	**210**	**260**	**310**

Sites of national importance: 210+.

Little Stint   *Calidris minuta*

A passage migrant, very scarce in spring, scarce in autumn and a very scarce winter visitor.

It was an average year with a minimum of 40 individuals recorded, although the presence of long-staying juveniles from mid September onwards makes it difficult to assess the true passage through the county.

In the first half of the year the only records were of two in the Black Point high tide roost on Jan 9th, one there on Mar 10th, 24th and 25th, one in a high tide roost with Dunlin and Ringed Plovers near Hurst Castle on Jan 10th and one at Normandy Lagoon on May 11th.

A protracted autumn passage involved at least 35 at seven sites between Aug 10th and Nov 1st. Records came from Pennington/Keyhaven Marshes (2, Aug 10th; up to 8, Sep 1st-Oct 2nd, max on Sep 20th); Farlington Marshes (1, Aug 23rd/24th; 1/2, Sep 4th-Oct 11th); Hayling Oysterbeds (1, Aug 25th), Titchfield Haven (1, Sep 2nd-6th; 1, Sep 11th; 1, Sep 25th), Black Point (1, Sep 14th/15th; 1, Oct 28th-30th; 2, Nov 1st) and inland at Ibsley Water (1, Aug 10th, 1/2, Sep 23rd-25th) and Tundry Pond (1, Sep 27th). The ten year totals of bird-days in autumn are as follows:

Autumn passage	1996	1997	1998	1999	2000	2001	2002	2003	2004	2005
Little Stint	400	28	175	64	29	113	48	34	72	35

Temminck's Stint   *Calidris temminckii*

A very scarce passage migrant (6,126,3).

Numbers returned to normal after the record total in 2004. There were three brief spring sightings of single birds. One flew north over Langstone Harbour and landed with Dunlin on South Binness Island on the evening of May 16th (JCr). Another, or the same, flew over Farlington Marshes at 1936 hrs on May 18th before landing briefly on North Binness Island and then being lost to sight over South Binness (JCr). Inland, one was at Ibsley Water on May 28th (LC). This is the sixth to be recorded in spring at that site, although the most recent was in 1994.

White-rumped Sandpiper   *Calidris melanotus*

A rare vagrant (0,16,1).

Following a day of heavy rain and cool south-easterlies, a summer plumaged adult was located in the high tide roost at Farlington Marshes at 1810 hrs on July 27th (JCr). At 1923 hrs, it flew off with two Dunlins and left south over Langstone Harbour. The record has been accepted by BBRC and is the 17th for Hampshire. Apart from single birds in 1963 and 1974, all have been since 1981 with the most recent at Testwood Lakes on Sep 15th/16th 2002.

Baird's Sandpiper
Calidris melanotus
A very rare vagrant (0,5,1).

A small sandpiper was found by GC & DLB at Fishtail Lagoon, Keyhaven Marsh at lunchtime on Sep 29th. It was then independently discovered by MC and identified as a juvenile Baird's Sandpiper by MPM later that day. It was subsequently seen by many observers until last seen on Oct 16th (photo). The record has been accepted by *BBRC* and is the sixth for Hampshire. All have occurred since 1986 with the most recent at Hayling Oysterbeds on May 8th 2000.

Baird's Sandpiper   © Peter Raby

Pectoral Sandpiper
Calidris melanotus
A rare passage migrant (0,70,3).

There were three records involving one at Park Shore, Needs Ore on Aug 2nd (GG, BG *et al*), a juvenile at Titchfield Haven from Sep 2nd-7th (KC *et al*, photo) and one which flew east over Hurst Beach at 0934 hrs on Sep 12th (MPM). The latter bird was subsequently reported on Keyhaven Marsh by unknown observers, and then re-located at Oxey Marsh before flying out to the Lymington River saltings (MPM, SW, MC).

Curlew Sandpiper
Calidris ferruginea
A scarce passage migrant, particularly so in spring. Rarely winters.

An average year with a minimum of 86 individuals being seen.

The only spring records were of singles at Titchfield Haven from Apr 27th-May 1st and on May 11th, Oxey Lake on May 4th, Keyhaven Marsh from May 11th-13th and inland at Winchester SF on May 28th.

The autumn passage began with one or two adults at Keyhaven/Pennington Marshes between July 29th and Aug 4th. A further influx started on Aug 21st, with the highest numbers from early September when passage of juveniles was at its heaviest. Records came from Keyhaven/Pennington Marshes (up to 13, Aug 21st-Oct 7th, max on Sep 12th), Farlington Marshes (up to 13, Aug 21st-Oct 23rd, max on Sep 4th), Titchfield Haven (1-5, Aug 21st-Sep 10th; 1, Oct 5th), Hayling Oysterbeds (1, Aug 25th; 1, Oct 28th), Needs Ore (4-7, Aug 26th-Sep 3rd), Ibsley Water (1, Aug 26th; 1, Sep 18th-25th) and Langdown, Hythe (1, Aug 28th). The ten year totals of bird-days in autumn are as follows:

Autumn passage	1996	1997	1998	1999	2000	2001	2002	2003	2004	2005
Curlew Sandpiper	190	60	105	135	95	89	138	40	150	78

Purple Sandpiper
Calidris maritima
A very scarce winter visitor and passage migrant to the coast [WCA 1, Amber].

It was another above average year with records from Barton on Sea now regular in winter.

In the first half of the year up to eight were noted at Barton on Sea between January 6th and April 1st (monthly maxima 7, 8, 4, 1). A single wandering bird (probably from Barton on Sea) was at Milford on Sea on Jan 8th and 15th. In the east of the county at Southsea Castle birds were present from January 22nd and lingered until May 18th (monthly maxima 1, 5, 11, 7, 3). Three seen there on May 18th were the latest ever there, surpassing the 1973 record by four days. In spring, birds were seen fairly regularly at Sandy Point, Hayling (often on the offshore tower) with up to five on nine dates between March 30th and May 4th. The latest ever was one at Hurst Beach on May 25th and 31st 1961.

The first returning bird was seen on the rocks off Hurst Castle on Sep 26th and the next flew past Milford on Sea on Oct 25th. At single was back at Barton on Sea from Nov 14th and was joined by a second on 18th. Both were present until Dec 27th at least. A single returned to Southsea Castle on November 19th and by the year's end three were present. Away from these prime sites, singles were noted at Park Shore, Needs Ore on Nov 1st, Black Point on Nov 8th and flying west at Sandy Point on Nov 11th. Two were at Hurst Castle on Nov 3rd. The records from Park Shore and Black Point are possibly the first ever from these sites.

Dunlin   *Calidris alpina*

An abundant winter visitor and common passage migrant; small numbers summer [SPEC 3, Amber, HBAP].

Overall, numbers remained at a stable level at both ends of the year. Internationally important wintering numbers were counted in Langstone Harbour in January, February, March and December. The count of 28,239 on Feb 12th was the highest for over a decade.

Elsewhere significant counts included 450 at Browndown on Dec 2nd and 250 at Titchfield Haven on Dec 31st.

Spring movement was only reported from Hurst where a total of 876 moved east between Apr 4th and May 28th.

Inland records were received from seven sites, not quite all for the passage periods; the exception was one seen at Casbrook Common on Nov 19th. The bulk of the records were from Ibsley Water, with up to three seen on seven dates between April 30th and May 28th, and a fairly regular passage in the autumn from July 16th until September 25th (monthly maxima 4, 17, 8, with the peak recorded on Aug 25th). Other records came from Winchester SF (1, Apr 26th; 1, May 16th; 2, Sep 10th-16th); Bourley Heath (6, Apr 30th); Woolmer Pond (1/6 on 9 dates, Apr 19th-May 13th); The Vyne (1, Mar 22nd; 1, Aug 13th; 1, Sep 30th-Oct 1st) and Testwood Lakes (2, July 31st).

Monthly maxima at the main high tide roosts are as follows:

	J	F	M	A	M	J	J	A	S	O	N	D
Chichester Harbour	8808	11404	7215	304	2038	2	150	506	447	1311	4296	5091
Verner Common	876	600	260	41					4	340	210	650
Black Point	3000	4500	2000	3	2		30	37	130	532	2000	2000
Langstone Harbour	25486	28239	17464	217	380	17	286	228	311	5026	8962	22356
Portsmouth Harbour	3933	1260	2481							44	217	1636
Hamble Estuary	865	750	1025	5			16	22	2	85	445	500
Langdown/Calshot	367	1150	895	22	12		5	7	11	150	450	858
Needs Ore/Beaulieu Estuary	766	1000	288	7	77		5	23	99	253	500	1000
Sowley/Lymington	900	620	500	250	220		20	50	15	150	600	1000
Lymington/Hurst	2800	2277	1047	160	335	4	145	200	250	459	2000	2000
WeBS count totals	36196	35692	22805	343	632		361	442	618	6530	11647	27605

Sites of international importance: 13,300+; national importance: 5600+.

Ruff
Philomachus pugnax

A scarce but regular passage migrant and winter visitor [ET, SPEC 2, WCA 1, Amber].

In the early year up to three were present on eight dates at Keyhaven/Pennington Marshes between January 3rd and February 13th. The only other record was of one at Titchfield Haven on January 23rd.

An above average spring passage took place between late February and mid May, involving about 49 birds, with records principally from the Keyhaven/Pennington Marshes (3 on Feb 26th, 4 Mar 4th then 5 on 5th, 12 on 12th with 5 last seen on 20th and a late record of 1 on May 15th/16th). Other records came from Lower Test Mashes (1, Apr 9th); Titchfield Haven (2, Mar 13th; 1, Mar 16th and a maximum of 3, May 12th-25th); Needs Ore (1, Apr 10th and 2, Apr 12th) and Farlington Marshes (1, Apr 14th and 1, Apr 16th).

Autumn passage was light and took place from July 9th until September 27th and involved approximately 25 birds. There were records from six sites with no site recording more than three on any one date. Inland records came from The Vyne (1, Sep 13th-20th) and Ibsley Water (1/2 on six dates between Aug 28th and Sep 10th).

Ruff   © Richard Ford

There were no records during the late winter period. The approximate monthly totals are tabulated below:

J	F	M	A	M	J	J	A	S	O	N	D
4	6	36	6	4	0	8	9	8	0	0	0

Bird-months	1996	1997	1998	1999	2000	2001	2002	2003	2004	2005
Ruff	164	112	120	119	76	61	197	79	75	81

Jack Snipe
Lymnocryptes minimus

A scarce but overlooked winter visitor and passage migrant [SPEC 3].

An above average year, with a minimum total of 156 recorded. Between Jan 1st and Apr 9th at least 112 were seen, with 47 at nine coastal sites and 65 at eleven inland sites.

Counts exceeding five were reported at Bourley Heath (10, Jan 2nd; 17, Jan 15th; 5, Jan 23rd and 9, Jan 26th), Farlington Marshes (9, Mar 12th) and Hook-with-Warsash (7, Feb 4th).

The first returning bird was seen flying around the Deeps at Farlington Marshes on September 24th. In the late year a minimum of 51 was seen, with 16 at seven coastal sites and 35 at nine inland locations. The only counts that exceeded five were from Bourley Heath (5, Oct 22nd; 6, Nov 6th and 5, Dec 19th).

The approximate monthly totals are tabulated below:

J	F	M	A	M	J	J	A	S	O	N	D
46	30	30	6	0	0	0	0	1	13	19	18

Common Snipe, Titchfield Haven © Peter Raby

Common Snipe
Gallinago gallingo

A formerly moderately common but currently declining breeder, common passage migrant and winter visitor [SPEC 3, Amber, HBAP].

Monthly maxima at sites where counts regularly exceeded 30 are tabulated below:

	J	F	M	A	M	J	J	A	S	O	N	D
Farlington Marshes	34	29	30	30	1	1	5	5	25	62	60	50
Titchfield Haven	40	30	10		1			10	9	20	20	33
Hook-with-Warsash	85	131	13	6				1	1	3	36	100
Lower Test Marshes	65	119	60	17				1	22	103	131	90

(continued)	J	F	M	A	M	J	J	A	S	O	N	D
Lymington/Hurst	75	30	44	10			1	16	6	14	51	8
Hucklesbrook Water Meadows	27		30						2	3	23	75
Itchen Valley CP	80	110	32	25			2	3	5	15	18	45
The Vyne Watermeadow	84	42	4	5				4	14	20	2	2
Hillside Marsh, Odiham	72	14	16								3	27
Woolmer Pond	205*	172	147	43	1				13		4	41

Away from these sites counts exceeding 20 were as follows: 80, Northney saltmarsh, Chichester Harbour, Jan 13th (flooded out by high spring tide); 37, Long Valley, Jan 23rd; 29, Emer Bog, Feb 6th; 34, Gosport Shore, Feb 12th; 39, Langstone South Moor, Mar 7th; 29, Ibsley Water, Sep 16th; 32, Winchester SF, Nov 22nd; 34, Fishlake Meadows, Dec 12th; 51, Keeping Marsh, Beaulieu Estuary, Dec 15th and 30, Ellingham Water Meadows, Dec 30th.

Reports in the breeding season were scarce. Nine were reported from the New Forest from widely scattered locations, including drumming display flight and birds defending territory. There were regular records throughout the summer of singles at Farlington Marshes, with two juveniles there on July 12th (but no indication of breeding). Elsewhere there were just three May records, all singles, at Titchfield Haven, Needs Ore, and Woolmer Pond.

Woodcock   *Scolopax rusticola*

A common resident and winter visitor [SPEC 3, Amber].

During the first winter period at least 40 were recorded, mostly in ones and twos; however, seven were seen in Roydon Woods on Feb 27th and four were flushed together at Longparish on Mar 26th.

In the breeding season, there were reports of 57 roding birds from 32 sites, including six at Havant Thicket and five at Warren Heath. This is unrepresentative of the county's breeding population.

There were only two records between July and October; one at Silchester Common on July 2nd and one at Moorgreen Farm on Oct 23rd.

In November and December at least 28 were reported including four seen at Botley Wood on Nov 27th.

There were no evident influxes during the year with just four passage singles seen at coastal sites in spring (Keyhaven Marsh, Mar 9th and Farlington Marshes, Mar 11th) and autumn (Calshot, Nov 14th and Sandy Point, Nov 20th).

Black-tailed Godwit   *Limosa limosa*

A common regular passage migrant and winter visitor. Small numbers summer [SPEC 2, WCA 1, Red, HBAP].

Counts in the first winter period indicated a county population of between 1491 and 2000. This range allows for movement between coastal sites.

Spring departures were noted from Langstone Harbour on four evenings between April 19th and 29th and totalled 88 birds. An unusually high percentage of the April flock were first-summers with, for example, 200 of this age in a flock of 265 at Farlington Marshes on April 25th, although by this date many of the adults present earlier in the month had left.

Autumn passage included peak year counts at Farlington Marshes of 750 on Aug 23rd and 742 on Sep 19th. Numbers fell away by November in Langstone Harbour as numbers in Portsmouth Harbour increased. The county population peaked at around 1600 in December – 10% of the British wintering population.

Monthly maxima at sites where counts regularly exceeded 100:

	J	F	M	A	M	J	J	A	S	O	N	D
Chichester Harbour	137	309	435	376	173	150	167	78	217	995	672	460
East Hayling	4	37	43				2	29		2	276	74
Langstone Harbour	110	96	450	450	198	114	325	750	742	689	153	73
Portsmouth Harbour	600	340	201	85				287	494	328	268
Titchfield Haven	150	140	167	36	156	82	64	45	18	32	139	368
Hamble Estuary	120	12	85	212	2		60	106	113	140	134	21
Lower Test/Eling/Bury Marshes	119	109	84	1				4	82	97	107	100
Needs Ore/Beaulieu Estuary	350	326	241	400	89	60	100	57	95	15	16	179
Lymington/Hurst	402	238	148	327	242	157	278	127	201	183	474	470
WeBS count totals	**1491**	**1271**	**1036**	**750**	**309**	**188**	**648**	**983**	**1086**	**1391**	**1504**	**1555**

Sites of international importance: 350+; national importance: 150+.

Inland records were received from seven sites. The bulk of them came from the Avon valley, where there were 26 at Sopley on Feb 7th, one at Ibsley Water on Mar 20th and May 11th, five there on June 13th, one from Aug 21st-Sep 2nd and on Sep 17th, and 40 at Hucklesbrook on Dec 6th. Other records came from four sites in the lower Itchen valley: Highwood Reservoir (3, Aug 30th; 1, Sep 1st; 1, Sep 19th-22nd), Itchen Valley CP (1, Aug 4th-18th; 3 N, Aug 31st), Lakeside (3 SE, Sep 2nd) and Moorgreen Farm (1 SW, Sep 25th), while other records were from Testwood Lakes (1, Jan 30th; 1, Apr 17th; 3, Aug 7th; 1, Aug 13th; 1, Oct 10th), the Vyne Watermeadow (1, Mar 12th/13th; 1, Aug 19th), and Winchester SF (7, Aug 5th).

Bar-tailed Godwit

Limosa lapponica

A moderately common passage migrant and winter visitor. Small numbers summer [ET, SPEC 3, Amber, HBAP].

Just three sites regularly held wintering birds during the year: East Hayling (peaks 711 on Feb 9th, 436 on Mar 3rd, 484 on Nov 5th and 600 on Dec 3rd); Langstone Harbour (peaks 400 on Jan 30th and 280 on Dec 14th) and the Lymington/Hurst area (peaks 26 on Feb

13th and 30 on Dec 11th). Elsewhere, winter records came from Eling (3, Dec 24th), Portchester (1, Feb 26th), Tipner Lake (1, Nov 26th) and Tanners Lane (8, Nov 7th).

Spring movements through the Solent involved a minimum total of 579 east between Apr 12th and June 1st with a peak day total of 120 at Hill Head on Apr 23rd. Individual sites totals included: Sandy Point (154 E, Apr 22nd-May 12th); Stokes Bay (128 E, Apr 22nd-30th); Hill Head (188 E, Apr 12th-May 17th) and Hurst Beach (307 E, Apr 13th-June 1st).

There were only two inland records, both from Ibsley Water, where there were singles on May 9th and 11th.

Monthly maxima at the main high tide roosts are as follows:

	J	F	M	A	M	J	J	A	S	O	N	D
Chichester Harbour	614	863	472	25	23	3	7	109	206	353	500	652
East Hayling	350	711	436	25	1		7	50	159	353	484	600
Langstone Harbour	400	200	90	70	60	15	109	175	200	254	215	280
Lymington/Hurst	25	26	20	5	37	1	2	2	10	16	25	30
WeBS count totals	90	475	131	5	1		119	11	177	161	517	628

Whimbrel
Numenius phaeopus

A common passage migrant. Rare in winter.

Up to five over-wintered in the early year, with birds being present in Chichester Harbour (1, Jan 15th-Mar 12th), Langstone Harbour (1, Jan 5th to at least Apr 9th), Bury Marshes (1-2, Jan 15th-Mar 12th) and Lymington/Hurst area (1, Feb 23rd-Mar 14th).

Easterly spring passage along the coast commenced on Apr 7th and continued until May 30th. Summing the maximum daily counts from any site gives a total of 896. Observations at some migration points were as follows: Sandy Point (237 E, Apr 17th-May 25th); Stokes Bay (128 E, Apr 17th-May 15th); Hill Head (427 E, Apr 7th-May 30th) and Hurst Beach (770 E, Apr 7th-May 16th). The peak movement occurred on April 22nd on which date 139 were logged at Hurst, 90 at Hill Head and 82 at Sandy Point.

Birds were reported from 14 inland locations, with counts exceeding six noted in the lower Itchen valley at Highwood Reservoir (34, Apr 27th; 43, May 1st), Itchen Valley CP (25, Apr 24th; 110, Apr 30th; 20, May 5th; 40, May 7th; 10, May 10th) and Moorgreen Farm (18, Apr 24th; 41, May 2nd), and elsewhere at Boldre, NF (9, Apr 30th) and near Pig Bush, NF (13, May 11th/12th).

Half monthly maxima during the spring and autumn passage periods are shown below:

	Apr 1-15	Apr 16-30	May 1-15	May 16-31	Jun 1-15	Jun 16-30	Jul 1-15	Jul 16-31	Aug 1-15	Aug 16-31	Sep 1-15	Sep 16-30	Oct 1-15	Oct 16-31
East Hayling		1	20		1		16	9	70	5	5	1		
Langstone Harbour	7	60	124	7	8	15	96	76	30	1	2	1	1	1
Portsmouth Harbour	1	73	13		4				1					
Titchfield Haven		33					1							
Hook Links/Bunny Meadows		5	25			1	1		1					
Curbridge		24	20				4							
Lwr Test/Eling/Bury Marshes	1	4					8	9	6			4		2
Langdown/Fawley	1	1	19	9			56	4			6			
Needs Ore		25	22			2	8	2	2					
Tanners Lane/Pitts Deep		23	30				2					1		
Lymington/Hurst		75	40	3		1	18	2	3			1	1	
Lower Itchen		34	110				1							

In June, up to two lingered on Langstone Harbour islands until 11th with returns noted from 25th. Inland, at least one was heard overhead at South Benyons Inclosure on 2nd.

In the late year up to seven over-wintered: Chichester Harbour (2, Dec 3rd); Langstone Harbour (1, Oct 4th-Dec 28th); Bury Marshes (1-3, Nov 26th-Dec 25th) and

Lymington/Hurst area (1, Dec 14th). The single in Langstone Harbour appeared to be same individual from previous years and is probably resident.

Curlew   © *Russell Wynn*

Curlew   *Numenius arquata*

A common passage migrant and winter visitor. Breeds in small numbers, mainly in the New Forest [SPEC 2, Amber].

Numbers wintering in the early and late year were about average (2000-2500). The highest count of the year was a high tide roost of 1811 at Langstone Harbour on Sep 17th - just above the nationally important threshold of 1% of the estimated British non-breeding population.

Monthly maxima at high tide roosts where counts exceeded 100 are as follows:

	J	F	M	A	M	J	J	A	S	O	N	D
Chichester Harbour	1255	869	724	370	45	768	1391	443	1296	1683	1498	849
East Hayling	511	246	242	140	35	334	470	290	611	744	370	243
Langstone Harbour	526	250	228	697	80	1000	1326	1405	1811	825	733	484
Portsmouth Harbour	411	215	296	106				334	356	136	462	
Hamble Estuary	108	84	74	18		4	14	21	32	50	76	60
Langdown/Fawley/Calshot	204	189	229	64	3	73	197	24	49	4	2	83
Needs Ore/Beaulieu Estuary	350	265	237	69			193	248	293	131	205	257
Lymington/Hurst	290	150	265	70	5	83	298	150	165	177	160	304
WeBS count totals	**2439**	**1437**	**1663**	**1122**	**110**	**1036**	**2577**	**2132**	**3341**	**2499**	**1982**	**2102**

Sites of national importance: 1500+.

Easterly passage past the traditional seawatching watch points was well below average, with the only significant counts being at Stokes Bay/Hill Head (35 E, on four dates with a peak of 12 on Apr 22nd) and Hurst Beach (49 E, Mar 9th-Apr 24th, max 22 on Apr 17th). However spring passage through, and departures from, Langstone Harbour was better recorded and, as usual, peaked in mid April. The WeBS count of 697 on Apr 9th probably corresponded with peak numbers. The only significant movements before that was on Apr 1st when 40 left high E, and Apr 3rd when 45 arrived from the west.

Subsequently a total of 362 left E or SE on four dates between Apr 14th and 24th and 101 arrived from the west or southwest on three dates between Apr 11th and 17th. The turnover of birds is probably significantly higher than any peak count.

During the breeding season no meaningful data were received for the New Forest, although casual records referred to 9-11 territories or pairs at seven sites. Otherwise, a single pair was recorded from Woolmer Pond.

Apart from the breeding records, inland reports came from the Avon valley at Burgate (1, Mar 6th), Hucklesbrook (up to 31, Mar 9th-Apr 9th, max on Mar 17th) and Ibsley Water (3, June 6th), Bitterne Park (4, Mar 10th), Itchen Valley CP (1, Jan 22nd-27th; 1, Feb 28th; 1, Mar 11th; 1, Aug 14th-18th; 1-2, Aug 22nd-26th; 1, Sep 1st), Moorgreen Farm (2, Jan 23rd; 1, Aug 14th/15th; 1, Aug 27th; 1, Aug 31st), Midanbury (1, Jan 16th), South Warnborough (1, Feb 18th), The Vyne (1, Apr 13th) and Testwood Lakes (1, Oct 30th; 2, Nov 13th).

Spotted Redshank   *Tringa erythropus*
A moderately common passage migrant and scarce winter visitor.

County totals were up on 2004 in the early winter period, with up to 15 present. As usual, sightings were mainly from the west of the county. For the months January to March the maxima were respectively: Lymington area (9, 9, 10) and Beaulieu Estuary (1, 2, 7). Reports on several dates from around Langstone Harbour probably all refer to the same individual.

Half monthly maxima during the spring and autumn passage periods are shown below for the main sites:

	Apr 1-15	Apr 16-30	May 1-15	May 16-31	Jun 1-15	Jun 16-30	Jul 1-15	Jul 16-31	Aug 1-15	Aug 16-31	Sep 1-15	Sep 16-30	Oct 1-15	Oct 16-31	Nov 1-15
Farlington Marshes	1	1				1			2	2	3	5	4	3	1
Needs Ore	1		1			3		4	7	8	13	7	3	1	7
Lymington/Hurst	8	4	1						1	1	1	1	1	1	2

Other records during this period came from East Hayling (1, Jun 26th); Ashlett Creek (1, Oct 25th and 30th) and Titchfield Haven (2, July 7th; 1, Aug 17th, 20th and 21st). The only inland record of the year was one at The Vyne on Sep 6th/7th.

Numbers at the end of the year were similar to 2004 and were again concentrated in the west. For the second half of November and December the maxima were respectively: Lymington area, six and seven and the Beaulieu estuary, seven and four. One was in Langstone Harbour regularly and another was at Titchfield Haven on Dec 3rd.

Redshank   *Tringa totanus*
A moderately common but declining breeder, common passage migrant and winter visitor [SPEC 2, Amber, HBAP].

The numbers in the eastern harbours in January were well down on last year; however there was an increase across the county in February and March. Counts in the second winter period were similar to 2004. Inland there was a count of 20 at The Vyne on Mar 22nd.

In the breeding season, pairs were located at the following sites: Keyhaven/Lymington (34 pairs); Langstone Harbour RSPB islands (20 pairs); Beaulieu area (8); Titchfield Haven (7); The Vyne (6), Hucklesbrook Water (3); Ibsley Water (2) and singles at Bishop's Dyke, NF and Testbourne Lake. There was limited reporting of breeding success, but three pairs successfully raised six young at The Vyne. Three juveniles were reported from Ibsley Water and at least two from Normandy. Up to three were reported from Woolmer Pond on many dates between Mar 20th and June 6th, but no breeding behaviour was observed. There were two other inland records: two at Headbourne Worthy CB on Apr 5th and a single at Beaulieu Heath on May 5th.

Reporting of autumn passage was limited to five east at Hurst on Sep 12th and singles at Highwood Reservoir on July 1st and 13th.

Monthly maxima at high tide roosts where counts exceeded 100 are as follows:

	J	F	M	A	M	J	J	A	S	O	N	D
Chichester Harbour	953	1260	1506	249	38	207	879	1073	1409	1596	1754	1647
East Hayling	301	275	338	10	8	5		35	183	375	354	146
Langstone Harbour	378	741	393	308	16	159	636	732	676	860	922	422
Portsmouth Harbour	260	340	448	154					686	247	619	602
Hamble Estuary	38	34	46	16			40	62	125	95	80	41
Southampton Water	117	80	92	10		2	34	68	41	76	101	125
Needs Ore/Beaulieu Estuary	56	5	53	38	11	16	49	73	112	141	106	207
Lymington/Hurst	162	225	134	70	30	47	144	200	200	250	270	192
WeBS count totals	**1003**	**1769**	**1562**	**614**	**68**	**213**	**860**	**1134**	**1964**	**2031**	**2525**	**1824**

Correction 2004: The combined Hampshire eastern harbour WeBS count on Jan 24th was 1341, not 1541.

Greenshank
Tringa nebularia

A moderately common passage migrant; scarce winter visitor [WCA 1].

A total of around 30 wintering birds was recorded in the first quarter from Langstone Mill House/Sweare Deep (1); Langstone Harbour north shore (1); Tipner Lake (1); Fareham Creek (up to 3); Upper Titchfield Haven (1, Feb 8th); Hook Links/Bunny Meadows (up to 3); Curbridge (up to 6 but maximum of 11 on Mar 5th, probably overlap with Hook); Hamble Common (1, Feb 12th); Lower Test Marshes (1); Needs Ore (2) and Lymington/Hurst area (monthly maxima 9, 8, 5).

Half monthly maxima during the spring and autumn passage periods are shown below:

	Apr 1-15	Apr 16-30	May 1-15	May 16-31	Jun 1-15	Jun 16-30	Jul 1-15	Jul 16-31	Aug 1-15	Aug 16-31	Sep 1-15	Sep 16-30	Oct 1-15	Oct 16-31	Nov 1-15
East Hayling								1		1	12		4	1	
Langstone Harbour	1	5	4	2	2		28	46	40	41	50	41	35	21	9
Portsmouth Harbour	1	1									3	2	2		
Titchfield Haven		2			1	1		1	2	2	1	2	2	1	1
Hook Links/Bunny Mdws							1	6	12	9	7	12	12	11	2
Curbridge	9	6					2	1	4	10	5		1		12
Lwr Test/Eling/ Bury Mshs	1	1						2	2	2		1	1	1	
Needs Ore	1	2	2		1	1	6	19	12	19	15	18	6	3	3
Tanners Lane/Pitts Deep							1	1		2			3	2	
Lymington/Hurst	5	7	1	1	1	1	4	10	16	24	18	16	16	17	11
The Vyne			4	3			1	3		1	1				

One summered at Normandy Marsh and two at Farlington Marshes. Other June records included those of single birds at Needs Ore on 9th and 21st and Titchfield Haven on 15th.

Other records came from Woolmer Pond (1, Apr 23rd); Lepe (1, May 4th, 6th and Aug 26th); Hurst Beach (1 E, May 12th); Sandy Point (2 W, July 8th; 2 E, Aug 15th; 1 E Aug 18th; 4 W, Aug 31st; 1 NW, Sep 4th); Itchen Valley CP (1 over July 9th; 2 over Sep 30th); Winchester SF (1, July 23rd; 2, Sep 16th); Ibsley Water (1/2 on 9 dates, July 30th-Sep 10th); Hamble (4, Aug 5th; 1, Oct 9th); Winchester SF (1, Aug 24th-28th; 2, Sep 16th-27th), Bisterne (1 S, Sep 24th) and Testwood Lakes (1, Oct 17th).

Records in late Nov-Dec came from Black Point (1), Langstone Harbour (2), Tipner Lake (2), Hook Links/Bunny Meadows (up to 8), Curbridge (up to 3), Chessel Bay (1), Needs Ore (up to 3), Tanners Lane (1) and the Lymington/Hurst area (9, Nov 20th; up to 5 in Dec).

Lesser Yellowlegs

Tringa flavipes

A rare passage migrant (0,11,1).

A juvenile moulting into first-winter plumage was found on the south scrape at Titchfield Haven at 1100 hrs on Oct 23rd (RKL). It remained until 1350 hrs, when all the birds on the scrape were flushed by a Sparrowhawk. It was again present from 1607-1613 hrs before flying off high to the west (KWM). On 25th, it was relocated on floods near Titchfield village (RM *et al*, photo), where it remained until 28th. It was also present briefly on the north scrape at Titchfield Haven on 26th. The record has been accepted by *BBRC* and is the 12th for the county. All have occurred since 1953, but this is only the third since 1977, with the most recent at Pennington Marsh from May 5th-7th 1999.

Green Sandpiper

Tringa ochropus

A passage migrant, scarce in spring and moderately common in autumn. Small numbers winter [WCA 1, Amber].

Reports were received from 26 sites in the early year. The highest monthly maxima from January to March were at Blashford Lakes (5, 3, 3) and the Lower Test Marshes (3, 3, 4). Elsewhere, in the early year, there were up to five in March at Dark Water, Lepe; up to three at Timsbury; two each at Langstone Harbour, North Somerley Lake and Alresford Pond, also up to two both at Testwood Lakes and in the Curbridge area. There were reports of singles from another 17 sites.

Half monthly maxima during the spring and autumn passage periods are shown below:

	Apr 1-15	Apr 16-30	May 1-15	May 16-31	Jun 1-15	Jun 16-30	Jul 1-15	Jul 16-31	Aug 1-15	Aug 16-31	Sep 1-15	Sep 16-30	Oct 1-15	Oct 16-31
Farlington Marshes	1				1	4	7	4	4	2				
Titchfield Haven					2	2	5	5	6	4	4	3	1	1
Hamble estuary							4	2	1					
Lwr Test Marshes	4	4	1	1	4	12	13	15	7	9	6	9	2	2
Dark Water, Lepe	4	4										1		
Needs Ore					1	3	7	2	3	3	3	1	2	
Ibsley Water						4	6	10	7	8	4		5	6
Testwood Lakes						3	1	1	4	4	4	2	1	2
The Vyne			1		1		2	1	4	5	1			
Tundry Pond					2	1		1	4	1	1	1	1	1

The last of spring was at Lower Test Marshes on May 8th. There were no further reports until June 8th, when the first of return passage was a single also at Lower Test Marshes. During the above period records of mostly ones and twos were received from a further 37 sites with three at Itchen Valley CP on July 10th.

Records for mid November to the end of the year were as follows: up to seven in the Blashford area; up to three both at Ellingham Meadows and Lower Test Marshes; two each at Bishops Waltham SF, Old Alresford CB, Bighton Lane CB, Testwood Lakes and Timsbury. Singles were reported from a further 15 sites, giving an approximate county total of 38. Approximate monthly totals were as follows:

	J	F	M	A	M	J	J	A	S	O	N	D
APPROX COUNTY TOTAL	23	20	31	11	1	11	60	72	55	30	31	30

Wood Sandpiper

Tringa glareola

A scarce passage migrant; more numerous in the autumn.

A below average year with a total of just 16 recorded. There was only one spring record of a single at Ibsley Water on May 28th.

There were approximately 15 in autumn as follows:

Ibsley Water: 1, July 28th; 1, Aug 8th; 2, Aug 9th; 1, Aug 10th; 1, Sep 17th; 1, Sep 25th.
The Vyne Watermeadow: 1, Jul 8th-15th; 1, Sep 2nd-7th.
Woolmer Pond: 1, Aug 8th-10th.
Tundry Pond: 1, Aug 19th.
Lower Test Marshes: 1, Aug 20th.
Farlington Marshes: 1, Aug 20th; 1, Aug 25th.
Normandy Lagoon: 2, Aug 31st.
Pennington Marshes: 1, Sep 25th.

Monthly totals for spring and autumn are shown below:

	Apr	May	June	July	Aug	Sep	Oct
Wood Sandpiper	0	1	0	1	10	4	0

The bird-months totals in autumn passage for 1996-2005 were:

	1996	1997	1998	1999	2000	2001	2002	2003	2004	2005
Wood Sandpiper	27	25	17	13	12	16	20	19	22	15

Common Sandpiper   *Actitis hypoleucos*

A moderately common passage migrant; a few regularly winter; has attempted breeding at least once [SPEC 3].

Records were received from four traditional wintering sites near the coast: Curbridge (monthly maxima, 2, 2, 3), Bedhampton/Broadmarsh (1, 1, 1), Riverside Park, Southampton (3, 4, 1) and Lower Test Marshes (2, 2, 2). Other winter reports were of singles at Sopley (Jan 16th and Mar 6th), Woodmill, Southampton (1, Jan 14th - possibly from Riverside) and Titchfield Haven (Mar 26th).

Half monthly maxima during the spring and autumn passage periods are shown below. The last bird of spring was at Wellington CP on June 1st. There were no further reports until June 15th, with one at Needs Ore.

	Apr 1-15	Apr 16-30	May 1-15	May 16-31	Jun 1-15	Jun 16-30	Jul 1-15	Jul 16-31	Aug 1-15	Aug 16-31	Sep 1-15	Sep 16-30	Oct 1-15	Oct 16-31
Hayling Oysterbeds			5	2			6	5	11		3			
Farlington Marshes		1	2	1		1	7	7	5	4	4	2	2	
Langstone Harbour		1	4	1		1	2	3	2	2	10	3	2	1
Portsmouth Harbour	2	1					3	3	10	4	5	9		
Titchfield Haven		1	1			2	1	3	2	3	11		1	
Hamble Estuary	1		2				6	10	10	3	5		2	
Curbridge	4	1	2					8	6	6	3		3	3
Riverside Park	10	9		1			1		1	2	5	1	1	
Lower Test Marshes	2	3	1			1	1	5	1	1	5	2	4	1
Testwood Lakes	1	5						4	4	4	3	2		
Lepe			2					8		3				
Needs Ore		1	5		1		1	5	1	4	3	1	1	
Lymington/Hurst	1	6	1	3		1	1	4	5	1	2			
Ibsley Water		1	1			1	3	6	2	6	4	2	1	
Lakeside, Eastleigh		5		1			1			1		2		
Winchester SF		3	1	1				1	1	1	5	1		
Woolmer Pond		1	3	1			1	4	5					

During the above period records were received from a further 40 sites including groups of two inland as follows: Black Dam, Basingstoke, Apr 24th; Fleet Pond, Apr 30th; Winnall Moors, July 4th; Highwood Reservoir 1/2, July 22nd–Sep 16th and The Vyne, Sep 7th.

Most of the reports for the second winter period came from the same regular sites as earlier in the year: Curbridge area (up to three); Bedhampton/Broadmarsh (up to 2); Hook Links (2, Dec 1st; 1, Dec 31st); Riverside Park, Southampton (1/2, with 3 on Dec 6th); Lower Test Marshes (2 on Nov 19th and 1 thereafter) and singles in Portsmouth Harbour on Dec 3rd and at Woodmill, Southampton on Dec 5th.

Monthly totals are shown below:

Bird-months	J	F	M	A	M	J	J	A	S	O	N	D
Common Sandpiper	10	9	9	56	43	10	101	105	108	22	10	13

Turnstone
Arenaria interpres

A moderately common passage migrant and winter visitor. Small numbers summer [Amber, HBAP].

The sum of January maximum counts was the same as for December 2004, although the birds were distributed differently around the coast. Maximum counts in the first winter period were substantially higher than in 2004. Maximum numbers in the autumn were also up on the previous year, but by the end of the year they were back to more normal levels. The count of 742 in Langstone Harbour on Oct 15th is thought to be the highest ever for the county.

Monthly peak counts at high tide roosts where counts exceeded 50 were:

	J	F	M	A	M	J	J	A	S	O	N	D
Chichester Harbour	153	142	228	83	72		21	48	41	197	93	288
East Hayling	10	7	62	50			21	48	20	48	58	19
Hayling Bay	150	50	81		3						80	63
Langstone Harbour	119	110	209	404	110	14	146	475	573	742	305	87
Portsmouth Harbour	50	3	19							4	69	93
Titchfield Haven/Hill Head	32	41	78	42	22	4	65	80	70	80	60	30
Hamble Estuary	119	104	102				5	80	113	117	40	111
Langdown/Fawley/Calshot	12	45	45	27		3	18	54	84	40	25	20
Beaulieu Estuary/Lepe	39	66	80	101	7	4	52	105		140	56	34
Tanners Lane/Pitts Deep		84	8	40	2		70	42	70	30	30	24
Lymington/Hurst	128	100	150	119	60	13	30	150	161	155	129	76
WeBS count totals	**439**	**385**	**295**	**524**	**138**	**16**	**200**	**718**	**899**	**1153**	**576**	**435**

Sites of national importance: 500+.

In addition to the above, there were reports of small groups from other localities along the Solent shore with maxima as follows: Browndown (25, Apr 3rd); Gilkicker Point (54, Dec 21st); Lee-on-the-Solent (13, Dec 28th); Brownwich (20, Mar 13th); Netley (26, Oct 14th) Weston (28, Dec 2nd) Cracknore Hard (20, Oct 15th) and Dibden Bay (1, Mar 13th).

The only reports of spring movements all came from Langstone Harbour, with a total of 82 on six dates between April 19th and May 17th, all heading inland. On July 27th, 11E were reported from Hayling Bay. There were inland records from Ibsley Water, with three on July 28th and 1 on Aug 21st.

Roosts are regularly seen on moored or wrecked boats in Langstone Harbour, including a flock of 209 on Mar 25th and 220 on Oct 18th.

Wilson's Phalarope
Phalaropus tricolor

A rare passage migrant (0,5,1).

A juvenile moulting to first-winter plumage was found at Farlington Marshes on Aug 21st and remained until 25th (HV *et al*, photo). The record has been accepted by *BBRC* and is the sixth for Hampshire. All have occurred since 1974, with the most recent at Pennington Marsh from Sep 9th-11th 1998.

Wilson's Phalarope, Farlington Marshes © Richard Ford

Red-necked Phalarope
Phalaropus lobatus

A rare passage migrant (3,36,3).

A juvenile was feeding at close range in the Beaulieu River off Needs Ore slipway at 1345 hrs on Oct 4th (MRa). There was an unconfirmed report of presumably the same bird on nearby Blackwater on Oct 8th. Another juvenile was found on a small gravel pit at Cherque, Gosport on Oct 7th (PNR, JAN, TFC *et al*, photo), remaining until the next day. It was colour-ringed and proved to have been ringed as a chick on Shetland in July. This is the first-ever sighting of a colour-ringed bird away from Shetland. Finally, one was seen feeding just offshore from Milford on Sea in rough weather from 1308-1328 hrs on Nov 11th (MPM). This is the latest ever for the county and the first for that month. The previous latest was at Farlington Marshes on Oct 28th 1987.

Grey Phalarope
Phalaropus fulicarius

A very scarce autumn and early winter visitor, usually occurring after gales. Very rare in December-February (?,244,9).

The first was at Ibsley Water on Jan 22nd (KTC). Sightings in winter are unusual and this is only the sixth county record in January. All other records for the year were in autumn.

The first of autumn was an exceptionally early juvenile/first-winter on Butts Pool, Pennington Marsh on Aug 25th/26th (RBW *et al*, photo) which was reported at nearby Oxey Marsh the next day. The previous earliest autumn record was at the same location on Aug 29th 1997. Subsequently, a small influx coincided with severe autumn gales. The first was a first-winter on the scrapes at Titchfield Haven on Oct 26th (BSD *et al*) to be followed by one at Hill Head on Nov 2nd which later moved onto the scrapes at Titchfield Haven (BSD). Further singles were off Weston Shore on Nov 3rd/4th (ARC *et al*), at Horsey Island, Portsmouth Harbour on Nov 5th (PJS), on a flood near Clarence Pier, Southsea on Nov 6th (*per* JMCk), flying west past Milford on Sea also on Nov 6th (EJW) and on a small pool at Burley GC, New Forest from Nov 12th-15th (SHS *et al*).

Pomarine Skua
Stercorarius pomarinus

A scarce passage migrant

An average year with 43 recorded during the spring passage period between late April and mid May and three in the autumn/winter period.

The largest flock of the spring occurred on Apr 24th when eleven (nine pale, one dark and one intermediate phase) were seen off Hurst spit (TP). They landed on the sea to the west and none entered The Solent. This conforms to the regular movement pattern observed from the Hurst area. It is thought that most movements from Hurst are south and then east along the Channel coast of the Isle of Wight. An exception was the individual seen passing Stokes Bay on Apr 28th at 1216 hrs and undoubtedly the same was seen 29 minutes later as it passed Sandy Point; on both occasions it was noted in the company of two Arctic Skuas. It seems probable that one observed passing Hurst earlier at 1150 hrs was the same.

In the late year a dark phase immature was off Hurst Castle on Sep 12th; the last was a first-winter seen well, down to 50m, at Hill Head on Dec 2nd - this is just the seventh county record in winter.

All records are summarised below:

Sandy Point, Hayling Bay: 1 E, Apr 24th; 1 E, Apr 26th; 1 E, Apr 28th; 1 E, May 1st.
Stokes Bay: 1 E, Apr 28th.
Hill Head: 1 E, Apr 29th; 2 E, May 17th; 1, Oct 19th; 1 first-winter, Dec 2nd.
Hurst/Milford on Sea: 1 E, Apr 23rd; 11, Apr 24th; 1, Apr 27th; 1 E, Apr 28th; 4 E, Apr 30th; 8, May

2nd; 8W, May 11th; 1 E, May 15th; 1 W, Sep 12th.

Arctic Skua   *Steracorarius parasticus*

A scarce passage migrant.

At least 192 were recorded during the year, of which 167 occurred during a record spring passage, including one inland. However, just 25 were seen in the autumn and winter periods. The earliest was a dark phase bird seen from Hayling Bay on Apr 4th. Eastward migration continued until June 18th. Spring records are summarised below:

Sandy Point/Hayling Bay: 40 E between Apr 4th and June 9th with peaks of 7 E, Apr 24th and 8 E, Apr 26th. Two dark phase birds seen on Apr 28th headed east with a single Pomarine Skua at 1245 hrs were earlier past Stokes Bay at 1216 hrs and off Hurst at 1142 hrs. A single pale phase on 1st May also accompanied a Pomarine Skua. Two loitering offshore for 20 minutes on Apr 24th and one N on June 9th have been included in the total.

Gilkicker to Hill Head: 37 E between Apr 19th and May 13th with peaks of 6 E, Apr 23rd, 5 E Apr 24th and 5 E, Apr 28th.

Lymington/Hurst/Milford on Sea: 103 E between Apr 18th and June 30th with peaks of 23 E, Apr 23rd, 9 E, Apr 26th and 18 E, May 2nd.

Phases of plumage were well recorded for the spring birds in the Hurst area, (61 dark phase, 41 pale phase and one unrecorded) and at Hill Head/Stokes Bay (20 dark, 12 pale, 5 unrecorded) but less so at Sandy Point/Hayling Bay where views were more frequently distant or in poor light (17 dark, 12 pale and 11 unrecorded). Overall, however, the ratio of dark/pale birds was very similar between sites.

Heavy rain and sea fog during the morning of Apr 23rd undoubtedly helped to produce the peak day count of the spring, 23 E at Hurst. Apart from example above there is remarkably little correlation between the timings of Arctic Skuas moving along The Solent; this raises questions about their behaviour during the spring passage along the county coast. It may well be that The Solent is used as a staging post as there is some evidence for overnight roosting and periods of feeding (attacking gulls and terns). Remarkably few seem to head non-stop between Hurst and Sandy Point.

There was an interesting inland record from Church Crookham, Fleet, on May 15th. This involved a pale phase heading north-east at 1330 hrs over the lucky observer's garden (CG).

The early autumn was much quieter, with a total of 24 from 17 records between July 17th and Oct 11th as follows:

Sandy Point: 2 E, July 17th; 1, Aug 14th; 1, Oct 6th.
Gilkicker - Hill Head: 4 offshore, Aug 14th; 1 W, Oct 23rd; 1 W, Oct 25th.
Needs Ore: 1, dark phase, July 19th; 1 W, July 29th; 1, Oct 11th.
Lymington/Hurst/Milford: 1, Aug 13th; 1, Aug 24th; 1, Aug 27th; 3, Sep 10th; 6 W, Sep 12th; 1 E, Oct 4th.

The largest movement was a flock of six which moved west out over Hurst spit on Sep 12th (five dark phase (not aged) and a light phase adult).

In the late autumn a dark phase individual (juvenile/first-winter) was seen in the eastern Solent area on Oct 25th, visiting Langstone Harbour, Southsea, Stokes Bay and Hill Head. Probably the same individual was in Langstone Harbour on Oct 29th and all afternoon on 31st. There were also records of a juvenile/first-winter dark phase in the Lymington/Hurst area on nine dates between Oct 27th and Nov 24th and was noted over Normandy, Pennington and Keyhaven Marshes. It is possible that just one individual was responsible for all the late autumn sightings.

Long-tailed Skua
Stercorarius longicaudus

A rare spring and autumn passage migrant also reported twice in winter (1,24,1).

One flew east past Milford on Sea at 1021 hrs on Oct 25th (MPM, SWd). A party of three was seen off Hengistbury Head, Dorset the same day. This species has been recorded in all but six years since 1990, the most recent being in 2003.

Great Skua
Stercorarius skua

A scarce passage migrant

The number of 'Bonxies' continues to increase with the highest ever annual total of 40 being reported; of these15 were seen from the eastern Solent and 25 from the west.

Spring passage provides the best opportunity to observe this species from Hampshire; about 35 were seen on eastward migration between Apr 16th and May 6th. Records are listed below:

Sandy Point/Hayling Bay: 1 E, Apr 22nd; 8 E (1,1,2,4), Apr 26th; 1 W, May 1st; 1, May 17th.
Stokes Bay/Hill Head: 2 E, Apr 23rd; 1 E, May 1st.
Hurst/Milford: 2 E, Apr 16th; at least 3 E, Apr 17th; 6 E (1,1,1,1,2), Apr 26th; 1 E, Apr 27th; 1 offshore, Apr 28th; 1 E, May 3rd; 1 E, May 6th; 2 S, May 20th.
Taddiford Gap/Barton on Sea: 1 E Apr 17th; 3 E, 1328 hrs Apr 23rd.

Eight east past Sandy Point on Apr 26th was the highest day total for the year and coincided with six east past Hurst. Of these only two together at Hurst at 1135 hrs possibly could be correlated with two past Sandy Point at midday. The individual recorded on May 1st observed flying west past Sandy Point was presumably the same first seen west past Hill Head and then returning eastwards.

The only summer record was of a single west past Milford on Sea on June 21st.

In the late year there were just four records coinciding with rough weather conditions: one west past Hurst Castle on Oct 8th; one south off Milford on Sea on 21st and another east there on Nov 11th, and one in Chichester Harbour entrance on Nov 6th.

Skua sp

Six records of unidentified skuas were reported. One early record on Apr 17th over Testwood Lakes was a possible Arctic Skua. Other sightings came from: Stokes Bay, Apr 30th; Sandy Point, May 12th; Hurst Spit, May 13th; Gilkicker, Oct 7th and Hayling Bay, Oct 28th (either an Arctic or Long-Tailed Skua).

The annual totals of bird-months 1996-2005 for all skua species are as follows:

	1996	1997	1998	1999	2000	2001	2002	2003	2004	2005
Pomarine Skua	33	119	9	25	61	45	22	30	63	46
Arctic Skua	169	122	101	120	115	139	151	221	189	192
Long-tailed Skua	2	0	1	0	1	1	1	1	0	1
Great Skua	10	15	27	15	32	38	24	31	35	40

Mediterranean Gull
Larus melanocephalus

A scarce but increasing visitor and scarce breeder (ET, WCA 1, Amber, HBAP).

This species was again enthusiastically recorded, with 787 records entered onto the database. Birds were seen on every part of the coast and at fourteen inland sites.

Around 56 were recorded during the first two months, with the regular spring build up starting in mid February. The highest counts were made in the eastern harbours in late March and early April, with 160 in Langstone Harbour (100 feeding in the channel NW of the Oysterbeds and 60 in from the east at dusk) on Mar 28th being the highest of four three figure counts. Away from the east, 28 were in the Lymington area on April 4th. Eastward, up channel movement was small and totalled only 19 at six sites between late March and early May.

In Langstone Harbour the number of breeding pairs almost doubled to 110, which raised a record total of 165 young (CC). Elsewhere, six pairs were found but there was no indication of successful breeding from any. Dispersal away from the colony on the Langstone Harbour islands is shown by counts of 165 juveniles there on July 15th but just three on 29th and, during the interim, a count of at least 33 juveniles which came to bathe at Farlington Marshes after a rain storm on 20th. Also, in the eastern harbours, an isolated peak of 21 (including 19 juveniles) in Chichester Harbour entrance channel on July 19th, a total of 14 (including 3 juveniles) at HMS Sultan playing fields on July 22nd and 12 at Titchfield Haven on 30th.

Numbers remained relatively low for the rest of the year but approximately 30 over-wintered. Tabulated below are the peak coastal day counts for each month and a list of all inland records:

	J	F	M	A	M	J	J	A	S	O	N	D
NW Chichester Hbr/E Hayling	8	38	82	80			21	4	4	5	3	1
Langstone Harbour	11	8	160	220	220	220+	385	4	1	2	2	1
Southsea	2	2	1	8	7	5	10	1	1	1	2	-
Tipner Lake, Portsmouth Hbr			3								16	5
Gosport area (incl Walpole Park and Haslar Creek)	11	3	8	4		3		7	1	2	12	6
Titchfield Haven / Stokes Bay	2	7	10	8	4	4	7	4	1	2	2	3
Lepe/Beaulieu Estuary		4	4	4	2			1	1			1
Lymington / Hurst area	2	4	14	28	9	4	6	1	1	2	4	6
Other coastal sites	8	18	20	21	3	4	28	6	5	11	8	2
Inland sites		3	2			17	29	7		3	1	1
Total	44	84	301	273	234	c230	c400	36	15	27	38	21

Inland records:

Alresford Pond: adult, Feb 10th.

Winchester SF: adult, Feb 10th; 2nd-winter, Feb 26th; juvenile, July 31st-Aug 2nd.

Meonstoke: 1, Mar 5th.

Old Winchester Hill: adult, Mar 26th; 17, June 4th: 13, June 26th; 10, July 6th. The two June records referred to birds feeding over recently planted maize fields.

Sway: adult, Apr 17th.

Waterlooville: At least four different birds (2 adults, 2nd-winter & 1st-winter) over the observers garden on 8 dates May 28th – Aug 17th. On July 9th one took a cheese roll from the garden. Not surprisingly this was the first time such behaviour had been recorded by the observer!

Teglease Down: 1, July 3rd.

Highwood Reservoir/Itchen Valley CP: juvenile, Aug 3rd & 10th; adult, Nov 26th.

Ibsley Water: juvenile, Aug 6th.

Testwood Lakes: juvenile, Aug 7th.

Newlands Farm, Fareham: 2nd-winter, Oct 8th.

Lakeside, Eastleigh: adult, Oct 15th.

Hambledon: 1, Nov 19th.

Moorgreen Farm: 2nd-winter, Dec 21st.

Laughing Gull   *Larus atricilla*

A very rare vagrant (0,0,1).

A first-winter was discovered at Haslar Creek, Gosport at 1300 hrs on Nov 5th (JAN, PNR, photo). It remained in the area until Nov 13th although it went missing at times. Other sightings of first-winters flying west off Hill Head at 1500 hrs on Nov 7th (MR), at Salterns Eight Acre Lake, Normandy on Nov 20th (SKW, JCa et al, photo), on a playing field at Henry Cort School, Fareham on Nov 22nd (TFC) and back at Haslar Creek on Dec 11th

(TFC) are assumed to relate to the same individual. Indeed, photographs of the Haslar and Salterns birds indicate that this is the case. The record has been accepted by *BBRC* and is the first for Hampshire. It occurred at a time of an unprecedented influx into Britain and Ireland involving in excess of 50 individuals.

Little Gull

Larus minutus

A scarce visitor, although sometimes moderately common, recorded in all months but most numerous in spring and autumn.

Approximately 300 were recorded during the year (*cf* 450 in 2004) and the pattern was similar to previous years. There were only three in the first three months but this total was followed by a strong spring passage, with around 180 moving through The Solent in April and May. Again, the peak was in the last ten days of April and movements had tailed off considerably by mid May. Thirty nine east past Sandy Point on Apr 22nd was the best of four double-figure counts. There were also two inland during this period.

Spring records from the main sea watching sites are summarised below:

Sandy Point/Hayling Bay: 68 E on 7 dates from Apr 19th to May 14th including 39 E on Apr 22nd and 18 E on Apr 26th.
Hill Head/Stokes Bay: 32 mostly E on 7 dates between Apr 19th and May 21st including 13 E on Apr 22nd.
Hurst Beach/Milford on Sea: 35 E on 11 dates from Mar 20th to May 17th including 10 E on Apr 22nd and 11 E on Apr 23rd.

One or two second-calendar year birds summered in the eastern harbours and were noted at Farlington Marshes between mid May and early August, Needs Ore on June 12th and Titchfield Haven on 20th (two). A small passage of mostly juveniles was then recorded through August and September until numbers increased from mid October. Around 90, mostly moving west, were noted in the four week period up to Nov 11th. The only double-figure counts were of 20 at Sandy Point and ten reported at Hurst Beach on Nov 11th. Unusually there were no later records.

Monthly maxima at the main coastal sites are tabulated below:

	Jan	Mar	Apr	May	Aug	Sep	Oct	Nov
Langstone Harbour			11	1	2	3	2	
Sandy Point/Hayling Bay	2		67	1		1	19	20
Hill Head/Stokes Bay			24	11	3	3	4	3
Lymington/Keyhaven				1	1		2	
Hurst Beach/Milford on Sea		1	25	9		1	5	18

All inland records were as follows:

Fleet Pond: 1, Apr 25th; 1, Sep 25th; 1, Oct 5th.
Ibsley Water: 1, May 1st.

Approximate bird month totals are given below:

J	F	M	A	M	J	J	A	S	O	N	D
2	0	1	139	42	3	0	6	8	38	51	0

Sabine's Gull   *Larus sabini*

A rare autumn passage migrant (1,144,2).

A juvenile was discovered at Sandy Point at 0850 hrs on Nov 11th (ACJ). It remained in the area until 1020 hrs before being relocated further west in Hayling Bay where it was present from 1200-1600 hrs at least (m.o.). It was seen briefly at Beachlands the next morning (KAT) and subsequently found dead at Sandy Point a week later (GSAS, photo). A second juvenile flew east at close range past Milford on Sea at 1221 hrs on the same day (MPM).

Black-headed Gull   *Larus ridibundus*

A numerous resident, passage migrant and winter visitor (Amber).

The number of breeding pairs at just over 10,000 was down slightly on 2004 but productivity was good in Langstone Harbour and the Keyhaven/Lymington area.

Selected counts of over 1000 are listed below, followed by the breeding records from the main sites:

Coastal sites
East Hayling, Chichester Harbour: 14,200 south to roost, Jan 6th. 10,000 (roost count), Jan 21st; 1000, July 22nd; 1500, Oct 15th.
Sandy Point: 1000, Aug 7th.
Farlington Marshes: 1000s hawking insects, Aug 29th.
Portsmouth Harbour/Paulsgrove Reclamation: 2119, Jan 15th; 1779, Feb 12th; 1606, Mar 12th; 1075, Sep 17th.
Dibden Bay: 1024, Jan 6th.
Lymington/Hurst area: 2000, Jan 26th.
Inland sites
Southleigh Farm: 3500, Jan 8th.
Waterlooville: 2000 hawking insects, Aug 17th.
Ibsley Water/Mockbeggar Lake (roost counts): 2150, Jan 9th; 4000, Feb 6th & 17th; 3200, Mar 6th; 2000, Sep 10th; 1410, Oct 29th; 4700, Nov 20th; 3500, Dec 18th.
Winchester SF: 1800+, Jan 28th; 1020, Feb 8th; 1000+, Dec 16th.
Breeding records:
Langstone Harbour islands: 4743 pairs raised 5200 young.
Hayling Oysterbeds: 27 pairs all failed probably due to rat predation.
Titchfield Haven: 110 pairs.
Needs Ore: 328 pairs, all unsuccessful.
Normandy Marsh: 765 pairs.
Normandy Lagoon: 20 pairs.
Pylewell Saltings: 3100 pairs. Productivity was regarded as good.
Boiler, Lymington: 1378 pairs. Productivity was regarded as good.
Keyhaven: 451 pairs.
Ibsley Water/Mockbeggar: 1 pair possibly nest building.

In the Lymington area, from Tanners Lane to Hurst, there were 5750 pairs estimated on the saltings, a decrease from the 7000 pairs in 2004.

Albino or leucistic individuals were recorded in Chichester Harbour (Jan 31st); Langstone Harbour (partial albino - Feb 13th; June 1st; returning leucistic bird – Feb 25th).

Ring-billed Gull   *Larus delawarensis*

A very scarce visitor seen annually since 1991 (0,34,1).

The adult at Walpole Park, Gosport remained until Mar 27th and returned for its third consecutive winter on Oct 5th, remaining into 2006 (PNR *et al*, photo). The only other record was of a second-winter at Gilkicker Lake on Aug 27th (IC, photo).

Common Gull   *Larus canus*

A common winter visitor and passage migrant; small numbers summer (SPEC 2, Amber).

In the early year counts in excess of 200 were made on eleven occasions, with 1350 at Corhampton on Jan 1st and 1500 through Farlington Marshes to roost on Mar 22nd the highest.

Evidence of spring passage was noted at several inland sites and, as in 2004, Old Winchester Hill provided the largest count with 700 on Mar 26th. Easterly up-channel movement was again recorded at only two coastal sites. At Sandy Point a total of 43 moved east on seven dates between Mar 15th and May 16th. At Hurst 227 were noted moving east between Jan 16th and May 15th.

During the breeding season a pair was present at one site, but the outcome of any nesting is unknown.

In the late year there were only three counts of over 200, with 1280 to roost through Farlington Marshes on Oct 27th the highest of these.

The leucistic/albino individual that was present at Farlington/Bedhampton in 2004 was again recorded at these sites in January, August and November and a further leucistic bird was at Hill Head on Feb 3rd.

Lesser Black-backed Gull   *Larus fuscus*

A common visitor, which occurs in all months; the first successful breeding attempt was in 2001. It is most numerous in autumn and increasing in winter (Amber).

A pair was present on South Binness Island during the summer, but they failed to raise any young. Inland, a pair bred on a factory roof in Andover, raising two young.

The highest counts were made inland. At Ibsley Water numbers peaked at 5098 on Oct 29th. At Blackbushe Airfield/Fox Lane GP, Eversley, numbers were much higher than in 2004 and reached a record 1785 to roost at Eversley on Oct 2nd. Only three counts were submitted for Eling/Redbridge, with the best being 40 on Sep 17th. Elsewhere, all of the three-figure counts were made at feeding sites on the chalk, the highest of which was 550 at Oakley, Basingstoke on Sep 1st. Monthly peak counts at the three main sites are tabulated below:

	J	F	M	A	M	J	J	A	S	O	N	D
Ibsley Water		250			13	75	200	30	1000	5098	2210	225
Eling								21	40	3		
Eversley/Blackbushe								1015	1350	1785	280	

Presumed spring passage was noted as follows:

Sandy Point: 15 E, Mar 21st; 7 E, Apr 22nd; 3, May 25th; 10, June 2nd.

Farlington Marshes/ Langstone Hbr: 88 moving (mainly E), on 28 dates between Feb 27th and May 20th, peak 10 E on Apr 3rd. Also 13, including 7 to roost at dusk, on Mar 23rd.
Hurst Beach: 18 NE, Mar 4th to Apr 4th.
Barton on Sea: 18 E, Apr 3rd.
Woolmer Pond: 14 N, May 1st.
Princes Marsh, Liss: 1 N, Mar 3rd; 11 N, Mar 7th.

Single birds of the race *intermedius* were recorded at Keyhaven/Pennington on Mar 20th with two there on Apr 2nd and at Walpole Park Lake on Nov 6th. There was also a probable in Langstone Harbour on Dec 3rd.

Herring Gull   *Larus argentatus*

A common winter visitor and passage migrant with a scarce but increasing breeding population; small numbers (mostly immatures) summer (Amber).

Peak numbers were recorded in the eastern harbours with four-figure counts being made at Sweare Deep, Chichester Harbour (1200, Aug 18th), and south down Langstone Channel to roost (2000, Oct 20th; 1850, Oct 22nd). Inland, the highest count was of 100 near Brockley Warren on Sep 27th but there were no other counts greater than 55.

In the breeding season a minimum of 34 pairs was located at three sites as listed below:

Burrfields, Portsmouth: 4 pairs.
Fawley Refinery: 12 pairs.
Langdown/Hythe MOD Base: 18 pairs.

There were no counts during the breeding season from the regular sites at Southampton Old Docks or Chickenhall Lane Industrial Estate, Eastleigh. Two juveniles at Dibden Purlieu in July were thought to have been raised locally and a pair at Alresford Pond were observed carrying sticks in April and sitting on an apparent nest in July.

The regular leucistic bird was around Southampton in April and July and was also recorded in the Hill Head/Stokes Bay area in April. A further leucistic bird flew south over Langstone Harbour on Dec 12th.

Yellow-legged Gull   *Larus michahellis*

Recently increased to moderately common but localised, mostly occurring in autumn.

Sightings were made at eight coastal and six inland areas – exactly the same as the previous two years. Double figure counts were made at Eling/Redbridge, Ibsley/Mockbeggar and Blackbushe/Eversley but numbers were down on previous years and Sweare Deep and Testwood Lakes were the only other sites to record more than two in a day.

Elsewhere a total of two adults and two juveniles was noted at Farlington Marshes from mid July to mid August and a further three birds were nearby at Hayling Island. Only three were recorded at Keyhaven/Pennington Marshes during the year. The regular adult at Hook-with-Warsash reappeared on July 5th and stayed until Sep 8th. More unusually, adults were recorded at Lakeside, Eastleigh on Jan 26th and Sep 4th.

Monthly totals for the county are summarised below:

	J	F	M	A	M	J	J	A	S	O	N	D
Eling/Redbridge	1	2	0	0	1	10	68	80	0	22	1	3
Ibsley Water/Mockbeggar	24	9	0	0	1	21	10	2	10	12	28	6
Blackbushe Airfield	0	0	0	0	0	0	0	4	10	6	1	0
Sweare Deep, Chichester Hbr	0	0	0	0	0	0	6	0	0	0	1	1
Others	1	0	2	4	1	1	7	2	1	2	1	3
APPROX COUNTY TOTAL	**26**	**11**	**2**	**4**	**3**	**32**	**91**	**88**	**21**	**42**	**32**	**13**

Caspian Gull

L.(a.) cachinnans

A very scarce autumn and winter visitor (0,31,2).

The only fully authenticated records were of first-winters at Black Point on Mar 30th (ASJ) and Fleet Pond on Sep 25th (GCS, JAE).

Iceland Gull

Larus glaucoides

A very scarce visitor, usually in winter, but recorded in every month except June (3,74,0).

The second-winter discovered at Walpole Park, Gosport on Dec 28th 2004 was recorded almost daily there until Jan 18th, with the only sighting away from there at Old Portsmouth on Jan 6th. Subsequently, it ranged more widely and was reported at Titchfield Haven on 22nd and 23rd, leaving west from Chichester Harbour entrance on 25th, at Walpole Park, Gosport on 28th then nearby at HMS Sultan playing field on Feb 2nd, Hill Head the next day, Walpole Park again on Mar 25th, 29th and Apr 8th, flying east over Farlington Marshes on Mar 28th, at Cams Hall, Fareham Creek the next day and finally in Haslar Creek on Apr 14th. Several observers noted that it showed distinct dark shading on the primaries, a feature which is consistent with the race *kumleini* or a *glaucoides/kumleini* intergrade. What was presumed to be the same individual returned to Hill Head on Dec 23rd (RJC).

Iceland Gull, Gosport   © Peter Raby

Glaucous Gull

Larus hyperboreus

A very scarce winter, usually in winter, but recorded in every month (3+,92,2).

There were two records involving a first-winter feeding with other gulls in the inclosures at Marwell Zoo on the afternoon of Feb 15th, which left west at 1530 hrs (SC), and a first-winter which flew east along the cliff top at Milford on Sea on Dec 2nd (RAC).

Great Black-backed Gull
Larus marinus

A moderately common winter visitor and passage migrant; small numbers (mostly immatures) summer; occasionally breeds.

During the breeding season a total of five pairs was located. Nesting was noted in Portsmouth (a pair raised one young) and Romsey (a pair raised two young) and in the Lymington area (three pairs raised three, two and two young respectively). One of the Lymington pairs was responsible for the predation of a large number of chicks from a Little Tern colony. A pair with a juvenile on Long Island, Langstone Harbour, in late July may have been the Portsmouth breeding pair or perhaps another from elsewhere locally.

Three-figure counts were reported on eleven occasions at sites in the east of the county as follows:

Chichester Harbour: 100, Oct 20th; 211, Oct 26th; 224, Dec 28th.
Sandy Point: 100, Nov 20th.
Langstone Harbour: 160, May 26th; 300, June 4th; 366, June 9th; 126, June 16th; 100, July 18th.
Southleigh Farm: 160, Jan 1st; 328, Jan 8th.

With the exception of the Southleigh Farm records above, no inland site recorded more than 30.

The only indication of passage was of three east in the Stokes Bay area in late March/early April and a single east at Ovington on Mar 16th.

Kittiwake
Rissa tridactyla

A passage migrant and winter visitor, usually scarce but sometimes occurring in large numbers after gales.

In many ways 2005 was an average year for Kittiwake in Hampshire, with sightings coming from five coastal sites. The exception, however, was a record - breaking count at one site in early April.

In the first ten weeks of the year, a scatter of individuals and single figure groups, moving mainly west, was recorded. Eighty-one off Hurst Castle on Jan 10th was the only count above seven at this time. The first indication of spring passage was one east past Hurst on Mar 13th and a further two there on 22nd. There were then records on 20 dates from Apr 2nd to June 24th and passage would have been unremarkable if it were not for the events of Apr 8th, when a staggering total of 829 moved west past Hurst Castle from 1545 – 1900hrs (MPM) following several days of wintry weather when successive weather fronts pushed eastwards before abating in the early morning. This totally eclipsed the previous record spring day count of 553 past Hill Head on Apr 7th 1979 and is not far behind the all-time Hampshire record of 900 west, also at Hurst, on Nov 8th 2001.

Spring passage east through The Solent is summarised below:

Sandy Point/Hayling Bay: 1, Mar 31st; 1, Apr 7th; 1 E, Apr 24th; 2, May 1st; 1 E, May 15th; 3 E, May 25th; 3, June 3rd; 1, June 11th; 1, June 14th.
Stokes Bay: 2, Apr 3rd; 2, Apr 19th.
Hurst Beach/Castle: 1 E, Mar 17th; 2 E, Mar 22nd; 19 E on 10 dates in April including 7E on 21st; 829 W, Apr 8th; 13 E & 2 S on 3 dates in May including 12 E on 15th; 1, June 12th; 1, June 24th.

After a lull from mid June to mid Aug, there was another pulse of records later in the month when 13 flew east at Hurst on 24th which constituted the only double-figure count since May. Another quiet period followed until late October/early November when birds appeared during unsettled weather. The year ended with singles at Hill Head and Sandy Point.

The monthly county totals were:

	J	F	M	A	M	J	J	A	S	O	N	D
Kittiwake	90	4	12	854	23	7	1	17	1	10	10	2

Little Tern
Sternula albifrons

A moderately common summer visitor and passage migrant; recorded once in winter.

The first was an early bird off Normandy Marsh from Apr 2nd-5th. Spring passage east through the Solent extended from Apr 6th-May 15th, with a minimum of 215 involved. Individual site totals were of 101 at Stokes Bay/Hill Head (peak 26, Apr 23rd), 94 at Hurst Beach (peaks 25, Apr 20th; 26, Apr 30th) and 84 at Sandy Point (peak 18, May 15th). The most noteworthy record was of six at Heath Pond, Petersfield on Apr 30th (ACS); this is the highest number inland since seven were at Broadlands Lake on June 30th 1978.

The breeding season was slightly improved on 2004, with 143 breeding pairs raising at least 23 young. In Langstone Harbour, 38 pairs raised 13 young on the islands while 45 pairs at the Oysterbeds all failed. Between Pitts Deep and Hurst, there were around 60 pairs at four sites, with at least ten young raised at one. Eight pairs at Normandy Lagoon in early May failed to nest and presumably attempted elsewhere in the Hurst area. The continuing predation from a variety of sources (including foxes, rats, crows, gulls and Kestrels), as well as the risk of tidal inundation, leaves this species in a precarious position in the county.

Post breeding gatherings in Langstone Harbour peaked at 160 on July 8th and 220 on July 28th. In August there was a maximum of 18 on 10th and the last a single bird on Sep 1st. Nearby, 101 flew west at Sandy Point on July 6th, presumably a local movement, and a maximum of 62 was in a day time high tide roost at Black Point on Aug 5th. The last was at Normandy Marsh on Sep 9th, apart from a late juvenile at Pennington/Keyhaven Marshes on Sep 30th and Oct 1st.

The table below summarises annual numbers of breeding pairs of tern species 1995-2005:

Breeding pairs	1996	1997	1998	1999	2000	2001	2002	2003	2004	2005
Sandwich Tern	219	246	250	334	356	271	402	244+	283	347
Common Tern	363	338	332	457	311	418	430	399	574	430
Little Tern	150	160	136	145	154	192	165	174	112	143

Whiskered Tern
Chlidonias hybrida

A rare vagrant (0,5,2).

An influx of around ten into England in May accounted for two individuals in Hampshire. One was present on the south scrape at Titchfield Haven from 1415-1430 hrs on May 20th (PD, RJC, photo). It left to the west and was probably the bird involved in an unconfirmed sighting at Testwood Lakes later that afternoon. Another was found with Common Terns at Tundry Pond at 1540 hrs the same day, remaining until 2006 hrs when it left north-east (BC, KBW *et al*). It was relocated at Fleet Pond, before leaving north with Common Terns at 2106 hrs, presumably to roost at Moor Green Lakes, Berkshire. It was located there at 0445 hrs the next morning (JMCk, BA), and subsequently returned to Fleet Pond between 0745 and 0915 hrs, then to Tundry Pond where it remained until 1735 hrs (m.o., photo). It flew into Moor Green at 1755 hrs. It roosted there overnight, again on May 22nd/23rd and was seen there on the afternoon of May 27th. However, it was not recorded in Hampshire during this time, but was present for long periods at King George VI and Staines Reservoirs, Surrey. The records have been accepted by *BBRC* and are the sixth and seventh for Hampshire. All have been since 1970, with the most recent of two at Mockbeggar Lake as long ago as May 14th 1988. The Tundry/Fleet bird was the first individual to linger in the county and was sought by many observers in consequence.

Little Tern © *Richard Ford*

Black Tern
Chlidonias niger

A scarce passage migrant.

Spring passage was light, with the first two east at Hurst Beach and one at Fleet Pond on Apr 23rd and then a further 29 evenly spread up to May 26th. All were coastal apart from single birds at Ibsley Water on Apr 24th and 30th.

The first return was one east at Hurst Beach on July 24th. Around 100 were recorded through until late September, with counts exceeding four at Titchfield Haven (11, Aug 17th), Needs Ore (5, Aug 19th), Black Point (5, Aug 19th) and Langstone Harbour (5, Aug 20th; 7, Sep 1st; 6, Sep 5th, all to roost at dusk; 9 (8 S), Sept 5th). Inland, the only records were of single birds at Ibsley Water on Aug 6th and Sep 4th, Fleet Pond on Aug 21st and Testwood Lakes on Aug 21st and 28th/29th. The last were two juveniles at Sandy Point on Sep 23rd.

The approximate monthly totals are tabulated below.

Apr	May	June	July	Aug	Sep
12	20	0	6	63	30

Sandwich Tern
Sterna sandvicensis

A moderately common summer visitor and passage migrant; small numbers now regular in winter.

Records in the period up to mid March came from Chichester Harbour (up to three adults and one first-winter throughout); Langstone Harbour (up to three adults and two first-winters throughout); between Sandy Point and Sinah (up to two adults and three first-winters between Jan 26th and Mar 11th) and Portsmouth Harbour (an adult on Jan 6th and an adult and a first-winter on Feb 13th). Study of the individual birds in Langstone and

Chichester Harbours revealed that at least nine and possibly 12 were wintering in the eastern harbours.

A summer-plumaged bird moving east off Sandy Point on Mar 15th was the first migrant detected. Eastward spring passage began in earnest from 20th, although was light until Apr 19th, when the first day total in excess of 25 was logged. In all, a minimum of 1199 moved through the Solent in the period up to May 18th, with individual totals of 657 logged from Hurst Beach/Milford on Sea, 769 from Stokes Bay/Hill Head and 682 from Sandy Point. The only day totals to reach three figures were of 300 east off Sandy Point on Apr 22nd and 117 east off Stokes Bay on May 2nd.

Breeding colonies were established in two areas, with 271 pairs raising at least 61 young in Langstone Harbour and 76 pairs raising at least ten young between Pitts Deep and Hurst. Feeding areas used by breeding birds were indicated by flocks of 38 at Titchfield Haven on June 1st and 70 at Sandy Point on June 14th.

Post-breeding flocks included 80 off Langdown, Hythe on July 23rd, 80 resting in a field at Eastney on Aug 2nd and 80 at Sandy Point on 4th. The nocturnal roost at the Kench, Langstone Harbour held 78 on July 26th and 80 on Aug 2nd. Numbers then fell until an influx later in the month, with 188 on 19th and 164 on 22nd. Subsequently, there were 80 on Aug 28th and 60 on Sep 5th. Evidence of departure was provided by westward movements of 38 off Barton on Sea on Aug 20th, 102 off Stokes Bay on Sep 26th and 35 off Milford on Sea on 30th. Small numbers remained in the eastern harbours in October and early November, with maxima of ten in Chichester Harbour on Oct 30th and 12 in Langstone Harbour on Nov 4th. Away from the east, the last were two at Keyhaven on Nov 9th.

Records in the period from mid November onwards came from Chichester Harbour (up to three adults and one first-winter throughout), Langstone Harbour (one up to Nov 18th, then three on Dec 8th, presumably Chichester Harbour birds) and Portsmouth Harbour (two off Hardway, Gosport on Dec 3rd).

2003 addition: one was at Ivy Lake, Blashford on May 30th (S Curson).

Common Tern   *Sterna hirundo*
A moderately common summer visitor and common passage migrant.

The trend of early arrivals evident in recent years continued, with the first at Fleet Pond on Mar 23rd/24th; two east at Hurst Castle on 29th; three at Sandy Point on 30th and singles in Stokes Bay and at Sinah GP on 31st. Eastward passage through the Solent continued until May 19th, with totals of 2406 at Hurst (peaks 700, Apr 26th; 633, Apr 30th); 2076 at Stokes Bay/Hill Head (peaks 292, Apr 22nd; 451, May 15th) and 1361 at Sandy Point (peaks 486, Apr 26th; 364, May 14th). Summing the daily maxima for any site during the period gives a minimum total of 3543. Off-passage gatherings included 250 at Fawley on May 4th and 200 at Hill Head on May 6th. Inland, double-figure counts were made at Ibsley Water (max. 16, May 14th), Fleet Pond (max. 15, Apr 24th) and Tundry Pond (max. 14, May 22nd and 28th). The Fleet and Tundry birds are mostly from the breeding colony on the Berkshire side of Eversley GP, where a peak of 66 was recorded on May 22nd.

A reduction in breeding numbers at all the main coastal colonies, including desertion of the Beaulieu Estuary, accounted for a decline from the record total of 574 pairs in 2004 to 430 this year. Between Pitts Deep and Hurst, 260 pairs bred in five colonies but success was again low. In Langstone Harbour, 151 pairs on the islands raised 31 young, but seven pairs nesting on Hayling Oysterbeds deserted on May 26th and have not been included in the total. Elsewhere, there were 12 pairs at Titchfield Haven, at least six pairs at Ibsley Water and one at nearby Spinnaker Lake.

Records during July included a maxima of 19 (4 juveniles) at Fleet Pond on 7th

(Eversley birds dispersing) and 37 (23 juveniles) at Ibsley Water on 16th. Post-breeding gatherings were prominent at Langstone Harbour and Titchfield Haven from the end of the month. In the latter area, a maximum of 500 was on the scrapes on July 31st, but offshore there were reports of 1000 feeding on the same date, 1400 on Aug 14th and 400 on Sep 2nd. In Langstone Harbour, there were 500 on July 28th and Aug 5th. Regular nocturnal roost counts there between mid August and early September produced: 1730 on Aug 13th; 2561 on 18th; 2410 on 19th, 2706 on 20th; 2690 on 21st; 2456 on 22nd; 2537 on 29th; 2401 on Sep 1st; 940 on 3rd and 851 on 8th. Noteworthy gatherings elsewhere included 250 off Mayflower Park, Southampton on Aug 19th/20th; 270 at Black Point on Aug 28th; 400 there on Sep 5th and 400 at Eastney on Sep 3rd. Later records included 100 at Black Point on Sep 26th (the last three-figure count) and five north-west past Hook-with-Warsash on Oct 1st. Numbers dwindled during October although there were still six at Calshot on 20th and five in Langstone Harbour on 22nd. The last were first-winters at Black Point and Hurst Castle on Nov 6th, and Langstone Harbour on 8th.

Roseate Tern   *Sterna dougallii*

A very scarce passage migrant which occasionally breeds.

In spring, the first was at Titchfield Haven on May 2nd. Subsequently, one or two were seen in that area until May 20th, and one on June 6th. Other spring records included two off Hurst Castle on May 6th; two east off Barton on Sea on 8th; two east off Hurst Beach on 11th, one east in Stokes Bay on 15th and one just inland; at Testwood Lakes on 28th (SSK *et al*). This is only the second ever inland record following one at Fleet Pond on May 21st 1997.

Post-breeding dispersal was evident from late June. In the Titchfield Haven/Hill Head area there were two adults on June 27th; one on June 29th, then three to five between July 17th and Aug 14th (max on July 31st and Aug 13th); two until Aug 21st and one until 31st. A juvenile was reported on Aug 29th. In Langstone Harbour, one adult was detected on five dates between July 8th and 26th, then on Aug 10th, five adults were in the nocturnal roost and a different adult was present during the day. Subsequent dusk counts included four on 18th, five on 19th, six on 22nd, two on 23rd and two on 31st. Other records came from Sandy Point (3 adults, July 6th); Barton on Sea (1 W, July 10th); Black Point (different adults, Aug 9th, 13th, 15th and Sep 5th); Langdown, Hythe (adult, Aug 14th) and Keyhaven (1, Sep 5th).

Arctic Tern   *Sterna paradisaea*

A scarce passage migrant.

Around 250 were recorded in spring, the most ever recorded in the county. The first were on Apr 22nd when 11 moved east off Hurst and 15 did likewise off Stokes Bay. The next day, at least 13 passed through Fleet Pond and 32 moved east or north over Langstone Harbour between 1130 and 1348 hrs. On 26th, flocks of 14 and 12 flew east off Sandy Point, while another group of 20 moved east off Hill Head. On May 2nd, 20 flew east in Stokes Bay, while on 4th, 16 flew north-east over Sandy Point and a flock of 68 arrived at Fleet Pond at 1925 hrs. Thirty-nine left NE at 1932 hrs but the remainder were still present at 1955 hrs (JMCk *et al*). This is the largest ever flock recorded in the county, and coincided with a heavy movement though inland sites in central England, involving in excess of 1000 birds. Four single birds were seen up to May 15th.

Autumn records involved a high total of around 50 individuals. These were mostly juveniles and included a late influx in late October and early November. The approximate half-monthly totals are shown below.

July 16-31	Aug 1-15	Aug 16-31	Sep 1-15	Sep 16-30	Oct 1-15	Oct 16-31	Nov 1-15	Nov 16-30
2	10	7	4	9	5	3	9	1

The first were adults at Ibsley Water on July 23rd and Sandy Point the next day. Apart from one at Ibsley Water on Aug 25th, all subsequent records were coastal, with five in the Langstone Harbour roost on Aug 13th and up to five at Sandy Point/Chichester Harbour from Sep 23rd-27th - the only counts to exceed two. November juveniles were recorded at Sandy Point (6th and 11th); Langstone Harbour (5th-12th); Tanners Lane (12th), Hook-with-Warsash (4th/5th and two on 12th/13th), Keyhaven/Hurst (one or two, 1st-13th); Weston Shore (14th) and Titchfield Haven/Hill Head/Brownwich (1st-19th – BSD *et al*). The Titchfield Haven bird is the latest ever for the county.

The table below summarises spring passage numbers (measured as the annual totals of bird-days for eastbound spring passage through the Solent) for all regularly occurring tern species:

Spring passage	1995	1996	1997	1998	1999	2000	2001	2002	2003	2004	2005
Sandwich Tern	940	516	292	269	923	819	1005	1059	1788	1208	1199
Roseate Tern	4	7	1	4	16	5	8	13	9	2	7
Common Tern	3440	2887	931	2398	5930	3916	4600	5600	2146	3463	3543
Arctic Tern	4	49	19	29	48	95	95	48	68	100	250
Little Tern	197	273	42	54	238	296	247	220	97	105	215
Black Tern	68	183	45	74	69	86	89	36	12	40	29

Guillemot   *Uria aalge*

A scarce but increasing visitor and passage migrant.

Numbers were well down on 2004 with only 53 sightings compared with 110. There was no repeat of the exceptional summer sightings and a mere five were seen between May and September, compared with 50 in 2004. Double figures were reported across the county in only two months.

Sandy Point/Black Point/Hayling Bay: 2, Jan 4th/5th; 1, Jan 20th; 3, Feb 27th; 1, Mar 18th; 1 on three dates, Oct 12th-28th: 1, Nov 19th-29th; up to 6, Dec 5th-31st.
Langstone Harbour: 1, June 19th; up to 2, Dec 15th-22nd.
Southsea: 1, Nov 30th; 1, Dec 20th.
Portsmouth Harbour: 1, Oct 3rd.
Stokes Bay: 1 E, May 18th; 1, Sep 10th.
Hill Head: 1, Jan 24th; 1 W, Feb 12th.
Southampton Water: 1, Feb 16th and 26th; 1, Oct 20th; 1, Nov 16th; 1, Dec 1st.
Needs Ore: 2, Jan 12th; 1, Sep 13th; 1, Dec 11th.
Hurst/Milford on Sea: 1, Jan 2nd; up to 4, Jan 10th-22nd; 1, Mar 9th; 1, Mar 16th; 1, Apr 22nd; 1, Apr 25th; 1, Apr 29th; 1, May 6th; 1, May 28th; 1 E, Oct 31st; 1, Nov 3rd; 1, Nov 12th; 1, Dec 13th.

There is no method of establishing how many different individuals were involved in the sightings, especially at Sandy Point. Monthly totals for the county have been conservatively calculated as follows:

J	F	M	A	M	J	J	A	S	O	N	D
10	5	3	4	3	1	0	0	1	8	5	13

Razorbill   *Alca torda*

A scarce but increasing passage migrant and winter visitor.

As with Guillemot, numbers were well down on the previous year with only 58 compared with 127 in 2004.

Sightings again peaked in January when birds were seen around Langstone Harbour and Hurst Beach. Off south-east Hayling Island there were regular sightings through to the end of March, with monthly peaks of four on Jan 31st, five on Feb 17th/18th, 21st and 28th and six on Mar 10th and 13th. It is impossible to know how many birds were present in this area where, interestingly, one of five was taken by a Common Seal on Mar 2nd but six were present on 10th! They have been recorded conservatively in the totals as only seven individuals.

Spring passage was noticeable at Hurst Beach where 15 were recorded on seven dates between Apr 3rd and May 13th, with five on Apr 26th the highest count. Nearby, five were off Taddiford Gap on May 2nd. There were no further records until one at Hurst Beach on Sep 9th. Numbers remained low with the highest of the late year counts being four at Hurst Beach on Sep 12th.

In the table below birds have been recorded only in the month in which they were first seen:

	J	F	M	A	M	J	J	A	S	O	N	D	Total
Number of records received	20	22	15	7	3	0	0	0	4	8	6	8	93
Probable number of birds	10	7	2	14	7	0	0	0	6	4	4	4	58

All records are summarised below:

Sandy Point/Black Point/Hayling Bay: Recorded on 35 dates between Jan 1st and Mar 25th. Monthly maxima as follows: Jan, 2 on 20th; Feb, 5 on 17th, 18th & 28th; Mar, 6 on 10th and 13th.
Langstone Harbour: 1 on six dates, Jan 31st-Mar 1st; 1, Nov 16th.
Eastney: 1, Feb 19th.
Portsmouth Harbour: 1, Dec 3rd.
Southsea: 1, Feb 6th & 8th.
Haslar creek: 1, Nov 17th/18th and Dec 2nd/3rd.
Gilkicker Point: 1, Sep 29th; 1, Dec 29th.
Hill Head: 1, Jan 22nd; 1 W, Oct 25th; 1, Nov 5th.
Hook-with-Warsash: 1, Jan 21st.
Southampton Water (Town Quay to Fawley): 1, Feb 5th/6th; 1, Oct 28th.
Lepe: 1, Jan 15th and 30th.
Pennington Marsh/Lymington: 1, Jan 15th; 1, Oct 10th; 1, Oct 15th.
Hurst/Milford Area: 1, Jan 2nd; 2, Jan 8th; 2, Jan 10th; 3, Jan 16th; 3, Jan 22nd; 2, Apr 3rd; 2 E, Apr22nd; 1 E, Apr 23rd; 1, Apr 24th; 5 (including 2 E), Apr 26th; 1, Apr 28th; 1, May 6th; 1, May 13th; 1, Sep 9th; 4, Sep 12th; 1W, Oct 8th; 1, Oct 26th; 1, Dec 22nd; 1, Dec 29th.
Taddiford Gap: 5, May 2nd; 1, Nov 17th.

Little Auk   *Alle alle*

A very scarce winter visitor, usually appearing following storms (10+,115+,2)

Singles were off Black Point, Hayling on Nov 15th (ACJ) and Hill Head on Dec 3rd (DH). The first of these attempted to join the high tide wader roost before teaming up with a Skylark and flying south!

Auk sp.   *Uria/Alca sp.*

A total of 64 were observed along the coast as follows:

Hayling Bay/Sandy Point: 2 E, Oct 28th; 14 (including 11 E), Feb 15th; 3, Mar 15th.
Milford on Sea/Hurst Beach: 2, Jan 10th; 7 E, Jan 16th; 1 W, Mar 11th; 12, Apr 23rd; 1 W, May 8th; 8, May 28th; 9, May 31st; 2, June 4th.
Taddiford Gap: 2, Nov 27th.

The table below summarises the annual totals of bird-months 1996-2005 of the four auk species:

Annual bird-months	1996	1997	1998	1999	2000	2001	2002	2003	2004	2005
Puffin	3	0	2	0	3	0	0	2	0	0
Guillemot	32	24	39	56	54	33	201	102	110	53
Razorbill	26	23	21	29	30	30	117	117	127	57
Little Auk	9	3	14	1	1	3	5	4	3	2

Feral Pigeon   *Columba livia*

A common resident.

No records were received this year.

Stock Dove   *Columba oenas*

A numerous resident and winter visitor [Amber].

Just two early year flocks reached three figures – 160 at Wade Court on Jan 4th, and 193 at Chilling Copse on Feb 23rd. Only nine other double figure counts were received during this period. In late spring, two gatherings of 37 and 40 were at Avon Causeway and Half Moon Common NF on May 22nd and 25th respectively. In addition, one flew in off the sea at Stokes Bay on May 12th.

BBS data produced 39% occupation of the 71 survey squares, at a mean of 2.0 per square. A scattering of other breeding season records was received but none was exceptional.

Autumn gatherings began to be noted in late July (25 at Gander Down on 31st), but no grounded flocks larger than 50 were recorded in the second half of the year. Given the increased interest in Wood Pigeon movements in recent autumns, it is perhaps no surprise that significant numbers of Stock Doves were also seen on the move in this season. Some 1500 diurnal passage birds were reported, although there may have been some duplication at the main sites in the far south-west of the county. A large majority flew west or south-west (e.g. 165 SSW at Hurst on Nov 1st), although the single biggest count was of 350 east on the same date at Broadmarsh.

Christchurch Harbour (Dorset) also recorded a heavy Stock Dove passage in autumn 2005, peaking at 1500 west on Nov 14th. Hampshire observers are encouraged to look for further evidence of autumn Stock Dove movements in amongst the larger numbers of Wood Pigeons.

Wood Pigeon   *Columba palumbus*

An abundant resident and winter visitor.

There were four early year reports of four figure flocks, on the chalk and the north-east, the largest being 3000 (minimum) at Bourley on Mar 1st, possibly indicating spring passage; in contrast to recent years, no movements suggesting spring migration as such were detected.

Once again, the BBS confirmed the abundance of the species during the breeding season. The species was present in every square, with an average of 27 birds per square.

The now expected autumn movement commenced early, with an unusual report of 1200 north over Pipers Wait in one hour on Aug 28th. However, as usual, the main movements came much later in the autumn. From mid-autumn, four figure counts began to be made along the coast, with a pronounced peak in the three weeks from Oct 23rd, especially on this first date and Nov 4th-7th, with another pulse around Nov 11th-14th. An astonishing (but no doubt under-representative) 141,000 birds were logged. The maximum count was of 23,544 SSW in less than three hours at Miles Hill near Fleet on Nov 5th, but many observers, both inland and on the coast, recorded four figure flocks in short spaces

Wood Pigeon © *Richard Ford*

of time. Unusually, the direction of movement was not overwhelmingly between south and west this year – many flocks were recorded heading south-east or east. The precise meaning of these movements remains poorly understood, but an impressive database of information now exists, and still further watching each autumn could further knowledge of the origins of these migrations/movements.

In neighbouring counties, and mirroring our records, 71,300 moved over five Sussex sites on Nov 4th, with 334,465 counted the next day (175,000 over Brighton alone); then 54,500 moving SW, with some heading SE, at another three Sussex sites on Nov 13th. To the west Christchurch Harbour also recorded high numbers, peaks of 35,000 and 41,000 south-west on Nov 4th and 14th respectively, while Portland recorded 74,120 (mostly south) in October and November, with a peak of 37,000 on Nov 1st.

Only two 1000+ flocks were reported in the late year, although an interesting late movement of 400 west was detected at Milford on Sea on Christmas Day.

Collared Dove   *Streptopelia decaocto*

A numerous resident and passage migrant.

Once more, the fragmentary nature of the data received means few conclusions can be drawn but BBS data indicate 54% of squares occupied, at a mean of 3.4 per occupied square which is 15% down on the previous ten-year average.

The peak garden count was of 60 at Petersfield on Jan 7th, but 175 near Fareham on Oct 4th and 340 at Penton Grafton on Nov 13th presumably relates to autumn concentrations around grain stores. A few reports of autumnal visible migration were received. For example, 15 moved high west or south-west over Broadmarsh on two dates in early November.

Turtle Dove
Streptopelia turtur

A moderately common, but declining, summer visitor and passage migrant [SPEC 3, Red, NBAP, HBAP].

One was present at Townhill Park, Southampton, from Christmas 2004 until at least Feb 6th (SI). This is about the sixth winter record for the county, and the first since early 2001.

Apart from this bird, a further increase in reporting saw 160 records entered into the database (up from 144 last year).

The first spring migrants were at Hurst and Farlington on Apr 25th, and a further seven were found before the end of that month. The main arrival was in early May, and while (as ever) it is very hard to estimate precise numbers of breeders, the species does appear to be holding its own (just) in the county, with 80-90 territories reported. A most welcome report was of 20 territories at Martin Down, an important site which has seen recent declines, and seven at Old Winchester Hill. No reports were received of local extirpations. After the usual sprinkling of dispersing birds in late summer, some 12 or so coastal migrants were detected, mostly in the eastern part of the county. Just one was noted in the latter half of Sep, and the last was one at Sandy Point on Oct 1st.

Ring-necked Parakeet
Psittacula krameri

A very scarce visitor, possibly resident.

Three reports of singles were received this year, from Locks Heath on Mar 13th, Hill Lane, Southampton on Aug 10th, and Dunmow Hill, Fleet on Sep 25th.

Cuckoo
Cuculus canorus

A moderately common summer visitor [Amber].

The first was a singing male at Faccombe on Mar 28th, and a further ten were recorded in the first ten days of April. A further 34 in the following ten days indicates the main arrival period, but passage and arrival took place at both coastal and inland sites well into May. The maximum spring migrant count was of seven at Farlington Marshes on May 4th.

Some 40-50 records of territorial males or proved breeding were submitted, allowing for duplication at well-watched sites, with three at Binswood and Titchfield Haven and four on the Beaulieu Estate. In addition 40% of the 75 BBS squares surveyed held Cuckoos, at an average of 1.7 per occupied square.

July records were received from just four sites, and most were juveniles, including singles on three well-scattered dates at Farlington Marshes which were probably all locally bred. Just eight probable migrants were recorded in late summer, three of them at Farlington Marshes, and seven of them juveniles. The last was an adult on the very early 'late' date of Aug 6th.

Barn Owl
Tyto alba

A moderately common resident [SPEC 3, WCA 1, Amber].

Reports came from about 80 sites, slightly higher than in recent years. The majority were from well-watched coastal areas and the main river valleys. Most were of single birds or pairs, the only exceptions being three seen at Titchfield Haven on Mar 30th and three at Lane End Down on Dec 17th. The latter record included two road-kills. Confirmation of breeding was obtained for 17 pairs, a much higher number than is normally reported. Many were in nest boxes underlining the importance of artificial nest sites for this species. One observer in the New Forest was fortunate to add Barn Owl to the list of species visiting his bird table.

Little Owl
Athene noctua

A common resident [SPEC 3].

As with the previous species, reports were received from about 80 sites mainly on the coast, the main river valleys and the north-east. Records from the centre of the county were sparse, probably reflecting the lack of coverage rather than the scarcity of the species. Most records were of one or two birds but there were three in the Hoe Cross area on Apr 3rd.

Tawny Owl
Strix aluco

A common resident.

Although this is the commonest of our owls, relatively few records are received and very little to indicate its population status. Even at the national level, the information obtained through the BBS is not statistically significant. It is, therefore, good to have results from the 2005 BTO Tawny Owl Survey which can be compared directly against one carried out in 1989. The 2005 survey, which was based on point counts in 109 randomly selected tetrads across the county, found a net increase in territories compared to the 1989 survey. Only the Wealden Heaths, in the extreme east of the county, showed a decrease. Full details of the Survey are described elsewhere in this Report.

Successful breeding was confirmed at 11 sites.

Long-eared Owl
Asio otus

A very scarce resident, passage migrant and winter visitor.

There were four records, all of single birds:

Beacon Hill, Warnford: Mar 1st (KWM).
Lasham: Feb 26th (GJSR).
Farlington Marshes: Aug 4th/5th (GHC, JCr).
South Stoneham: Nov 11th (J Tamblyn per JMC).

Although this was an improvement on 2004, the run of poor years continued. The last proven breeding record for this species in the county was in 2000.

Short-eared Owl
Asio flammeus

A scarce but regular winter visitor and passage migrant that occasionally breeds [ET, SPEC 3, Amber].

Numbers in the first winter period were boosted by counts from a roost site in the centre of the county which peaked at nine on Feb 16th. There were also counts of up to three from the New Forest and from two other sites, one in the south and one in the north of the county, and two from Langstone Harbour islands and another northern site. Elsewhere there were records of singles at Mottisfont on Mar 10th and Firgo Farm, Longparish on 31st.

Likely passage birds were seen at Chilbolton on Apr 21st, Eversley GP on Apr 25th, at a site in the west of the county on May 4th and the last at Keyhaven/Pennington Marshes on May 7th.

In autumn, the first were not seen until Oct 15th when singles were at Sandy Point and Woolmer Pond. Thereafter most records were from coastal sites including: Tournerbury (1, Oct 23rd; 2, Nov 19th); Fort Nelson (1, Nov 2nd); Titchfield Haven (1, Nov 4th and 26th); Sandy Point (1, Nov 12th and 19th); Langstone Harbour/Farlington Marshes (1, Nov 12th, 17th and 19th); Sinah GP (1, Nov 13th) and Keyhaven/Pennington Marshes (1, Dec 16th and 19th; 2, Dec 31st).

Numbers at the central roost site built up to three by Dec 22nd. The only other inland records in the late year were of singles at Martin Down on Dec 20th, Kingsclere on Dec 27th and Kingsley on Dec 28th.

Approximate monthly totals are tabulated below:

J	F	M	A	M	J	J	A	S	O	N	D
13	19	14	9	2	0	0	0	0	4	7	7

2004 addition: one was flushed by a Marsh Harrier from the south scrape at Titchfield Haven at 1610 hrs on Aug 30th; it was probably the bird seen at Hook-with-Warsash the next day.

Nightjar   *Caprimulgus europaeus*

A moderately common summer visitor and passage migrant [ET, SPEC 2, Red, NBAP, HBAP].

The first were heard at Millyford Bridge, NF and Bourley & Long Valley in the north-east on May 1st. Thereafter all records were of churring males at probable breeding sites.

Following the higher-than-usual effort put into counting this species in 2004 as part of the national Nightjar Survey, the 2005 effort was much reduced. Available counts of territorial males at sites outside the New Forest are tabulated below:

Thames Basin Heaths		Wealden Heaths	
Bourley North/Long Valley	12	Alice Holt, Abbotts Wood Inclosure	2
Bramshill Plantation	11	Broxhead Common	2
Bourley South/Bricksbury Hill	7	Woolmer Forest	24
Eelmoor Marsh/Pyestock Wood	2	Longmoor Inclosure	19
Eversley Common/Castle Bottom	3	**South-central sites**	
Hazeley Heath	5	Ampfield Wood	3
Tweseldown	3	Embley Wood	2
Velmead Common	1	Great Covert Wood	6
Warren Heath/Heath Warren	16	**South-eastern sites**	
Yateley Heath Wood	4	Botley Wood	3
Other Northern sites		Havant Thicket	3
Basing Forest	2	West Walk	7
Benyons Inclosure	4	**Elsewhere**	
Silchester Common	6	Burton Common	6
Tadley Common	3	Chawton Park Wood	3
		Martin Down	1

A further 36 territories were counted in the New Forest, including a notable nine at Badminston Common.

The last, and only record after mid July, was one at Embley Wood on Sep 1st.

2004 addition: one was photographed perched on the window sill of a house at Barton on Sea on Oct 7th (V Sullivan). This is the first in October since one was found dead on a road near Titchfield Haven on Oct 1st 1977. The latest ever were at Dur Hill Down on Oct 13th 1974 and Farlington Marshes on Nov 23rd 1958.

Swift   *Apus apus*

A numerous summer visitor and passage migrant.

The first was at Farlington Marshes on Apr 15th followed by reports of singles at Hurst Beach and Woolmer Pond on 16th, and Lower Test Marshes and Testwood Lakes on 17th. The next were not until 23rd when up to three were seen at eight sites. The first double-figure counts were made on Apr 24th with 18 at Farlington Marshes and 59 at Testwood Lakes and the first three figure counts next day, with 110 at Farlington Marshes and 100 at Titchfield Haven. The highest count of the spring was 365 moving north at Lymington/Hurst on May 9th.

Breeding information was, as usual, limited but the general impression is of a continuing slow decline in the number of breeding pairs. Of the very few records received, ten pairs bred on Whale Island in Portsmouth Harbour and three pairs on a housing estate in Stubbington.

The first birds reported moving south were 150 over Winchester on July 26th. Return passage was unspectacular with a high count of 240 moving south-east at Itchen Valley CP on July 31st. Numbers dropped rapidly in August with a high count of just 33 at Farlington Marshes on Aug 6th and no double-figure counts after that date. There were five September and four October records, with the last birds reported by *Birdguides* at Lymington/Hurst on Oct 29th and Chandlers Ford on Oct 30th.

Alpine Swift   *Apus melba*
A rare vagrant (3,6,1).

One flew east over Brownwich on Oct 30th (GO). The record has been accepted by *BBRC* and is the tenth for Hampshire. This is a typical flyover sighting and follows the three in 2004. What was presumably the Brownwich bird was reported flying east over Sandy Point shortly afterwards, but there are inadequate notes to submit the record.

Kingfisher   *Alcedo atthis*
A moderately common resident whose numbers may be severely depleted during harsh winters [ET, SPEC 3, WCA 1, Amber].

Records were submitted from around 130 sites. The majority were from the main river valleys, the north-east GPs and, in autumn and winter, from the coast. Breeding season records (March-July) were received from almost 50 sites although few successful nests were recorded. Counts of territories included five at Somerley Park in the Avon valley and two at Longstock.

Most records were of ones or twos but there were eight at Yateley GP on Sep 26th, seven on Oct 19th and six on Nov 9th. There were also six in the Lymington/Hurst area on Sep 25th. Regular counts at well-watched sites confirmed post-breeding dispersal to the coast. This was demonstrated most clearly at Normandy and Titchfield Haven, where monthly bird-day totals were:

	J	F	M	A	M	J	J	A	S	O	N	D
Normandy	34	17	5	4	0	1	28	58	62	64	58	56
Titchfield Haven	19	13	4	1	0	2	14	22	19	30	26	22
Black Dam	5	1	0	0	0	2	2	6	2	3	2	2

The same pattern was also observed on inland waters as demonstrated by bird-day totals for Black Dam NR, Basingstoke in the above table.

Bee-eater   *Merops apiaster*
A rare vagrant (0,10,3).

Two circled with Swifts over the scrapes at Titchfield Haven at 1130 hrs on June 3rd before leaving north (BG, RP). These were probably the two present at Shoreham, West Sussex on June 1st/2nd. An adult was watched over the northern edge of Sandy Point NR between 1804 and 1811 hrs on Aug 3rd before drifting off west (ACJ).

Hoopoe
Upupa epops

A very scarce passage migrant; bred on eight occasions during 1953-59 but not since (?,207,3).

The only spring record was of one at Cams Bay on Apr 9th (DIB). In autumn, one was reported at Broughton on Oct 6th and 7th (per SKW) and a very late bird was seen at Sandy Point from 0720-0735 hrs on Nov 14th (ACJ). This was subsequently seen flying west over Beachlands (RHM) and was reported, later that day, from Walpole Park, Gosport (per JRDS).

Wryneck
Jynx torquilla

A very scarce passage migrant; formerly bred.

Wryneck   © George Spraggs

Following an above average year in 2004, this year was even better with records from 13 sites involving at least 15 individuals. All records were of singles in autumn: Farlington Marshes, Aug 25th-Sep 1st, Sept 12th and Sep 17th-29th; Alton, Aug 31st-Sep 2nd; Milkham Bottom, New Forest, Sep 1st; Needs Ore, Sep 3rd; Mill Rythe, Sep 6th; Abbottswood Romsey, Sep 9th/10th; Titchfield Haven, Sep 11th-18th; IBM Lake, Sep 18th; Sandy Point, Sep 21st-29th; Keyhaven, Sep 26th; Backley Plain, NF, Oct 9th; Langstone Harbour north-west shore, Oct 15th and Stubbington, Oct 23rd.

Green Woodpecker
Picus viridis

A common resident [SPEC 2, Amber].

Results of the 2005 BBS show that the population of Green Woodpeckers in south-east England has increased by 28% since the survey began in 1994. It is a common bird in

Hampshire, particularly on heathland where densities can be high. Examples of counts on the north-east heaths included nine territories on Warren Heath and Heath Warren, 20 on Bourley & Long Valley including Tweseldown and Velmead Common, and 21 in Longmoor Inclosure. Elsewhere, on non-heathland sites, there were six territories in the Lymington/Hurst area, seven on the Lower Test NR CBC plot and six at Brownwich/Chilling.

At Farlington Marshes, a non-breeding site, bird-day counts showed a clear autumn peak, indicative of dispersal, probably from nearby breeding areas:

	J	F	M	A	M	J	J	A	S	O	N	D
Farlington Marshes	4	0	8	3	2	0	2	1	18	24	5	9

Records from other coastal non-breeding sites showed a similar distribution with singles at Paulsgrove Reclamation on Jan 15th; Milton Reclamation on five dates between Aug 18th and Oct 8th; Great Salterns Quay, Langstone Harbour on Oct 4th and Taddiford Gap on Nov 14th.

Great Spotted Woodpecker   *Dendrocopos major*

A common resident.

This species is faring even better than Green Woodpecker. BBS results show that its population in south-east England has more than doubled between 1994 and 2005. Counts on the north-east heaths included 20 territories on Bourley & Long Valley/Tweseldown/Velmead Common, seven on Heath Warren/Warren Heath and 20 in Longmoor Inclosure. Elsewhere there were nine successful territories in Roydon Woods NF.

Records from coastal non-breeding sites peaked in late summer/early autumn, indicative of either post-breeding dispersal or longer distance migration. Monthly bird-day totals from well-watched sites were as follows:

	J	F	M	A	M	J	J	A	S	O	N	D
Normandy	0	0	2	2	6	8	16	20	8	6	9	6
Farlington Marshes	1	1	1	0	0	1	1	3	2	6	1	7
IBM Lake	1	0	0	0	0	4	0	0	7	6[†]		0
Sandy Point	0	0	2	0	0	1	7	3	5	2	2	0

[†] total for October and November.

The Farlington records included one high west over the reserve on Sep 28th. Probable migrants were also seen at several other coastal sites including: Keyhaven, four on Sep 10th; Hook Spit, one in tamarisk on Sep 24th; Brownwich, one west on Sep 24th and Broadmarsh/Bedhampton, one west on Nov 1st and one north-west on Nov 14th.

Lesser Spotted Woodpecker   *Dendrocopos minor*

Once moderately common, perhaps now a scarce resident [Red, HBAP].

Although the UK population fell by 73% over the period 1970-2003, recent indications in Hampshire suggest that the local population is now recovering. The 2005 total of 152 records from 92 sites represents an almost 50% increase over 2004, and maintains the upward trend first noted in HBR 2003. Of the 92 sites, 39 were in the New Forest, with 13 in both the Test Valley and in the north-east.

Lesser Spotted Woodpeckers are essentially sedentary so most of the records would indicate birds on, or close to, their breeding sites. Even so, direct evidence of successful breeding was sparse. Young were seen at Bordon, Fleet Pond, and Brinken Wood in the New Forest. A juvenile was trapped at Southampton Common on Aug 15th.

Lesser-spotted Woodpecker © *Nigel Jones*

Woodlark
Lullula arborea

A moderately common but local resident and passage migrant [ET, SPEC 2, WCA 1, Red, NBAP, HBAP].

Following another mild winter, there were several early records of singing males at known breeding sites, including singles at Warren Heath on Jan 1st, Woolmer Forest on Jan 2nd, Winter Down on Jan 3rd and three at Bourley and Long Valley on Jan 11th. Consistent with an early return to breeding sites, there were few reports of wintering flocks, although 26 at Wickham Hundred Acres on Jan 5th was the largest gathering of the year. Other flocks during the first winter period included six at Princes Marsh also on Jan 5th, and 11 at Farnborough airfield on Feb 22nd. The only record indicative of spring passage was of a bird moving north at Lower Test Marshes on Feb 2nd.

During the breeding season 121 pairs/singing males were located in the north-east (Thames Basin and Wealden Heaths plus other north-east sites). This was an increase on 2004 (114 pairs/singing males). About 36 pairs/singing males were reported from the New Forest.

The expansion from traditional heathland sites to agricultural areas, first noted in HBR 2000, continued. Records of territorial birds came from farmland areas across the county. In the following table, these are grouped by river valley to give an idea of their geographical spread. Singing birds were found in a variety of habitats including set-aside and a newly sown bean field. Such habitats remain vulnerable to farming activities. For example, at Winters Down up to six territories were destroyed when 150 acres of fallow were ploughed in early April.

Counts of pairs/singing males outside the New Forest are tabulated below:

Thames Basin Heaths		The Warren, Oakhanger/Blackmoor GC	1
Bourley North/Long Valley	9	Woolmer Forest	17
Bourley South/Bricksbury Hill	3	**Other North-east**	
Bramshill Plantation	8	Bentley	1
Eversley Common/Castle Bottom/Blackbushe	5	Stodham Park	2
Eelmoor Marsh/Pyestock Wood	2	**Test valley**	
Hawley Common	2	Casbrook Common	1
Hazeley Heath	2	Mottisfont area	3
Minley	1	Romsey area	2
Tadley & Silchester Commons/Benyons Inc.	5	Other Test Valley	5
Tweseldown	3	**Itchen valley**	
Velmead Common	2	Twyford	1
Warren Heath/Heath Warren	16	**Meon valley**	
Yateley Common	6	Brookwood Copse	1
Yateley Heath Wood and Lowmans Wood	3	Shirrell Heath	1
Wealden Heaths		West Meon	2
Bramshott Chase/Bramshott Common	1	Wickham Hundred Acres	1
Broxhead Common	5	**Other sites**	
Hammer Common	1	Ampfield	1
Liss Nurseries/Prince's Marsh/Lyss Place Fm	4	Breamore Down	1
Longmoor Inclosure	19	Hinton Ampner	1
Ludshott Common	2	Parnholt Wood	1
Shortheath Common	2		

Passage birds were noted in October and November at several non-breeding sites: Fleet Pond (1, Oct 2nd); Gilkicker Point (3 E, Oct 3rd); New Milton (1 E, Oct 7th); Hambledon (1 E, Oct 8th); Tanners Lane (1 E, Oct 10th); Sandy Point (5 E, Oct 15th, 1 W, Nov 4th); Old Winchester Hill (1 SW, Oct 15th); Browndown (1, Oct 15th/16th); Budds Farm SF (3 E, Oct 16th); Sinah Common, (2 N with finches, Oct 16th); Milford on Sea (2 S, Oct 23rd); Broadmarsh (1 W, Oct 26th and 1 SE, Nov 4th); Lower Test Marshes (1 S, Oct 29th) and Hook-with-Warsash (4 N, Nov 13th).

Notable post-breeding flocks were reported from both breeding and non-breeding sites. The former included 14 at Woolmer Pond on Sep 15th, 12 nearby at Woolmer Forest on Oct 23rd and 10 at Princes Marsh, Liss on Dec 22nd. The wintering flock at Wickham Hundred Acres held 14 on Oct 29th and remained at this level until the end of the year. Flocks at non-breeding sites included seven at Preshaw on Oct 22nd and eight in stubble at Hartley Wintney on Dec 8th.

Skylark *Alauda arvensis*

A numerous resident, passage migrant and winter visitor [SPEC 3, Red, NBAP, HBAP].

The largest flock reported during the first winter period was 300 at Brownwich on Jan 15th. The count of 150 at Sleaford Malthouse Farm on Jan 13th was the highest there for several years but the only other three-figure count was 125 at Gander Down on Jan 23rd.

The only evidence of spring passage came from Hurst Castle, with one over on Mar 19th, and Sandy Point, with seven bird days recorded in March (including three east over the sea on 21st), one in April and three in May.

Breeding season counts included: 16 territories in the Lymington/Hurst area; 32 at Brownwich/Chilling; 13 on the Langstone Harbour islands; 11 on Windmill Down and 23 on half of Butser Hill. At Bourley and Long Valley SSSI where the species has re-colonised the area following the creation of suitable grassland habitat, the number of territories continued to increase from four in 2003, to eight in 2004 and at least nine in 2005.

In autumn, diurnal movement was noted at both coastal and inland sites between Sep

20th and Nov 22nd, peaking in the second half of October. Half-monthly totals of movements and settled flocks are tabulated below:

Skylark	Sep 15-30	Oct 1-15	Oct 16-31	Nov 1-15	Nov 1-15
Movements	12	327	596	214	26
Flocks on the ground	9	629	741	590	678

The largest flock during the autumn period was 300 at Hillside on Oct 16th. Other notable counts included 210 in a stubble field at Stoke Charity on Nov 14th and 214 at Upham on Nov 17th.

Sand Martin   *Riparia riparia*

A common breeding summer visitor and numerous passage migrant [SPEC 3, Amber].

There was no early arrival in 2005, with the first birds being two at Tundry Pond on Mar 18th. A scattering of single figure counts over the next few days preceded the first major arrival of 84 at Testwood Lakes on Mar 24th. This site also held the largest spring gathering, of 200 on Apr 10th. Arrivals continued, albeit with mostly low double figure counts, well into May, with the last spring migrants at Farlington Marshes on May 21st.

Some 800 active nest holes were counted at just six colonies, with the largest count of 543 at Kimbridge (down from 722 in 2004). A roost of 700 was noted at Lower Test Marshes as early as June 25th presumably including many juveniles. Records recommenced at Farlington Marshes just two days later.

Autumn migration proper was certainly under way by mid July, with many reports of small parties on the move along the coast, and significant gatherings at sites such as Ibsley Water (maximum 400 on Aug 6th). A distinct easterly component was evident in late Aug movements, with some 150 birds noted on the move at a variety of locations. A total of 131 were recorded in Sep, and a further 13 in Oct, including the last (10) east at Gilkicker Point on Oct 16th.

Swallow   *Hirundo rustica*

A numerous summer visitor and abundant passage migrant [SPEC 3, Amber].

After an early arrival at Testwood Lakes on Mar 13th, the next was not until 24th, with a further 48 reported until the end of the month, and a peak of six at Titchfield Haven on 31st. The first double figure count did not come until Apr 10th, with 34 at Testwood Lakes. The main arrival was underway on Apr 16th, when a large passage occurred over Keyhaven/Pennington Marshes (over 500 estimated moving north in just 2½ hours) and smaller peaks were noted at Fleet Pond, Tundry Pond and Farlington Marshes that day. Another major pulse of arrival occurred on Apr 24th, with over 800 birds detected, including 249 north at Sandy Point and 324 north-west at Testwood Lakes. Still more appeared next day in classic spring fall weather conditions with 121 arriving over Sandy Point, 100 north-west over Broadmarsh and at least 200 feeding at Farlington Marshes. The largest spring count was made on Apr 29th with 710 north at Pennington Marsh. Smaller scale arrivals continued throughout May and daily recording at Sandy Point detected movement up to June 10th.

BBS data suggested no change in breeding fortunes – Swallows were found in 68% of squares, with almost five birds per occupied square on average.

Autumn roost data was received only from Farlington Marshes this year, where numbers were rather lower than usual. The count peaked at 350 on Aug 7th, and during the traditional peak period in late August, just 80 or so were present.

Visible autumn migration began about Aug 9th, and an impressive series of three figure counts (43 in all) was made between Aug 31st and Oct 14th. Almost 20,000 Swallows were counted over this period, mostly on the coast and 62% between Sep 10th and 25th. The highest counts were of 1342 south at Gilkicker Point on Sep 3rd, and a stunning 5730 east at Barton on Sea in just 2½ hours on the morning of Sep 18th. Numbers quickly tailed off in mid October, with the last double figure count being of 30 at Calshot on 20th, and only 22 (mostly singles) seen after that. The last certain migrant was one at Tanners Lane on Nov 7th (one of only two in that month). One at Mengham (Chichester Harbour) on Dec 8th might conceivably have been a wintering bird rather than a late migrant.

2004 addition: A BTO Swallow Feeding Survey was conducted at four points in each of 46 randomly selected tetrads across the county between May 23rd and Aug 11th 2004. Swallows were seen foraging at 46% of the184 points surveyed. Other species noted at the same survey points included: Sparrowhawk - 4%, Kestrel - 8%, Hobby - 2%, Swift - 11%, House Martin - 26% and Sand Martin - 2%.

House Martin *Delichon urbicum*

A numerous summer visitor and abundant passage migrant [SPEC 3, Amber].

Just five were found in late March, the first being one at Testwood Lakes on 27th. Numbers remained low in early Apr, with just 25 seen in the first ten days. The first main arrival was closely correlated with the second large pulse of arriving Swallows in late April, with peaks of 53 north at Sandy Point on Apr 24th/25th and 100 gathered at Fleet Pond on 28th. Arrivals were poorly detected along the coast in May and numbers gathered inland were generally low with just four counts over 100. The spring maximum was 225 at Fleet Pond on May 17th. Low scale migration continued into June with passage detected at Farlington Marshes as late as Jun 17th.

No data relating to breeding were submitted this year. BBS data do not track this species well, but no change in breeding figures was suggested by the available data.

A post-breeding gathering of 300 on Binstead church roof (80% juveniles) on Aug 6th presaged the autumn migration, and by the end of the month, many small parties were on the move. Almost 15,000 migrants were logged through the autumn, well down on last year's exceptional totals, but some impressive movements were recorded nonetheless. While numerous three figure flocks were seen in mid September all along the coastal strip, and also inland, the biggest counts were made in the far south-west, especially at Barton on Sea. A total of 1181 flew east on Sep 10th, and 2170 did the same on the 18th. Several large flocks were seen in early Oct, with a major movement of c.1500 at Hurst on Oct 7th. Thereafter, numbers rapidly dwindled, with 35 east at Farlington Marshes on Oct 16th being the last record there. Just eight were seen subsequently in the county, with only one of them in November, at Titchfield Haven on 8th.

Red-rumped Swallow *Cecropis daurica*

A very rare vagrant (0,2,2).

One was watched for 15-20 seconds at close range as it migrated over Farlington Marshes in association with a House Martin at 0913 hrs on May 10th (JCr). This was one of 20 recorded in England during the spring. In autumn, one was watched for one minute with a large flock of hirundines over Keyhaven Harbour and reedbed at 1710 hrs on Oct 10th (TP). This was one of just three records in autumn in Britain.

Both records have been accepted by *BBRC* and are the third and fourth for the county. The only earlier records were of single birds at Farlington Marshes on Nov 14th 1987 and at Pitts Wood Inclosure on July 10th 2002.

Richard's Pipit
Anthus novaeseelandiae

A rare passage migrant (0,24,2).

One was heard and seen as it flew east over Butts Lagoon, Pennington Marsh at 0830 hrs on Sep 17th. It landed 100 meters to the east and was observed briefly before flying off north-east, apparently landing again in the vicinity of Oxey Marsh (RBW). This is the earliest autumn record for the county, the previous earliest being on Sep 19th 1999 nearby at Pennington Marsh. At least six were reported in southern England on Sep 17th/18th. Over two months later, another individual was located in fields adjacent to Lower Pennington Lane at 1015 hrs on Nov 20th (SKW).

Tawny Pipit
Anthus campestris

A rare passage migrant (0,14,1).

One flew over calling at Sandy Point at 0625 hrs on May 10th. It landed on the beach where good views were obtained but was then lost to view as it moved off north (ACJ). This is the first ever spring record for the county.

Tree Pipit
Anthus trivialis

A moderately common summer visitor and passage migrant (Amber).

The first arrival was at Barton on Sea GC on Mar 16th. The next reported was one on Mar 27th at Bricksbury Hill and another on 30th at Sinah Common. These two sightings preceded a steady influx over the next four weeks. The first singing male was heard at the Canadian Memorial in the New Forest on Apr 5th. Passage continued throughout April and a group of nine was found grounded by thick fog at Hurst Castle on Apr 25th.

In the north-east of the county, on the Thames Basin and Wealden Heaths, a total of 112 territories were established over 18 sites. No reports were received from Hawley Common and Bramshott Common where one and four pairs, respectively, established territories last year. At Warren Heath, numbers were halved from ten territories last year to five.

The New Forest attracted slightly better coverage this year resulting in 22 singing males/territories being located at 19 sites. Away from the strongholds of the north-east and the New Forest, a total of 45 singing males heard from 16 different sites was a significant and welcome increase over previous years. A total of ten singing males was heard at Medstead and seven each at Chawton Park Wood and in Ringwood Forest. Counts of pairs/singing males at sites away from the New Forest are shown in the table below.

Thames Basin Heaths				
Bramshill Plantation	5	Conford Moor	-
Bourley South/Bricksbury Hill	12	Cranmer Heath	-
Bourley North/Long Valley	6	Longmoor Inclosure	27
Castle Bottom	3	Ludshott Common	-
Eelmoor Marsh/Pyestock Wood	2	Passfield Common	-
Hawley Common	-	Shortheath Common	5
Hazeley Heath	6	The Warren, Oakhanger	-
Heath Warren	1	Woolmer Forest	20
Holt Pound Inclosure	1	**Other sites**	
Lichett Plain/Lowman's Wood	1	Abbotts Wood	2
Pyestock Heath	-	Ashmansworth	2
Tweseldown	3	Bentley	4
Velmead Common	4	Bossington	1
Warren Heath	5	Broughton Down	1
Yateley Common	3	Chawton Park Wood	7
Yateley Heath Wood	4	Damerham	1
Wealden Heaths		Four Marks	2
Bramshott Common	-	Lasham Airfield	1
Broxhead Common	4	Liphook	1
		Lower Froyle	1

Other sites (continued)		Ringwood Forest	7
Medstead	10	Spearywell Wood	1
Newton Valence	3	Upper Froyle	1
Noar Hill, Selborne	1		

Passage migrants were first seen on Aug 1st when singles were seen at Sandy Point and in the Lymington/Hurst area. A total of 185 was seen at 35 sites during the autumn, with numbers peaking between late August and mid September. Maximum counts occurred on Sep 12th when 14 were seen at Sandy Point and 15 in the Hurst Beach/Milford on Sea area. There were two records in October, of two heard over Old Winchester Hill on 1st and a very late bird moving west over Lakeside, Eastleigh on 24th.

Meadow Pipit   *Anthus pratensis*

A locally common resident, numerous passage migrant and winter visitor (Amber).

Early in the year, flocks of 20 or more were recorded from 16 different sites up to mid March, amounting to a total of 752 individuals across the county. Counts from the *New Forest wintering survey* totalled 128 on Jan 23rd and 109 on Feb 20th. Outside the New Forest, maximum counts of 55 were recorded from the Lymington/Hurst area on Jan 15th and 105 from Porton Down on Feb 15th.

Spring passage was noted between Mar 4th and May 6th, with a total of 5124 seen heading mainly north at 15 different locations. At the main coastal watch points, totals of 1428 were seen heading north or north-east at Hurst Beach/Milford on Sea between Mar 4th and Apr 25th and 605 at Sandy Point between Mar 11th and May 6th. A flock of 50, grounded by fog at Sinah Common on Mar 19th, was the only record of birds on the ground during spring. Overhead passage peaked between Mar 18th and Apr 2nd. On Mar 20th there were counts of 508 from the Lymington/Hurst area and 308 (in three hours) from Barton on Sea GC. On Mar 26th, peak counts of 359 were recorded from Hurst Beach/Milford on Sea and 455 from the Lymington/Hurst area. The spring peak at Farlington Marshes was 300 moving north-east on Apr 1st. Away from the coast, maximum totals of 143 were recorded flying north-west at Testwood Lakes on Mar 13th, 72 heading north at Long Down on Mar 18th, 72 heading north-west at Testwood Lakes on Mar 20th and 74 moving north at Long Down on Apr 15th.

Evidence of breeding was recorded at 16 sites around the county. A total of 22 singing males/territories was noted in the Lymington/Hurst area and 26 territories were established on the Langstone Harbour islands. In the north-east, a total of 17 territories was noted from five different sites. Elsewhere, 26 territories were recorded from 9 sites, the maximum concentration being 11 territories at Butser Hill. These figures show that there is a stable breeding population away from the New Forest where the species is most numerous.

A total of 13,140 was seen on passage in the autumn. Large movements were noted on Sep 12th when 1741 moved south-east at Hurst Beach/Milford on Sea, 1630 moved north-east at Farlington Marshes and 702 moved north-east at Sandy Point. Another large movement occurred on Oct 14th when 1220 moved north-east at Farlington Marshes and 737 headed east at Needs Ore/Beaulieu Estuary. A further 800 flew east at Sandy Point on Oct 16th. Major counts received from inland sites included: 250 at Nether Wallop on Sep 28th; 225 at Fleet Pond on Oct 1st, 645 at Testwood Lakes (during a full day of observation) and 200 at Old Winchester Hill on Oct 2nd.

During the winter months, counts of flocks exceeding 20 were received from 16 sites. Peak counts exceeding 50 individuals comprised: 60 at Eastleigh SF on Nov 20th; 60 at Tundry Pond on Nov 21st; 60 at Hordle Cliff on Nov 27th and 85 at Newlands Farm on Dec 10th. Additionally, a total of 97 was seen across 22 locations in the New Forest on Dec 17th.

Rock Pipit   *Anthus petrosus*

A very scarce resident, scarce passage migrant and winter visitor.

In the first three months of the year, 129 were recorded from 25 sites, with peak sightings of six at Barton on Sea on Feb 27th. Four singing males were heard there on Mar 20th but there was no evidence of breeding on this early date. At least nine were at Farlington Marshes on Mar 18th.

Five territories were established at Hurst Castle during the breeding season. Possible breeding was noted from ten other sites, amounting to a minimum total of 13 breeding pairs in the county. A pair was observed feeding young in the nest in the walls of Southsea Castle during May. An adult was seen carrying food at Barton on Sea on June 26th and two juveniles were seen fledged there on July 7th - the only report of successful breeding at any site. There were further sightings at Barton on Sea of up to three on six dates, with the last report of two on Oct 16th. The breeding report from Southsea represents the first known instance of breeding between Calshot and Seaford, East Sussex.

The earliest arrival in autumn record was one at Bunny Meadows, Hamble estuary, on Sep 19th - the first at Farlington Marshes on 22nd and at Sandy Point on 27th - with a total of 27 bird-days there to Nov 20th. During the last three months of the year, a minimum of 103 was seen at 31 sites. Peak gatherings during this period were a maximum of 19 in the Hurst Beach/Milford on Sea area on Nov 5th, eight at Langdon/Hythe on Dec 12th and 18th and eight on the Langstone Harbour islands on Dec 11th.

Scandinavian Rock Pipit *(A. p.littoralis)*

A very scarce passage migrant and winter visitor.

Records of birds showing characteristics of this race were received from Langstone Harbour and the saltings between Lymington and Hurst Castle.

At least some Rock Pipits acquire distinctive summer plumage in late winter/early spring when it is possible to distinguish this form from nominate *petrosus*. A single was noted at Farlington Marshes on three dates between Feb 22nd and Mar 30th (JCr) and at least two individuals in the Lymington/Hurst area on six dates between Mar 12th and Apr 4th (TP, KP, MPM, MR and M&ZW). As over-winterers may well move out of the county before commencing their moult into summer plumage, it is currently impossible to say what proportion of the county winter population of Rock Pipits are of this form.

Water Pipit   *Anthus spinoletta*

A scarce winter visitor and passage migrant.

In the first three months of the year, a minimum of 49 was seen at six different localities. There were peak monthly counts of 21, 26 and 29 at Lower Test Marshes. Elsewhere, an unusually high count of nine was at Keyhaven on Mar 28th, suggesting passage birds.

At least eleven individuals lingered into April, with maximum counts of three at Keyhaven and two at Dark Water, Lepe, both on 3rd; two at Needs Ore on 7th and three at Farlington Marshes on 9th - the last was at Lower Test Marshes on 16th.

The first autumn arrivals were one over Pennington Marsh on Oct 7th and another at Titchfield Haven on Oct 9th. Following these, singles were seen at Farlington Marshes on Oct 17th, Lower Test Marshes on Oct 21st and in the Lymington/Hurst area on Oct 22nd. A further two were observed at Dark Water, Lepe on Oct 23rd. A minimum of 26 was reported from seven separate localities over the winter period. The table below shows the monthly maxima at the main sites and approximate monthly totals for the whole county.

	J	F	M	A	M	J	J	A	S	O	N	D
Langstone Harbour	2	3	2	3						1	3	3
Lower Test Marshes	21	26	29	17						2	4	5
Lymington/Hurst Area	1	3	9	1						2	2	1
Titchfield Haven										1		
APPROX COUNTY TOTAL	**24**	**32**	**41**	**23**	**1**					**6**	**10**	**10**

Yellow Wagtail

Motacilla flava

Formerly a summer visitor and common passage migrant now rare in summer but still moderately common on passage (Amber, HBAP).

One was seen and heard flying north over Farlington Marshes on Mar 18th (JCr), the earliest for the county apart from one at Gilkicker Point on Mar 10th 1968. This was followed more than a week later by one flying east at Hurst Beach on Sea on 30th and one flying north over Keyhaven on 31st. There was then steady passage from mid April through May, with a total of 181 migrants seen at 19 coastal sites. The best site proved to be Farlington Marshes, where 67 bird-days was recorded during the spring up to May 17th. Additionally, 14 birds were seen at 6 inland sites, the majority of records coming from Woolmer Pond. The last migrant was one seen at Lakeside, Eastleigh on May 22nd. Peak counts during this period were 17 at Stokes Bay on Apr 24th, 32 grounded at Farlington Marshes on Apr 25th, and 16 at Gilkicker Point on Apr 30th. Half-monthly totals of bird-days are tabulated below:

Mar 15-31	Apr 1-15	Apr 16-30	May 1-15	May 16-31
2	4	152	36	2

The only records during the breeding season were of one flying into a reedbed at Farlington Marshes on June 8th and another at Needs Ore on July 20th. There have now been no positive breeding records for this species since 2002.

Post-breeding dispersal was noted from the beginning of August, with one at Testwood Lakes on Aug 7th and another the following day at Barton on Sea GC. Numbers peaked during the latter half of August, with 275 seen in the Lymington/Hurst area between Aug 19th and Oct 10th and 258 at Sandy Point between Aug 18th and Sep 26th. High numbers were recorded leaving the Barton on Sea GC roost with 129 and later another 94 west on Aug 28th and 84 on Sep 3rd. Counts of 25 or more are summarised below, followed by the minimum half-monthly totals.

Coastal sites
Sandy Point: 36, Aug 31st; 35, Sep 2nd; 40, Sep 12th.
Farlington Marshes: 25, Sep 2nd; 25 (roosting), Sep 13th.
Gilkicker Point: 39, Sep 3rd; 56, Sep 10th.
Browndown: 50, Aug 29th.
Titchfield Haven: 51 (roosting), Sep 27th.
Brownwich: 25, Sep 11th.
Hook-with-Warsash: 35, Aug 29th; 38, Aug 30th; 52, Aug 31st; 30, Sep 28th; 35, Sep 29th.
Lymington/Hurst: 31, Aug 31st.
Hurst Beach: 32, Sep 3rd; 46, Sep 9th, 34, Sep 12th.
Barton on Sea: 229, Aug 28th; 125, Aug 31st; 127, Sep 3rd; 109, Sep 11th.
Hordle Cliff: 26, Sep 12th.
Inland sites
Ibsley Water/Mockbeggar Lakes: 25, Aug 31st.
Abbottswood, Romsey: 45, Sep 15th.
Westend Down: 50, Sep 7th.

Half-monthly totals of bird-days are tabulated below:

Aug 1-15	Aug 16-31	Sep 1-15	Sep 16-30	Oct 1–15	Oct 16-31
6	1083	1348	333	32	0

The last for the year were singles at Farlington Marshes and Walpole Park, Gosport, on

Oct 8th.

Blue-headed Wagtail *(M. f. flava)*

A very scarce passage migrant, which has bred (n/r,n/r,2).

Records of birds showing characteristics of this race were received from two sites.

A male was at the Deeps on Farlington Marshes on May 1st (JCr *et al*). Up to four males and a female were seen at Abbottswood, Romsey between Sep 10th and Sep 15th (DT, KL, KP, PL). Both of these records involve birds found feeding with Yellow Wagtails *M.f.flavissima*.

Grey Wagtail   *Motacilla cinerea*

A moderately common resident, passage migrant and winter visitor (Amber).

During January and February, a minimum of 100 was reported from 56 sites, with a maximum of 15 at Chilbolton SF by Mar 2nd. A pair was reported nest building at Ellingham Meadows, Southampton on the early date of Feb 9th.

Probable migration was noted during the spring: two on the rocks at Barton on Sea on Feb 16th; one at Sandy Point on Mar 6th and two there on Mar 10th; singles flying north at Farlington Marshes and Sinah GP on Mar 11th and one flying north at Sandy Point on Apr 12th.

During the breeding season, a total of 34 pairs was confirmed as breeding at 28 different sites. Additionally, 21 pairs/singing males were noted at 18 sites.

Strong autumn passage was observed at most coastal sites between mid August and mid September; however, it is not known how many sightings are just post-breeding dispersal as birds were seen flying to all points of the compass. A minimum total of 331 was seen at 19 coastal sites. Peak counts were all in September, with 13 moving north or east at Sandy Point on 1st and 11 there on 10th, ten flying west at Barton on Sea GC on 7th, eight moving south at Hurst Beach on 12th and nine heading north-east at Farlington Marshes on 13th.

During November and December, a total of 101 was reported from 75 sites with a maximum gathering of 12 at Eastleigh SF on Dec 19th.

2004 additions: a roost at King's Pond, Alton held three on Sep 25th, 10-12 on Oct 21st, 13-14 on Nov 13th and 20 on Nov 27th.

Pied Wagtail   *Motacilla alba*

A numerous resident, abundant passage migrant and winter visitor.

Three-figure counts in the early part of the year, some involving winter roosts, were recorded as follows:

Coastal sites
Havant: 150, Feb 4th.
Calshot: 230, Jan 24th.
Inland sites
Bishops Waltham SF: 190, Jan 29th; 230 Feb 5th.
Kings Pond, Alton: 120, Feb 22nd.
Camp Farm SF: 150, Jan 18th; 200, Feb 16th; 150, Mar 14th.

Spring migration was noted at four coastal sites: two were seen heading north-west at Eastney on Mar 16th; 11 flew west at Barton on Sea on Mar 13th; three flew north-east at Farlington Marshes in March, one on 16th and two on 20th; and a total of 26 were seen flying north at Sandy Point between Mar 11th and May 1st. Additionally there were records of "alba" wagtails moving at four sites, including a spring total of 42 north at Hurst Beach/Milford on Sea between Mar 9th and Apr 15th which included at least 16 definite Pieds but also several Whites (which see separate account), and nine at Farlington Marshes.

A minimum total of 1838 was recorded on autumn passage at 11 coastal sites; undoubtedly a percentage of these were White Wagtails *M.a.alba* but were unable to be recorded as such. In autumn visible migration is better ascribed to "alba" wagtails rather than either of the two regularly occurring forms. An overall total of 463, mainly flying eastwards, was recorded at Sandy Point between Sep 10th and Nov 1st. At Farlington Marshes/Langstone Harbour, double-figure movements were made on five dates in October including 42 heading north-east on 14th. Peak movements were observed at Barton on Sea totalling 209 east on Oct 5th and 268 east on Oct 16th. An apparent hybrid male Pied Wagtail/White Wagtail was present at Farlington Marshes between Nov 7th and 17th.

During the latter part of the year, congregations of 100 or more birds were recorded at the sites listed below:

Coastal sites
Portsmouth: 200, Oct 28th.
Gilkicker Point: 155, Oct 16th.
Barton on Sea: 209, Oct 5th; 268 Oct 16th.
Inland sites
Ibsley North Lake: 105, Aug 8th.
Ibsley Water/Mockbeggar: 200, Sep 10th.
Barton Stacey SF: 140, Dec 6th.
Eastleigh SF: 150, Nov 16th; 200, Nov 20th; 300 Nov 26th.
Bishops Waltham SF: 100, Nov 26th; 110, Dec 28th.
Basingstoke: 370, Oct 13th; 159, Oct 14th.
Hook: 250, Sep 26th; 300, Dec 8th.
Darby Green SF: 150, Dec 11th; 250, Dec 18th.
Camp Farm SF: 175, Dec 28th.

White Wagtail *(M.a.alba)*

A scarce spring and autumn passage migrant.

A total of 47 birds showing characteristics of this race was recorded in spring and 40 during the autumn.

The earliest record was of one at Appleshaw, Andover on Mar 5th. Other March records were one at Hurst Beach/Milford on Sea on Mar 19th; one at Farlington Marshes on 26th, and three on Beaulieu estuary also on Mar 26th, with one on 27th. A minimum total of 47 was seen in spring, with a peak count of nine in the Lymington/Hurst area on Apr 10th.

During the autumn, passage was noted at five coastal sites - the majority of records coming from the well-watched Lymington/Hurst area where 12 were seen between Sep 17th and Oct 25th. All records are listed below.

Coastal sites
Farlington Marshes: 1, Mar 26th; 1, Apr 16th; 1, Apr 19th; 1, Apr 21st; 3, Apr 30th.
Stokes Bay: 1, Oct 25th.
Titchfield Haven: 1, May 1st.
Brownwich: 3, Apr 29th.
Hook-with-Warsash: 2, Apr 1st; 1, Apr 2nd; 1, Apr 3rd; 1, Oct 16th; 3, Oct 17th.
Needs Ore/Beaulieu Estate: 3, Mar 26th; 1, Mar 27th; 1, Sep 20th.
Lymington/Hurst area: 9, Apr 10th; 2, Apr 11th; 2, Apr 16th; 3, Sep 17th; 1, Sep 25th; 1, Sep 30th; 1, Oct 5th; 1, Oct 16th; 2, Oct 18th; 2, Oct 23rd; 1, Oct 25th.

Hurst Beach/Milford on Sea: 1, Mar 19th; 1, Apr 22nd; 1, Apr 24th.
Hurst Castle: 1, Apr 4th; 2, Apr 25th.
Barton on Sea GC: 1, Apr 10th; 1, Apr 24th; 1, Sep 17th.
Inland sites
Appleshaw, Andover: 1, Mar 5th.
Testwood Lakes: 1, Apr 25th.
Westend Down: 1, Apr 30th.
Woolmer Pond: 1, Apr 21st; 1, Apr 22nd; 2, Apr 25th; 1, Sep 2nd; 3, Sep 5th; 10, Sep 9th; 6, Sep 12th; 1, Sep 14th.

Waxwing   *Bombycilla garrulus*

A rare winter visitor, sometimes occurring in large numbers.

The early year saw the largest ever invasion of Waxwings into the county. Estimation of numbers is extremely difficult but there was certainly in excess of 1000 present at the peak of the influx in February, when the largest flock was of 352 at the Ordnance Survey, Southampton on Feb 13th. Flocks also exceeded 100 in ten other distinct areas. Birds were present in dwindling numbers through April with the final reports of eight at St Mary's, Southampton on May 1st, 18 at Church Crookham the next day and one in a Lymington garden on May 8th (MPM), the latest ever for the county. Full details of the invasion are given in a paper elsewhere in this Report.

Small numbers again reached England in the late year, but the only report in Hampshire was of six in the car park at Langstone Technology Park, Havant on Dec 15th.

Wren   *Troglodytes troglodytes*

An abundant resident, passage migrant and winter visitor.

Wren   © Richard Ford

All 15 records received related to counts of breeding territories, including: 169 at Lower Test Marshes (157 in 2004 and 162 in 2003); 113 at Titchfield Haven (97 in 2004, 115 in

2003); 109 at Longmoor Inclosure; 54 at Brownwich/Chilling and 44 at Deadwater Valley. The BBS fieldwork detected a mean of 9.5 per square, slightly above the ten-year average but dropping one place in the rankings to the seventh most abundant species encountered.

Dunnock   *Prunella modularis*
An abundant resident, passage migrant and winter visitor (Amber).

Of the 17 records received, 14 related to counts of breeding territories including: 61 at Lower Test Marshes (56 in 2004, 44 in 2003); 29 at Longmoor Inclosure; 22 at Titchfield Haven NNR (20 in 2004); 20 at Brownwich/Chilling and 11 at Deadwater Valley. BBS fieldwork detected a mean of 3.7 per square, slightly above the ten-year average, and placing it as the 18th most abundant species recorded during BBS. Many grounded birds at Normandy from Oct 12th to 16th were suggestive of an influx.

Robin   *Erithacus rubecula*
An abundant resident, passage migrant and winter visitor.

During the breeding season counts of territories included: 116 at Longmoor Inclosure; 58 at Brownwich/Chilling; 95 at Lower Test Marshes (75 in 2004, 76 in 2003); 44 at Deadwater Valley and 43 at Titchfield Haven NNR (34 in 2004 and 2003). BBS statistics shows a detection rate average of 8.6 per square. 16 of the 20 records received related to counts of singing/territorial birds. Evidence of migratory movement was a single in off the sea at Hurst Beach Mar 20th and a bird flying north over Hurst Castle on 29th.

Nightingale   *Luscinia megarhynchos*
A scarce and declining summer visitor, previously moderately common (Amber, HBAP).

In spring there were 61 singing males reported from 32 widespread sites, almost unchanged from 63 at 30 sites in 2004, but significantly higher than the low point of 30 at 21 sites in 2003. Botley Wood remains the most important breeding site. Less encouragingly, there was just one record from Martin Down which held ten territories as recently as 1999.

The first was a singing male at Ashlett Creek on Apr 4th, the next at Casbrook Common on 7th. The main arrival was from Apr 21st, with singing males reported from 15 sites by the end of the month. The maximum count at any one site was six at Botley Wood, this figure rising to at least nine territories when taking into account adjoining areas. Other areas/sites which held territories included: four each at Dunbridge and Fishlake Meadows and three at Ashlett Creek/Calshot. There was a total of 11 records in June with the last two singing males of the year heard at Botley Wood on 19th. The only autumn record was one seen at Bushfield Camp on Aug 21st.

Bluethroat   *Luscinia svecica*
A rare passage migrant (2,26,1).

One was present for several days in a garden in Liss Forest in late March. An account of the bird's occurrence was published in the *Bordon Herald* (per MFW, photo). Five were recorded between 1996 and 2002 but all other occurrences were before 1982.

Black Redstart
Phoenicurus ochruros

A scarce passage migrant and winter visitor; occasionally breeds (WCA 1, Amber).

A total of six was present during the first winter period, all at coastal sites. Spring passage was slightly stronger than usual, commencing with one at Hook-with-Warsash on Mar 10th. Three records came from inland sites, all singles at: Stockbridge Down on Mar 21st, Lockerley on Mar 27th and Rownhams on Apr 1st. Coastal records, all singles unless stated otherwise, included: three migrants at Sandy Point in March (with two overlapping with a wintering bird); up to two at Fawley Power Station on Mar 31st; singles at Hurst Castle on Mar 22nd and Apr 1st; Hill Head on Mar 30th; in a Warblington garden on Mar 31st and Southsea Castle on Apr 25th.

One pair bred in the north-east of the county and successfully raised two broods. Other breeding season records were one at Southampton Airport on June 4th and one at Eastney on July 7th.

Autumn passage began early with a bird at Barton on Sea on Aug 8th. Passage peaked in the first half of November with two at Titchfield Haven on Nov 5th and three at Sandy Point on Nov 13th. Five remained on the coast into December, with the last reported at Workhouse Lake on 23rd.

Coast and immediate hinterland
Sandy Point, Hayling Island: 1 on many dates, Jan 17th-Mar 24th (remaining from 2004); additional migrants on Mar 16th/17th, 21st/22nd and 26th/27th; 1 on 11 dates, Nov 5th-Nov 24th with 3, Nov 13th; 1 on five dates, Dec 9th-Dec 21st.
Mengham: 1, Apr 4th.
Beachlands: 1, Feb 7th; 1, Oct 30th.
Warblington: 1, Mar 31st.
Farlington Marshes: 1, Nov 14th.
Southsea Castle: 1, Apr 25th; 1, Dec 6th.
Fort Cumberland: 1, Nov 13th; 2, Nov 16th.
Long Meadow, Portsmouth: male, Jan 16th-Feb 7th; 1, Dec 10th-Dec 24th.
Eastney: 1, July 7th.
Newlands Farm, Fareham: 1, Oct 23rd.
Gosport: 1, Nov 12th; 1, Dec 23rd.
Hill Head: 1, Mar 30th; 2, Nov 5th.
Hook-with-Warsash: 1, Mar 10th; 1, Mar 12th; 1, Mar 14th.
Lower Test Marshes: 1, Oct 17th.
Calshot: 1, Feb 4th; 1, Mar 10th; 1, Nov 4th; 1, Nov 7th; 1, Dec 4th.
Fawley Power Station: 1, Mar 9th; 1, Mar 31st; 1, Nov 9th; 1, Nov 15th.
Eastleigh Sewage Works: 1, Feb 25th; 1, Nov 26th.
Southampton Airport: 1, June 4th.
Needs Ore/Beaulieu Estate: 1, Nov 1st; 1, Nov 8th; 1, Nov 29th.
Romsey: 1, Nov 13th.
Lymington/Hurst area: female, Mar 22nd; 1, Apr 1st; 1, Oct 21st; 1, Oct 30th-Oct 31st; 1, Dec 22nd.
Barton on Sea: male, Jan 3rd and 15th; male, Feb 1st-17th; male on four dates, Mar 4th-18th; 1, Aug 8th-Sep 10th; 1, Nov 12th.
Inland Sites
Woolmer Pond: 1, Sep 24th, 1, Oct 23rd.
Chineham: 1, Nov 4th.
Whiteley Pastures: 1, Oct 28th.
Blackfield: 1, Oct 15th.
Rownhams: 1, Apr 1st.
Stockbridge Down: 1, Mar 21st.
Lockerley: 1, Mar 27th.

Monthly county totals were:

J	F	M	A	M	J	J	A	S	O	N	D
3	4	9	4	0	2	1	3	2	10	15	5

Redstart
Phoenicurus phoenicurus

A locally common summer visitor (mostly to the New Forest) and passage migrant (SPEC 2, Amber).

Denny Wood, May 2005 © Marcus Ward

The first arrival was at Sandy Point on Mar 24th and the first at a breeding site was at Woolmer Pond on Mar 26th. Subsequent records of migrants at non-breeding sites involved 28 birds at 11 coastal sites up to Apr 30th including 12 at Farlington Marshes (max 5 on Apr 16th) and seven at Sandy Point, and a further seven birds at seven inland sites to early May.

Breeding season coverage in the New Forest produced counts of at least 178 singing/territorial males, including sites holding eight or more as follows: 12 at Mallard Wood; 11 at both Highland Water and Sloden Inclosures; at least ten at Mark Ash Wood; eight at each of four sites - Bolderwood, Cadnam's Pool, Denny Wood and Wooson's Hill Inclosure.

On the north-east heaths, a total of 46 pairs/singing males was located at seven sites as follows: 20 at Longmoor Inclosure; 20 at Woolmer Forest; singles at Birch Piece, Bourley North, Heath Warren, Liphook, Waggoners Wells and Warren Heath; this compares with 36 pairs/singing males at seven sites in 2004. The last bird recorded within this area was at Longmoor Inclosure on Sep 19th.

Two singing males, at least one of which was paired, were located at Faccombe in the north west of the county on June 7th.

The first evidence of post-breeding dispersal was of two birds at Nursling GP on July 26th. It was a poor year for autumn passage with the first record on Aug 9th at Hayling Oysterbeds. The approximate half-monthly totals for the autumn passage period are tabulated below. A comparison with the same figures in 2004 indicates that the total autumn passage recorded in 2005 was down by 56%.

	Aug 1-15	Aug 16-31	Sep 1-15	Sep 16-30	Oct 1-15	Oct 16-31	Total
2005	3	28	30	15	3	0	79
2004	6	54	78	30	7	2	177

At Old Winchester Hill coverage during this period produced a total of 29 bird-days recorded over 15 dates between Aug 15th and Oct 1st (compared to 78 bird days in 2004), with maximum counts of five on Sep 1st and four on Sep 16th/17th. Elsewhere, the highest counts were ten at Martin Down on Aug 30th, three at both Sandy Point and at Farlington Marshes on Sep 2nd and three at Northney on Sep 9th. The last record of the year was a single at Normandy on Oct 13th.

Reported spring and autumn totals for 1996-2005:

Bird-months	1996	1997	1998	1999	2000	2001	2002	2003	2004	2005
Spring	110	90	16	21	17	18	12	9	69	23
Autumn	94	184	109	57	121	136	114	209	177	79

Whinchat   *Saxicola rubetra*

A very scarce summer visitor and common passage migrant (HBAP).

The first of the year was at Hurst Beach on Apr 24th. The following day there were three at both Lepe and Hurst. Subsequent spring passage was typically light with a total of 18 recorded in April and 12 in May. The peak day passage was on Apr 30th with 11 at seven sites. The last of the spring was at Testbourne Lake on May 25th.

There was no evidence on breeding, with just one record of a singing male at a site in the New Forest on May 26th. The first evidence of post-breeding dispersal was provided by a male and female at Farlington Marshes on July 8th.

Autumn passage peaked in the first half of September. The table shows half-monthly maxima for Farlington Marshes and the minimum half-monthly total for all other sites:

	Aug 1-15	Aug 16-31	Sep 1-15	Sep 16-30	Oct 1-15	Oct 16-31
Farlington Marshes	1	15	26	11	3	0
Other sites	8	62	101	89	21	3

As in previous years heaviest passage was recorded at Farlington Marshes, with 77 bird-days in August (peak 15 on 18th), 219 bird-days in September (max. 26 on 13th) and 14 bird-days in October (max. 3 on 2nd). Counts from elsewhere included: seven at Hook-with-Warsash on Aug 21st; in September eight at Hambledon on 3rd, seven at Chilling on 6th, ten at Lymington/Hurst on 13th and 17th, nine at Winchester SF on 16th and eight at Barton on Sea GC on 17th. The last for the year was one at Keyhaven Marsh on Oct 23rd.

Spring and autumn totals of bird-months for 1996-2005 follow:

	1996	1997	1998	1999	2000	2001	2002	2003	2004	2005
Spring	70	45	41	25	27	21	55	24	58	39
Autumn	324	181	284	349	330	304	794	919	642	799

Stonechat   *Saxicola torquatus*

A moderately common but local resident and partial migrant (SPEC 3, Amber).

Winterers in the early year totalled at least 194 at 62 sites. The *New Forest Wintering Survey* found 29 on Jan 21st/22nd. Maxima at other sites were: 11 at Titchfield Haven on Jan 24th; eight at Hook-with-Warsash on Jan 16th; seven at Lymington/Hurst on Jan 15th and six each at six other sites – Winchester SF (up to 6 regularly), Broadlands Estate on Jan 16th and Feb 13th, Itchen Valley CP on Jan 22nd, Titchfield Haven on Jan 31st, Calshot on Feb 4th and Barton on Sea on Feb 8th.

Spring passage was noted from mid February onwards and continued into late May.

Peak counts during this period were 12 at Toyd Down on Feb 24th and 11 at Barton on Sea GC on Mar 20th.

During the breeding season coverage of the Thames Basin and Wealden Heaths was again notable, with a total of 131 pairs/singing males located and comparable to the 132 in 2004. Elsewhere, coverage was improved with several new sites recording pairs/singing males for the first time. Data from the New Forest remained scant with no records of pairs/singing birds received. Observers are requested to submit all records of breeding Stonechats. BBS data show 12% of squares occupied, with a mean detection of 0.5 per square, both marginally higher than the ten-year average. All records of pairs/singing males away from the New Forest are tabulated below.

Thames Basin Heaths		**Tadley Common**	1
Blackbushe Airfield	4	Tweseldown	6
Bourley Heath/Long Valley/Bricksbury Hill	28	Velmead Common	1
Bramshill Plantation	6	Warren Heath	4
Bramshott	1	Yateley Common (HCC)	10
Eelmoor Marsh	7	**Wealden Heaths**	
Eversley Common	3	Broxhead Common	4
Hawley Common	2	Longmoor Inclosure	17
Hazeley Heath	4	Woolmer Forest	32
Shortheath Common	1	**Coastal sites**	
Silchester Common	1	Sandy Point	1

Coastal sites (continued)		Other Sites	
Barton on Sea	3	Bishopstoke	2
Browndown	3	Cheesefoot Head	1
Eastney	1	Crondall	1
Hook-with-Warsash	1	Fishlake Meadows	1
Hordle Cliffs	2	Houghton	1
Lymington/Hurst	4	Ibsley Lake North	1
Lower Test Marshes	6	Itchen Valley CP	1
Milton on Sea	2	Porton Down (Hampshire)	2
Sinah Common	2	Romsey	1
Titchfield Haven	1	Winchester SF	4

Minimum monthly totals reported in autumn were: August, 48 at 14 sites; September, 186 at 33 sites and October, 240 at 51 sites. Notable peak counts included: 22 at both Hook-with-Warsash on Oct 2nd and in the Lymington/Hurst area on Oct 22nd; 19 at Farlington Marshes on Oct 2nd; 15 at Barton on Sea on Sep 24th; 12 at both Crondall on Aug 13th and Titchfield Haven on Oct 1st; 11 at Needs Ore on Oct 11th and Calshot on Oct 30th.

In November and December a total of 327 was recorded at 75 sites (198 at 80 in 2004). The *New Forest Wintering Survey* found 103 on Nov 19th/20th. Other maxima included 14 at Hordle Cliff on Nov 14th; 12 both at Calshot on Nov 4th and Lymington/Hurst on Dec 22nd; 11 both at Farlington Marshes on Nov 17th and Itchen Valley CP on Nov 19th.

Wheatear   *Oenanthe oenanthe*

A scarce summer visitor and common passage migrant.

The first arrival was a single at Bighton Lane CB on Mar 14th. The first arrival in double-figures was on Mar 17th when 16 were at Farlington Marshes, followed by a further 136 bird-days there until May 21st. This was significantly lower than the 380 bird-days recorded in 2004 and was a return to the more usual levels recorded in 2003 and 2002 (118 and 209 bird-days respectively). Peak spring counts were of 60 in the Lymington/Hurst area on Apr 25th and 21 at Farlington Marshes on Apr 28th. Inland, the spring total was 94 (105 in 2004), with no site attaining double-figure counts. The minimum half-monthly totals are tabulated below:

Year	Mar 1-15	Mar 16-31	Apr 1-15	Apr 16-30	May 1-15	May 16-31	Total
2004	7	197	128	520	109	10	971
2005	6	114	38	186	56	10	406

One pair nested in the south of the county and successfully raised one young.

Two were recorded at coastal sites in July, at Needs Ore between 17th and 20th and at Hook-with-Warsash on 19th. Minimum half-monthly bird-day totals are tabulated below and compared to the highest passage over the last decade which took place in 2004:

Year	July 1-15	July 16-31	Aug 1-15	Aug 16-31	Sep 1-15	Sep 16-30	Oct 1-15	Oct 16-31	Nov 1-15	Nov 16-30
2004	2	7	103	629	366	132	66	16	2	1
2005	0	2	93	107	217	148	55	29	10	0

Autumn records from coastal sites with double-figure counts are summarised below:

Sandy Point: 98 bird-days between Aug 3rd and Oct 16th with 11, Sep 24th.
Farlington Marshes: 206 bird-days between Aug 3rd and Nov 1st with 12, Sep 2nd; 12, Sep 12th; 18, Sep 13th; 10, Sep 17th.
Gilkicker Point: 14, Sep 3rd; 17, Sep 12th.

139

Chilling: 22, Sep 12th.
Newlands Farm, Fareham: 10, Sep 4th.
Needs Ore/ Beaulieu Estuary: 10, Sep 6th; 12, Sep 17th; 11, Sep 20th.
Lymington/Hurst area: 10, Aug 23rd; 21, Sep 17th; 11, Sep 18th.
Hordle Cliff: 22, Sep 12th.
Barton on Sea: 60, Aug 20th; 11, Aug 28th; 15, Sep 12th; 15, Sep 17th; 12, Sep 24th.

Notable peak counts at inland sites included six rising to ten at Woolmer Pond from Sep 14th to 19th, six at Abbottswood, Romsey on Sep 12th, and five at Battle Green, Hamble-don on Aug 14th.

Fledgling Wheatear © *Trevor Hewson*

There were two records for December. A first-winter male was at Beachlands, South Hayling on 13th and a male was at Langstone South Moor on 22nd. There have been six records for that month since 1979 including the latest ever at Chilling from Dec 30th 1994 into 1995.

Reported spring and autumn bird-day totals for 1996-2005 follow:

Totals	1996	1997	1998	1999	2000	2001	2002	2003	2004	2005
Spring	798	525	441	308	385	554	555	198	971	406
Autumn	368	429	419	555	399	551	551	936	1324	658

Desert Wheatear
Oenanthe deserti

A very rare vagrant (0,1,1).

A first-winter male was located on the beach just west of Sandy Point at 1010 hrs on Nov 13th (ACJ, SKW). It disappeared at 1023 hrs but fortunately was relocated at Beachlands, about two miles to the west, at 1235 hrs, where it remained until dusk and was enjoyed by many observers (TAL, PAG *et al*, photo). The record has been accepted by *BBRC* and is the second for Hampshire. The first was a female at Farlington Marshes from Nov 4th-19th 1961.

Ring Ouzel
Turdus torquatus

A scarce passage migrant; has wintered.

At least 100 were reported in 104 records submitted. As usual, spring records were very scarce with just five. Numbers in autumn, however, were exceptional: the heaviest passage ever recorded in the county with 20 at Old Winchester Hill on Oct 22nd the highest ever site count (KWM). Other counties in south-east England recorded even greater numbers at the same time: on Oct 20th, there were at least 60 at Cissbury Ring, East Sussex and over 50 in the Portland Bill area.

In spring five were noted between Apr 2nd and Apr 30th.

Basing, The Millfield: male, Apr 2nd.

Old Winchester Hill: male, Apr 23rd.
Sandy Point, Hayling: male, 2 females, Apr 25th; female, Apr 26th-30th.

The first autumnal return was on Oct 6th, with the last bird on Nov 12th. Autumn passage was concentrated in the period Oct 16th to 23rd, with a total of 186 bird-days. The peak passage was 49 on Oct 22nd: a male, three females and two first-winters were on Old Winchester Hill with a single and a flock of 13 noted arriving from the east; nine in the Lymington/Hurst area; seven at Leaden Hall, NF; two females and two first-winters at Fort Widley, Portsdown Hill; three at Marchwood and ones and twos at four other sites. All records are tabulated below:

Sandy Point: 2, Oct 6th; 1, Oct 16th-22nd, with 2, Oct 18th and Oct 21st; 1, Oct 29th; 1, Oct 31st and Nov 1st.
Ibsley Water: 1, Oct 7th.
Old Winchester Hill: 1, Oct 7th; 5, Oct 19th; 1, Oct 20th-21st; 20, Oct 22nd; 6, Oct 23rd; 1, Oct 24th-26th; 3, Oct 29th; 1, Oct 31st; 1, Nov 5th; 1, Nov 12th.
Burr Bushes, NF: male, Oct 14th.
Lymington/Hurst: male, Oct 15th; 2, Oct 18th; 4, Oct 21st; 9, Oct 22nd; 3, Oct 23rd; 1, Oct 26th-27th; 3, Oct 29th; 5, Oct 31st.
Bishop Waltham, West Hoe Farm: male, Oct 17th.
Winchester: 1, Oct 17th.
Leaden Hall/Black Gutter Bottom, NF: 3 males, Oct 17th; 6, Oct 20th; 1, Oct 21st; 7, Oct 22nd; 1, Oct 27th.
Farlington Marshes; male, Oct 18th-21st.
Ockley Plain: 4, Oct 18th.
Portsdown Hill: 1, Oct 18th.
Calshot: 10, Oct 19th.
Westend Down: 1, Oct 19th.
Brownwich: 1, Oct 20th; 1, Oct 31st.
Fort Widley: 4, Oct 22nd.
Itchen Valley Country Park: 2, Oct 22nd; 1, Oct 27th-31st; 1, Nov 9th.
Marchwood: 3, Oct 22nd-23rd; 1, Oct 25th; 2, Oct 29th.
Browndown: 2, Oct 22nd-23rd; 1, Oct 25th; 2 Oct 26th.
The Vyne Watermeadow: 1, Oct 22nd.
Ibsley North Lake: 1, Oct 23rd.
Gander Down: 1, Oct 23rd.
Moorgreen Farm, Romsey: 1, Oct 23rd.
Butser Hill: 4, Oct 23rd.
Grandfather Combe: 7, Oct 23rd.
Titchfield Haven: 1, Oct 23rd; 1, Nov 7th.
Magdalen Hill: 3, Oct 29th.
Ocknell Plain, NF: 2, Nov 1st; 1, Nov 4th.
Ashley Warren: 1, Nov 4th.
Fawley Power Station: 1, Nov 9th.

2000 addition: one was at Moorgreen Farm on Apr 20th (GH-D).

Blackbird   *Turdus merula*

An abundant resident, passage migrant and winter visitor.

As in previous years, significant winter numbers were recorded at Petersfield. One observer noted early and late year maxima in his garden, with 44 there on Feb 25th (corresponding to a cold weather period) and 49 on Dec 29th (present at dawn, after leaving a large local roost).

Breeding records included: 75 territories at Lower Test Marshes, (64 in 2004 and 76 in 2003); 53 at Longmoor Inclosure; 22 at Deadwater Valley, Bordon; 39 at Brownwich/Chilling and 29 at Titchfield Haven (21 in 2004 and 27 in 2003).

BBS data show a mean of 16.8 detections per square, above the ten year average (1996-2005) of 14.4. It was the third most frequently recorded species, dropping one place from 2004.

The largest count of the year was 50 at Milton Reclamation on Oct 6th which included a partial albino bird. A flock of 25 was reported from Abbottswood, Romsey on Oct 20th, with influxes of 47 at Itchen Valley CP on Oct 22nd and 30 at Normandy on Nov 12th, followed by a flock of 40 at Broadlands Estate on Dec 5th.

Fieldfare   *Turdus pilaris*

A numerous to abundant winter visitor and passage migrant (Schedule 1, Amber).

Numbers were again above average in both the early and late year. The approximate monthly totals are tabulated below:

J	F	M	A	M	J	J	A	S	O	N	D
5300	4770	2100	9	0	0	0	0	1	235	1280	5560

Flocks in excess of 100 were reported from 31 sites in the early winter months. In January the largest flocks were: 540 at Chattis Hill on the 12th; 400 at Black Gutter, NF on the 3rd; 335 at Chidden on the 3rd and 350 at Ellingham Meadows on the 9th. In February the largest count was 600 at Kingsley Sand Pit on 20th. Seven sites hosted three-figure flocks in March, including 450 at Wheely Down on 13th and 300 at Farley Mount on 3rd. There was an early departure with numbers rapidly declining in the last week of March and just three April records including: six at Tundry Pond on 2nd and the last of the early year, two at Mottisfont on 12th.

The first autumn arrivals were one at Crondall on Sep 29th, followed by one at Old Winchester Hill on Oct 1st and 50 at Danebury on Oct 2nd. There was then a lull until Oct 16th when 27 arrived including 20 north-east at Mottisfont but the main autumn movement commenced on Oct 29th when 130 moved east over Old Winchester Hill. Visible migration continued until Nov 14th with a peak day movement recorded on 5th on which date a total of 989 was recorded including 300 west or south-west at Itchen Valley CP and 321 west at Broadmarsh – the largest recorded movement of the year. In November there were just five grounded flocks in three-figures, the largest being 200 at South Warnborough on Nov 27th. In December numbers appeared to be above average with 14 three-figure flocks reported including: 400 at Temple Manor and 500 at Lane End Down on 11th; 400 at Hartley Mauditt on 16th and the largest, an impressive 2000, at Swanmore Apple Farm on 31st.

Song Thrush   *Turdus philomelus*

A numerous resident, passage migrant and winter visitor (Red, NBAP).

No records of significant flocks were received for the early part of the year.

Counts of breeding territories included 20 at Lower Test Marshes (18 in 2004, 28 in 2003 and 18 in 2002), 12 at Brownwich/Chilling, seven at Longmoor Inclosure six at Deadwater Valley and four at Titchfield Haven. The mean detection of 3.5 per square was in line with the ten year average (1996-2005).

The first evidence of autumn arrival was eight at Farlington Marshes on Sep 20th, rising to an autumn maximum of 30 there on Oct 6th; on the same day an influx of 25 was reported at Sandy Point and 15 were at Milton Reclamation. The highest count of the autumn was 58 at Old Winchester Hill on Oct 9th which was comprised mostly of new arrivals. Other notable autumn influxes included 41 at Itchen Valley CP on Oct 22nd and 40 at Appleshaw, Andover between Oct 22nd and 25th and again on Nov 8th.

Redwing
Turdus iliacus

A numerous to abundant winter visitor and passage migrant (Schedule 1, Amber).

Early year numbers were again higher than in recent years but below the exceptional numbers of December 2004. In the late year, numbers built up steadily through October, reaching a peak in mid November, but then dropped off following a heavy diurnal passage in the second week of November and remained at a low level for the rest of the year.

J	F	M	A	M	J	J	A	S	O	N	D
5200	3900	2200	12	0	0	0	0	0	11,600	10,130	950

In the early year, flocks in excess of 100 were seen at 42 sites. The largest gatherings were: 433 at Black Dam, Basingstoke on Jan 30th; 400 at Pitts Wood, NF on Jan 2nd; 370 at Tundry Pond on Feb 26th; 340 at Harley Mauditt on Feb 24th; 320 at Fleet Pond on Jan 23rd; 260 at Weston on Jan 26th and 250 at Dogmersfield Park on Jan16th. Migration in the first part of the year included 300 south over Portsmouth at night on Jan 1st and 90 north over Appleshaw at dusk on Jan 10th. The last spring records were one at Cove on Apr 8th and one at Brocks Hill on Apr 17th.

The first late year returns were six at Bassett and 161 west over Lakeside on Oct 1st. On Oct 16th there was a very heavy passage, notably: 7372 north-west over Testwood Lakes; 1000 north over Deadman Hill, NF; 250 north-east over Mottisfont; 183 west over Lakeside; 130 north over Regents Park and 100 north over Gander Down. The period Nov 2nd-10th saw another heavy passage: significant counts included 200 east over Sandy Point and 186 south-west over Black Dam; Basingstoke on 2nd; then 135 south-west over Fleet Pond on 4th; followed by the peak on 5th, with 1720 south/south-west over Itchen Valley CP, 1250 west over Broadmarsh; 560 west over Waterlooville; 277 north-west over Hook-with-Warsash; 232 west over Old Winchester Hill; 227 west over Miles Hill and 140 west over Riverside Park. A few days later there were 340 west/south-west over Itchen Valley CP and 250 west over Casbrook Common on 9th; then 832 over Black Dam, Basingstoke on 10th. The only large count, after the November passage, was 500 at Swanmore Apple Farm on Dec 31st.

Mistle Thrush   *Turdus viscivorus*
A numerous resident and passage migrant (Amber).

There were no flocks reported in the early year. Counts of breeding territories received included 15 at Longmoor Inclosure, three at Browning/Chilling and two at Deadwater Valley. BBS showed that the species was encountered in 61% of squares, a 5% decrease on 2004, but 3% above the average for the ten-year period 1996-2005.

Double-figure post-breeding flocks were noted from six sites between June 21st and Aug 29th. The highest counts were: 20 at Lakeside CP on June 20th; 20 at Lyss Place Farm, Liss on June 21st; 16 at Stubbington on July 23rd; 50 at Roydon Woods, NF on Aug 18th; 16 at Princes Marsh, Liss on Aug 21st and 47 at Hill Barn, Bullington on Aug 29th. Movements in this period included 43 west over Island Thorns Inclosure, NF on Aug 29th, 25 west at Abbottswood, Romsey on Sep 9th and 11 over Iley Point on Nov 5th.

The only late year concentration was 18 at Pitts Wood Inclosure on Oct 18th.

Cetti's Warbler   *Cettia cetti*
A moderately common resident; first bred in 1979 [WCA 1, HBAP].

The number of sites at which singing males were recorded in the breeding season increased from 23 in 2004 to 37 in 2005. However, the total number of singing males remained the same at around 180, meaning that the county population has probably stabilised at around 200 pairs. There was a noticeable decrease at Titchfield Haven from 40 territories in 2004 to 28 in 2005 (38 individuals were ringed) and numbers at Lower Test Marshes had also reduced from 31 in 2004 to 26 this year. Other counts of singing males included Farlington Marshes, 12; Lymington/Hurst, 11; Somerley Park, 11; Longstock, 9; and Itchen Valley CP, 9. At least three pairs were known to be successful at this last site.

As usual, singing was recorded at many sites outside of the breeding season. For example, eight were singing along the River Avon above Fordingbridge on Sep 25th, and five at Titchfield Haven on Oct 9th. At Farlington Marshes a total of 25 individuals, including 13 first-winters, was ringed between May 1st and Oct 29th.

Records away from the coast and main river valleys continue to be scarce. One was recorded at the Millfield, Basing on five dates between Feb 17th and Apr 30th, while a singing male at Woolmer Pond on Apr 10th was the first ever there. Singles were at Bransbury Common on Feb 16th and Oct 7th, and an unmated male at Winchester College Meadows on several dates from Apr 16th into June was the first for several years at this site.

2004 addition: one ringed at the Hawthorns, Southampton Common on May 10th was the first for the site.

Grasshopper Warbler   *Locustella naevia*
A scarce summer visitor which has declined considerably since 1970 [Red, HBAP].

As noted in the 2004 report, the optimism connected to this species fortunes in the county continued in 2005, with an increase to at least 17 reeling males at 15 locations between Apr 12th and May 24th. Although the majority of these reports were on one day only, there does seem to be a genuine upward trend. The records are listed as follows:

Lakeside, Eastleigh: 1, Apr 12th.
Titchfield Haven: 1, Apr 15th-June 18th, with 2, Apr 30th-May 1st.
Roydon Woods: 1, Apr 15th.
Itchen Valley CP: 1, Apr 17th.
Avon Water, Keyhaven: 1, Apr 17th.
The Millfield, Basing: 1, Apr 20th-July 9th.
Calshot: 1, Apr 23rd-June 4th.
Sandy Point: 1, Apr 25th.
Keyhaven/Pennington Marshes: 1, Apr 25th.
Farlington Marshes: 1, Apr 25th.
Tweseldown: 1, Apr 26th-28th.
Wildhern: 1, Apr 27th.
The Kench: 1, Apr 30th.
Hook-with-Warsash: 1, May 4th.
Fishlake Meadows: 1, May 24th-July 20th (with 2 heard on some dates)

The only confirmed breeding was at Fishlake Meadows, but it can also be presumed likely from the presence of long-staying males that nesting was at least attempted at Titchfield Haven, Calshot, and the Millfield, Basing. One singing at Winnall Moors on July 3rd may also indicate breeding.

In autumn an incredible 329 were ringed at Titchfield Haven between July 23rd and Oct 8th. This exceeds last year's record figure of 237 trapped at this site. Despite the increase in numbers being ringed, there seems to be a lack of any returns that would indicate from where these birds, the vast majority of which are first-winters, originate from. A total of 19 was also trapped at Farlington Marshes between Aug 17th and Sep 10th. The increase in numbers mist-netted at the Haven over the last few years can be seen in the following table:

2000	2001	2002	2003	2004	2005
05/8 – 25/9	28/7 – 03/10	27/7 – 25/9	01/8 – 28/9	24/7 – 05/10	23/7 – 08/10
13	84	232	88	237	324

One reason for the greater catch figures is the use of overnight/early morning tape luring. One at Old Winchester Hill on Oct 9th was the second-latest for the county, the latest being one at Farlington Marshes on Oct 20th 2001. The highlights for Titchfield Haven and Farlington Marshes, plus all other records, are listed below:

Titchfield Haven (all mist-netted): total of 324, including 29, Aug 21st; 31, Aug 28th; 21, Sep 13th.
Farlington Marshes (all mist-netted): total of 19, including 4, Aug 17th; 6, Aug 28th.
Keyhaven Marshes: adult, Aug 19th.
Gilkicker Point: 1, Aug 27th; 1, Sep 6th.

Milkham Bottom, NF: 1, Aug 28th.
Stokes Bay: 1, Aug 28th.
Noar Hill, Selborne: 1, Sep 7th.
Sandy Point: 1, Sep 8th; 1, Sep 15th.
Lower Test Marshes: 1, Sep 17th.
Old Winchester Hill: 1, Oct 9th.

Aquatic Warbler   *Acrocephalus paludicola*

A rare autumn passage migrant (1,74,3).

Three were trapped and ringed at Farlington Marshes during August, a first-winter and an adult on 7th (PJL, DAB, photos) and a first-winter on 11th (DAB, photo). The vast majority of the 78 recorded in the county have been trapped, and of these, only three were adults.

Sedge Warbler   *Acrocephalus schoenobaenus*

A common summer visitor and passage migrant.

The first of the year was one at Titchfield Haven on Mar 18th (BSD). Apart from one at Stratfield Saye on Mar 17th 1963, this is the earliest record for the county. This individual may account for further sightings at Titchfield Haven on Mar 26th and Apr 1st, as the main arrival did not commence until Apr 7th, when there were singing males at the Millfield, Basing and Lakeside, Eastleigh. The highest reported counts on spring migration were 12 at Bishopstoke on Apr 24th and 12 at Keyhaven/Pennington Marshes on the following day. There were several records of singing males on passage away from typical habitat; e.g. one at Hartley Wintney GC on Apr 26th and May 9th, and one at Deadwater Valley, Bordon on May 1st.

Records of presumed breeding involved around 135 singing males at 19 sites. The highest counts were 36 at Lower Test Marshes (49 in 2004); 35 at Titchfield Haven; 23 at Longstock (32 in 2004), and eight at Stratfield Saye – the latter being a good count from the north-east of the county where it is a scarce breeder.

The first autumn migrant was a juvenile at Sandy Point on July 21st. Reports came from a further 20, mostly coastal, sites. A total of 767 was ringed at Titchfield Haven, while the highest visual counts were of 12 at Milton Reclamation on Aug 9th and 25 at Farlington Marshes on Sep 13th (a late date for double-figure field sightings, most of these birds were seen away from the main reedbed, and corresponded with a large overnight arrival of other migrants). The last for the year was inland, at the Millfield, Basing on Oct 6th.

Reed Warbler   *Acrocephalus scirpaceus*

A common but local summer visitor and passage migrant [HBAP].

The first of the year was at Fleet Pond on Apr 2nd, only one day later than the earliest ever. Possibly the same individual was also heard there on 4th and 10th, but the main arrival was noted from Apr 11th onwards, with one at Farlington Marshes and four at Titchfield Haven. Spring passage was generally unremarkable, but late migrants at unusual sites included singing males on Breamore Down on May 29th, and at Alverstoke on June 19th.

Breeding season records amounted to approximately 400 pairs/singing males at 25 sites. Counts included: 112 at Titchfield Haven; 94 at Lower Test Marshes (98 in 2004); 32 in a partial survey of the Lymington/Hurst area; 22 at Fleet Pond; 20 at IBM Lake; and 18 at Longstock (14 in 2004).

As usual, very few were reported in autumn but the Titchfield Haven ringing total of 409 indicates a substantial passage. A count of 20 at Milton Reclamation on Aug 9th, presumably, included breeding birds and their young. Otherwise, apart from two trapped on Farlington Marshes on Oct 2nd, there were only eight records of single birds at seven sites between Sep 3rd and Oct 2nd. The last for the year was one at Keyhaven/Pennington Marshes on Oct 22nd.

Melodious Warbler   *Hippolais polyglotta*
A rare autumn passage migrant (0,13,1).

One was watched at close range in bushes adjacent to IBM Lake, Cosham between 0850 and 0910 hrs on Sep 17th (TMJD). It then moved 50 metres to another group of bushes but could not be relocated. Recent sightings were made in late August 2002 and 2003, but prior to that just 11 were recorded between 1961 and 1992.

Blackcap   *Sylvia atricapilla*
A numerous summer visitor and passage migrant; moderately common in winter.

In the first two months of the year there were at least 58 reported from 37 locations. As usual, the majority of these records were in observers' gardens. Good numbers continued to be reported in March, with the highest count being seven in Winchester on 17th.

As ever, the arrival date of the first spring migrants was difficult to assess due to the presence of wintering birds. However, the first at Farlington Marshes was on Mar 25th, and at Titchfield Haven and Lower Test Marshes the following day. In April, double-figure counts included 17 at Farlington Marshes on 3rd, and ten at Sandy Point and at Keyhaven/Pennington Marshes on 25th. Breeding season counts included: 51 at Lower Test Marshes (46 in 2004); 41 at Itchen Valley CP; 20 at Titchfield Haven; 18 at Brownwich/Chilling; and 13 in Straits Inclosure, Alice Holt Forest. BBS data indicated that 80% of squares were occupied in 2005, the same as in 2004.

Ringing activities at Titchfield Haven produced a total of 543 trapped between July 22nd and Oct 9th, with the peak being 112 on Sep 17th. Several double-figure counts were made at Farlington Marshes: 65 on Sep 13th (*cf* Sedge Warbler); 55 on 18th (including 16 trapped north of the A27); and 30 on Oct 3rd. Elsewhere, there were 35 at Old Winchester Hill on Aug 28th, with 25 there on Sep 11th; 20 at Sandy Point on Sep 18th; and 16 at Milton Reclamation on Sep 20th. There were only three records in the second half of October, including the last for Farlington Marshes on 29th. The return to gardens was from Nov 4th onwards. In the last two months of the year a minimum total of 59 was reported at 43 sites, including four at IBM Lake, Cosham on Nov 5th, and four at Stubbington in late December.

Garden Warbler   *Sylvia borin*
A common summer visitor and passage migrant.

The first for the year was one at Titchfield Haven on Apr 11th, followed by one at Timsbury on Apr 16th. Small falls of five at Sandy Point, and six at Farlington marshes, both on Apr 30th, were the highest counts noted on spring passage.

Although much under-recorded as a breeding species, at least 57 singing males were reported from 24 sites. Highest concentrations included eight at Camp Farm GP on May 15th (which probably included some migrants), six at Itchen Valley CP (an increase from only two in 2004), and five at Binswood. BBS data indicated that 21% of 1 km-squares were occupied in 2005, a marked improvement on the 14% recorded in 2004.

In autumn a total of 45 was ringed at Titchfield Haven between July 23rd and Sep 20th (52 in 2004). At Farlington Marshes there were only 19 bird-days between July 23rd and

147

Sep 18th, included two trapped during ringing activities, and a maximum of three on Aug 20th. Elsewhere, there was one migrant reported in July (in a Fareham garden on 26th), eleven from six sites in August (including four at Oxley's Coppice, Fareham on 7th), and six at four sites in September. The last for the year was one at Lower Test Marshes on Oct 4th.

Barred Warbler   *Sylvia nisoria*

A rare autumn passage migrant (0,8,1).

A first-winter was seen briefly but well in the observer's garden at Solent Court on Sep 18th (PMP). Apart from the first in 1965, all previous records were between 1986 and 2003.

Lesser Whitethroat   *Sylvia curruca*

A moderately common but declining summer visitor and passage migrant; recorded four times in winter.

A total of 190 reports was received, an increase on last year, and a response by observers who submitted all records due to the relative scarcity of this species. The first of the year was one in a Chineham garden on Apr 16th, followed by one at Sandy Point on 19th. There were reports of up to three birds from a further 19 sites during April. The species was widely reported in small numbers throughout May, and it is difficult to separate singing males on territory from those passing through. For example, at Lakeside, Eastleigh, there were 19 bird-days between Apr 26th and July 29th, but no breeding on site.

Including both reported territories and singing males from mid May onwards, there was an approximate total of 79 pairs at 44 sites, a notable improvement over the 47 pairs at 27 sites in 2004. Highest counts included eight singing males at Itchen Valley CP (four in 2004), six at Farlington Marshes, and six on Portsdown Hill. The species is still very scarce in the well-watched north-east of the county, where there were only five reports throughout the year (all in spring).

Autumn passage was reported from 22 locations. Highest counts came from Farlington Marshes with seven on both Aug 25th and Sep 18th, and Testwood lakes with six on Aug 21st. The last inland was at Old Winchester Hill on 2nd. There were few records after mid September and just three in October, the latest being at Farlington Marshes on 8th.

However, the last for the year was a probable first-winter feeding on fat, apple, bread, and peanuts in a Fareham garden from Dec 7th-31st (DFB). There are at least five previous instances of over-wintering in Hampshire, although this individual appears to be the first known to be present at the year-end. Inevitably, it becomes the latest record for the county.

Whitethroat   *Sylvia communis*

A numerous summer visitor and passage migrant; recorded three times in winter.

The first of the year was a male at Sandy Point on Apr 3rd and 5th, followed by one at Lower Pennington Lane on 7th. The first at Farlington Marshes was on Apr 11th and had increased to nine singing males by 19th, with five at Sandy Point the following day. Peak counts in spring included 20 at Keyhaven/Pennington Marshes on Apr 25th, 26 at Farlington Marshes on May 1st and 13 at Stokes Bay on May 9th.

Breeding reports came from most parts of the county. Noteworthy counts of pairs/singing males included 45 at Itchen Valley CP (34 in 2004); 36 in a partial survey of the Lymington/Hurst area; 34 at Longmoor Inclosure (55 in 2004), 24 at Brownwich/Chilling and at least 24 in the Crondall/Roke Farm/Well area. One territory was also located on the Langstone Harbour Islands. BBS data gave conflicting information: there was a decrease in 1-km square occupancy from 51% in 2004 to only 40% in 2005, but the actual total number of territories in these squares was the highest ever at 102.

In autumn, a total of 30 was ringed at Titchfield Haven between Aug 6th and Sep 17th, well down on the 90 trapped there in 2004. Double-figure counts of migrants included 25 at Milton Reclamation on Aug 9th; 25 at Farlington Marshes on Aug 25th and Sep 13th; 22 at Oxleys Coppice, Fareham on Aug 7th; 20 at Stokes Bay on Aug 29th; and 15 at Gilkicker Point on Aug 28th and Sep 6th. There were four reports in October, with the last being at Farlington Marshes on 11th.

Whitethroat   © Richard Ford

Dartford Warbler   *Sylvia undata*

A moderately common resident, largely confined to the heaths of the New Forest and north-east but with small numbers in coastal scrub [ET SPEC 2, Schedule 1, Amber, HBAP].

Survey work on the Thames Basin and Wealden Heaths produced a total of 223 pairs/singing males. Allowing for the lack of coverage at Ludshott Common which held 49 territories in 2004, and incomplete coverage at a few minor sites, the overall total is similar to the record count of 279 in 2003. Coverage of coastal sites was incomplete, but improved on 2004 with a total of 20/21 pairs/singing males reported, including the first confirmed breeding at Eastney and Barton on Sea. In the New Forest, coverage of one area produced a total of 58 territories compared with 59 in 2003 and 78-85 in 1992. However, the area of suitable habitat in that area has declined since 1992; the national survey in 2006 should provide a new base line for the level of the New Forest population, which was around 600 pairs in 1994 and approaching 650 pairs in 1999.

Counts of pairs/singing males away from the New Forest are tabulated below:

Thames Basin Heaths		Ludshott Common	nc
Blackbushe Airfield	7	The Warren, Oakhanger	0
Bourley Heath/Long Valley	11	Woolmer Forest	66
Bramshill Plantation	11	**Coastal sites**	
Bramshott Heath	0	Sandy Point, Hayling Island	1
Bricksbury Hill	19	Sinah Common	2
Eversley Common/Castle Bottom	4	Eastney	2
Hawley Common	1	Gilkicker Point	2
Hazeley Heath	7	Browndown	3
Pyestock Heath	4	Hook-with-Warsash	1
Silchester Common	2	Hamble Common	nc
Tadley Common	2	Dibden Bay	nc
Tweseldown	3	Fawley Refinery	nc
Warren Heath	1	Calshot	2
Velmead Common	0	Needs Ore	nc
Yateley Common (HCC)	1	Lymington/Hurst	4/5
Yateley Common South (MOD)	9	Milford on Sea	nc
Yateley Heath Wood	0	Barton on Sea	3
Wealden Heaths		**Other Sites**	nc
Bramshott Common	1	Burton and Poors Commons	nc
Broxhead Common	8	Ringwood Forest	1
Cranmer Heath	nc		
Longmoor Inclosure	66		

Records from presumed non-breeding localities are summarised below.

Roydon Woods: 1-4, Jan/Feb;1-2, Aug/Dec.
Hamble Old Airfield: 1, Jan 6th-Feb 21st.
Fawley Refinery: 1, Jan 6th and Feb 18th.
Dibden Bay: 2, Jan 7th;1-2, Mar 13th/14th; 2, Nov 30th.
Paulsgrove Reclamation: 1, Jan 15th and Feb 12th.
Northney, Hayling Island: 1, Feb 26th and Sep 6th.
Needs Ore: 2, Mar 13th; 1 on 5 dates, Aug 21st-Dec 4th
Titchfield Haven: 1, Aug 8th; 2, Oct 2nd-Dec 31st.
Lepe: 1, Aug 21st, Nov 18th and Dec 11th.
Farlington Marshes: 1-4, Oct 4th-Nov 12th; 4-6, Nov 14th-17th; 1, Nov 19th-Dec 29th.
*Brownwich/Chilling:*1-2, Oct 8th-Nov 14th.
Eastoke: 1, Oct 23rd.
Newlands Farm, Fareham: juvenile, Oct 23rd.
Blendworth Common: 1, Dec 13th.

Also of note were counts from the *New Forest Wintering Bird Survey* which produced 41 at 21 sites on Jan 22nd/23rd, 15 at nine sites on Feb 19th/20th, 111 at 19 sites on Nov 19th/20th and 53 at 22 sites on Dec 17th/18th.

Pallas's Warbler   *Phylloscopus proregulus*
A very rare vagrant (0,3,1).

One was seen at close range in the observers' garden at Lymington at 1355 hrs on Jan 15th (DE, GG). It departed before other local birders could arrive and unfortunately was not relocated. This individual presumably arrived in southern England the previous autumn and remained to winter. Previous records were in 1974, 1988 and 2004 and all during the period Oct 26th-Nov 23rd.

Yellow-browed Warbler   *Phylloscopus inornatus*
A rare autumn passage migrant (0,26,3).

A good year, with a total of three recorded as follows: Titchfield Haven on Oct 4th (BSD, DP); Wittering Road, Sandy Point from Oct 6th-10th (ACJ *et al*) and IBM Lake on Oct 16th (TMJD). The last was a year to the day since another was found at the same site by the same observer.

Within the illustration the following handwritten annotations appear:

OLIVE GREEN / UNDER PARTS

HORN COLOURED BILL

PALE CROWN STRIPE

LONG SUPERCILIUM

DISTINCT EYE STRIPE

WHITE UNDERSIDE

VERY ACTIVE. FLITTED

TWO CLEARLY MARKED WING BARS

OFF YELLOW LEGS

TITCHFIELD HAVEN OCT 4th '05

EAST SIDE

CALLED FREQUENTLY OBVIOUS HIGH PITCHED

Phyllosc

YELLOW-BROWED WARBLER

DAN POWELL, ROSIE POWELL, BARRY DUFFIN

Yellow-browed Warbler, Titchfield Haven, Oct 4th   © Dan Powell

Radde's Warbler   *Phylloscopus schwarzi*

A very rare vagrant (0,0,1).

One was discovered in an area of scrub at Stokes Lane near Sherborne St John at midday on Dec 7th (AP, NM *et al*, photo). The bird remained until dusk but was not relocated the following day. The record has been accepted by *BBRC* and is the first for Hampshire.

Wood Warbler   *Phylloscopus sibilatrix*

Formerly moderately common but now scarce and local summer visitor and scarce passage migrant [SPEC 2, Amber].

The first for the year was one at Bratley Wood, NF on Apr 9th (HJV). This equals the earliest ever for Hampshire, the other being at Horsebridge in 1988. There was then a gap of twelve days until the next arrivals at the Millfield, Basing, and Salisbury Trench, NF on Apr 21st. Other presumed migrants were singing males at Titchfield Haven on Apr 30th; Tweseldown on May 2nd; Deadwater Valley, Bordon and IBM Lake on May 7th; Longmoor Inclosure on May 11th and Marchwood on May 29th.

During May and June a minimum total of 71 singing males was recorded from 42 sites in the New Forest, an increase from 60 at 31 locations in 2004. However, the only site that held more than four was Highland Water Inclosure, where there were seven singing males on May 17th. Several of those recorded were found in the course of observers taking part in the BTO Woodland Bird Survey. The only record of potential breeding outside the New Forest involved an apparently unmated male at Woolmer Forest between May 30th and June 16th.

There were just two autumn records: one in a Fareham garden on Aug 4th, and the last for the year at Acres Down on Aug 9th.

Chiffchaff   *Phylloscopus collybita*

A numerous summer visitor and passage migrant; moderately common in winter.

In January and February a minimum of 105 was reported from 48 locations. At Eastleigh SF a total of 22 was trapped between Jan 1st and Mar 1st, with 14 on Jan 1st being the highest number seen on one day. Other early year counts included eight at Barton Stacey SF on Jan 31st; seven at Chilbolton SF on Feb 16th; and six at Fullerton SF on Feb 22nd. The attraction of sewage-farms for this species as food sites during the winter months is well established.

The first migrants started to appear from about Mar 10th, with the first double-figure count being 12 at Bishopstoke on 20th. At Farlington Marshes the peak was 13 on Apr 1st, with 13 at Keyhaven/Pennington Marshes the next day. At Blashford Lakes 20 were present on Apr 3rd, while the peak spring count at the Millfield, Basing was 23 on Apr 21st. Counts of territories/singing males during the breeding season included 67 at Itchen Valley CP (54 in 2004); 28 at Lower Test Marshes (40 in 2004); 22 at Brownwich/Chilling; 21 at Titchfield Haven; and 14 at Longmoor Inclosure (25 in 2004). BBS data indicated 82% square occupancy, the same as in 2004.

A singing male at Longparish from Apr 9th-16th showed some characteristics of the recently split species Iberian Chiffchaff *Phylloscopus ibericus*, but its identity was not confirmed.

During the autumn a total of 318 was

Chiffchaff, Longparish   © Peter Raby

ringed at Titchfield Haven between Aug 30th and Oct 9th, with a maximum of 57 on Sep 17th. Other peak counts included 35 at Old Winchester Hill on Aug 28th; 20 at Abbottswood, Romsey on Sep 17th; 30 at Keyhaven/Pennington Marshes on Sep 18th; and 25 at Barton on Sea GC on Sep 24th. In October highest numbers passed through on 8th, when 39 were ringed at Titchfield Haven, and 16 were at Brownwich. The autumn peak at Farlington Marshes was only ten on Oct 3rd.

In the last two months of the year an impressive 128 were ringed at Eastleigh SF, including 70 new birds and 58 re-traps. Elsewhere, numbers during November and December totalled approximately 101 at 51 sites, with the highest count being six at Titchfield Haven on Nov 30th.

Scandinavian/Siberian Chiffchaff *(P. c. abietinus/tristis)*

The majority of reports of Chiffchaffs showing characters of these eastern races came from Eastleigh SF. Three, one new bird and two re-traps, were caught between Jan 1st and Mar 1st, with one also being reported on five dates between Mar 7th and 23rd. In the late year a total of 17 new and two re-traps was ringed, perhaps indicating that these races are commoner during the winter months than previously thought (PC *et al*). None of these birds was definitely assigned to the race *P.c.tristis*. Elsewhere, single *P.c.abietinus* were seen at Lower Test Marshes on Mar 12th (MJP) and Barton on Sea GC on Sep 17th (SGK) while a probable *P.c.tristis* was seen well at Broadmarsh on Nov 29th but was not heard to call (JCr).

Willow Warbler   *Phylloscopus trochilus*

An abundant summer visitor and passage migrant [Amber].

The first was one at Oxey Marshes on Mar 18th, with the main arrival being from 24th onwards. The recent trend of this species arriving earlier continued, with March records from 34 sites – as recently as 2001 March records came from only eight locations. The highest count during this period was ten at Sinah GP on 31st. Good numbers were noted on passage at this site in April, with 35 on 1st, 25 on 4th and 50 on 14th. At Farlington Marshes there were 78 bird-days between Mar 25th and May 2nd including two double figure counts – 10 on Apr 11th and 12 on 16th. Other double-figure counts included 21 at Sandy Point and 18 at Keyhaven/Pennington Marshes, both on Apr 12th. One was noted coming in off the sea at Hurst Castle on Apr 25th.

During the breeding season noteworthy numbers of singing males/territories included 38 at Longmoor Inclosure (51 in 2004 – the decline mainly due to the clearance of birch), 15 at Leckford, and ten at Queen Elizabeth CP. BBS data showed that 43% of squares were occupied, a very small decline compared to the 44% in 2004.

The first autumn migrant was one on North Binness Island, Langstone Harbour on July 16th. At Titchfield Haven a total of 122 was ringed between July 23rd and Sep 17th, with a peak of 24 on Aug 7th. At Old Winchester Hill there were 53 on Aug 7th and 28 on Aug 9th. There were 17 at Sandy Point on Aug 4th, followed by 14 at Keyhaven Marshes on Aug 7th, and the maximum count of the autumn at Farlington Marshes of 11 on Aug 30th. Records came from nine sites during September, with peaks of four at Old Winchester Hill on 2nd and at Sinah GP on 10th. There were two October sightings of singles at Old Winchester Hill, on 14th and 20th.

Goldcrest
Regulus regulus

A numerous resident, passage migrant and winter visitor [Amber].

The annual breeding survey at Longmoor Inclosure indicated an increase with 32 singing males compared to 22 in 2004, and 39 in 2003. BBS data, however, showed that 60% of squares were occupied, a small decrease compared to 2004.

As usual, spring records at coastal sites peaked in March, with 15 at Sandy Point on 20th and 13 at Farlington Marshes on 24th. There were also 30 at Eastleigh SF on Mar 13th. Autumn passage was virtually non-existent, with the highest count being only eight at Farlington Marshes on Nov 15th.

Firecrest
Regulus ignicapilla

A scarce resident, passage migrant and winter visitor [WCA 1, Amber, HBAP].

Records outside the breeding season came from both coastal and inland sites, but showed clear peaks in March and October when most were from the coast. The approximate monthly totals of non-breeding birds together with the number of sites are shown in the table below:

Firecrest	J	F	M	A	M	J	J	A	S	O	N	D
Birds	11	9	21	0	0	0	0	0	5	28	13	8
Sites	9	8	16	0	0	0	0	0	3	16	8	6

During the first winter period, most records were of singles but two over-wintered at Eastleigh SF. There were two at Sowley Pond on Jan 12th, two at Lakeside, Eastleigh on Mar 15th and at least three at Sandy Point (where only one wintered) throughout March. It was not always possible to distinguish between wintering and migrant birds, but a first-winter male trapped on Jan 4th on Southampton Common, and re-trapped on Feb 14th, had originally been ringed on Nov 11th 2004 and had clearly over-wintered. In contrast, one in the east seawall bushes at Farlington Marshes on Mar 25th and one that landed on rocks at Hurst Castle on Mar 29th, were certainly migrants.

The first record of a bird singing on territory was from the New Forest on Mar 15th. Counts of singing males there totalled 63, including one site with 11 and two with nine each. In the north-east, 26 singing males were located at three sites, one of which held 24. Elsewhere, an additional nine singing males were located at five sites, bringing the county total to 98 territories - 10% lower than in 2004. Annual totals of territories for the ten years 1996-2005 are given in the table below.

Firecrest	1996	1997	1998	1999	2000	2001	2002	2003	2004	2005
Territories	17	17	18	30	50	47	53	109	109	98

Successful breeding was confirmed in the New Forest and at a site in the east of the county.

The first autumn record from a non-breeding site was one at Sandy Point on Sep 1st/2nd. Three first winter males were trapped in a Fareham garden during September, but numbers peaked in October with seven trapped on Southampton Common during the month. There were three at Old Winchester Hill on Oct 12th; two at Sinah on Oct 22nd and 29th; and up to three at Titchfield Haven from Oct 27th to the end of the year. Most records in November and December were from coastal sites, with multiple sightings at Keyhaven/Pennington Marshes, Southampton Common, where four were trapped in November, Hook-with-Warsash, Titchfield Haven and Sinah. There were also a few inland records during this period, including one in a garden at Upper Shirley on Nov 14th and 20th, and another in a garden at Cove on Dec 26th.

The information given above about birds trapped on Southampton Common is particularly interesting. In total 12 birds were trapped, one in January, which was re-trapped in February, seven in October and four in November. One of the November birds, first trapped on 10th was re-trapped on 29th. Of the 12, eight were first winter birds. It is clear that at least one over-wintered and possible that others did as well. While we can speculate that these may be local breeding birds, consistent with recent early returns to breeding sites in spring, we cannot rule out that they are autumn immigrants.

2004 additions: ten individuals were trapped at the Hawthorns, Southampton Common, as follows: a first-winter male on Feb 13th previously trapped there on Nov 5th 2003, a first year male on Sep 14th, a female, probably adult, on Oct 8th, a first-winter male on Oct 19th, a first-winter male and female, and a male, probably adult, on Nov 1st, an adult male on Nov 2nd previously trapped there on Dec 9th 2003, and first-winter males on Nov 3rd and 15th.

Spotted Flycatcher   *Muscicapa striata*

A moderately common summer visitor and passage migrant [SPEC 3, Red, NBAP, HBAP].

Over 230 records were received, a response by observers to send in all records due to the decline of this species. The first spring record was for Binswood on Apr 27th, followed by one at Steep Marsh on 29th, and singles at Basing, Titchfield Haven and Lakeside, Eastleigh the following day. Due to the problem of differentiating between territorial and migrant birds, passage during May was noted mainly at coastal sites. The maximum number reported was of just three at Northney on May 25th, and the only spring record at Farlington Marshes was on May 23rd.

There were 32 reports of likely breeding (nests, pairs or family parties), with presence during the breeding season coming from an additional 47 locations around the county. The species continues to be (presumably) under-recorded in the New Forest, from where it was reported from just 15 sites (including five pairs in Church Place Inclosure). The Test Valley appears to be a stronghold, being reported from at least 13 localities.

As usual, the species began to be observed more frequently in August. Some of the sightings referred to family parties, but the majority were migrants scattered throughout the county. Double-figure counts at this time included 15 at Beacon Hill, Warnford on 18th; ten at Northney on 20th; and ten at Bur Bushes, NF on 28th. Numbers through September were below average; although the species was observed at many inland and coastal sites, the highest counts were only six at Farley Mount on 2nd and four at Cams Bay on 17th. Only one was trapped at Southampton Common, Sep 12th (*cf* additions below). The last sighting for the year was a single in Kiln Road, Fareham on Oct 1st.

Minimum half-monthly totals of all autumn records for the past six years are tabulated below and show that passage was reasonable in August but poor in September:

	Aug 1-15	Aug 16-31	Sep 1-15	Sep 16-30	Oct 1-15
2000	8	81	25	8	2
2001	35	330	101	25	2
2002	23	148	107	16	3
2003	40	173	166	40	0
2004	15	97	89	38	1
2005	59	151	57	25	1

2003 additions: a total of 12 was trapped at the Hawthorns, Southampton Common between Aug 30th and Sep 15th, the most ever.

2004 addition: one was trapped at Southampton Common on Sep 14th.

Pied Flycatcher

Ficedula hypoleuca

A scarce passage migrant; rarely breeds.

The first of the year was a very early record of a male at Whitten Pond on Mar 26th (PMS). This equals the earliest ever for Hampshire, the previous individual being at Sandy Point in 1996. However, it was a poor spring passage with only another five reports between Apr 8th and 30th. All records are listed below:

Whitten Bottom, NF: male, Mar 26th, flew off W.
Stoney Cross, NF: male, Apr 8th.
Wilverley Plain, NF: male Apr 9th.
The Millfield, Basing: 1, Apr 21st; female, Apr 30th.
Sandy Point: female, Apr 30th.

Autumn passage was below average, with a minimum of 21 well-scattered reports. It also started later than in 2004 when there were several records in the first half of August. The first was a female/first-winter feeding on elderberries in the observer's garden in Petersfield on Aug 17th and the last was one at IBM Lake on Sep 20th. The maximum was three at Northney Paddocks on Aug 30th. All records are as follows:

Petersfield: 1, Aug 17th.
Northney Paddocks: 1, Aug 18th; 1, Aug 23rd; 1, Aug 29th; 3, Aug 30th, 1, Sep 13th.
Farlington Marshes: 1, Aug 20th.
Hayling Island Lifeboat Stn: 1, Aug 23rd.
Titchfield Haven: 1, Aug 23rd; 1, Sep 18th.
Waterlooville: 1, Aug 23rd.
Milton Reclamation: 1, Aug 30th; 1, Sep 3rd.
Martin Down: 1, Aug 30th.
Woolmer Pond: 1, Aug 30th.
Southampton Common: 1, trapped, Sep 2nd.
Sandy Point: 1, Sep 13th.
Baddesley Common: 1, Sep 14th.
Chandlers Ford: 1, Sep 15th.
IBM Lake, Cosham: 2, Sep 17th; 1, Sep 18th-20th.

2003 additions: ten first-winters were trapped at the Hawthorns, Southampton Common on Aug 21st, 30th (3), Sep 5th (2), 9th, 12th (2) and 23rd

Bearded Tit

Panurus biarmicus

A scarce resident, passage migrant and winter visitor [Schedule 1, Amber, HBAP].

Nearly 100 records came from eight coastal sites, with no inland records. All the sites are listed below.

Reports of breeding birds and successful outcomes were sparse. In the Keyhaven area two were holding territory on June 1st. Juveniles were reported at Farlington Marshes on June 22nd and 27th. At least four pairs bred at Titchfield Haven, where newly fledged birds were reported in May, June and August.

Monthly maxima at breeding sites are summarised below:

	J	F	M	A	M	J	J	A	S	O	N	D
Farlington Marshes	10	4	2	8		16		1	25	20	9	1
Titchfield Haven	6	3	2	8	8	8	8		20	20	7	12
Fawley Refinery		1	2			9			1			
Keyhaven	4		5	5		2	10		16		7	13

Elsewhere, records were as follows: 13, Lower Test Marshes, Mar 26th; 3, Needs Ore, Apr 2nd, with 7, Nov 22nd and 2, Dec 20th; 2, Curbridge, Oct 22nd; 2, Normandy Lagoon, Nov 18th-Dec 26th; 1, Calshot, Nov 28th and Dec 24th.

Long-tailed Tit
Aegithalos caudatus

A numerous resident.

The larger groups reported were: 36 at the Millfield, Basing on Oct 6th; 30 at Firgo Farm, Longparish on Dec 6th; 25 at Lakeside, Eastleigh on Nov 14th and 24 at Fishlake Meadows on Nov 21st. A total of 12 breeding territories was reported from Lower Test Marshes NR and five from Longmoor Inclosure. The BBS showed 48% of squares occupied compared with an average of 50% in the previous ten years.

Long-tailed Tit   © Nigel Jones

Blue Tit
Cyanistes caeruleus

An abundant resident and passage migrant.

No records were received of large winter parties. In the breeding season there were 49 territories reported at Lower Test Marshes NR, 48 at Longmoor Inclosure, 40 at Brownwich/Chilling, 34 at Titchfield Haven and 12 at Deadwater Valley, Bordon. There were no other reports exceeding ten.

The BBS showed a total of 97% of squares occupied, slightly higher than the previous ten year average, at a mean detection of 10.8 per square compared with an average of 11.3 in the previous ten years.

Great Tit
Parus major

An abundant resident.

No winter records were received. During the breeding season the higher counts were: 42 territories at Lower Test Marshes NR; 41 at Longmoor Inclosure; 32 at Brownwich/Chilling and 27 at Titchfield Haven. The BBS showed a total of 92% of squares occupied at a mean of 6.5 per square compared with an average of 91% and 5.5 in the previous ten years.

There has been a modest increase in the population which was first observed in 2000.

Coal Tit   *Periparus ater*
A numerous resident.

As last year there was a report of a large number of breeding territories at Longmoor Inclosure. This year there were 102, compared with 73 in 2004 and 66 in 2003. The BBS showed 36% of squares to be occupied compared with an average of 34% in the previous ten years. The mean number detected was 1.2 per square compared with an average of 1.1 in the previous ten years.

There were six reports of eight individuals exhibiting characters of the continental form *ater* along the coast in October: three at Taddeford Gap on 9th (TP), one at Sinah Common on 11th/12th (TAL, PAG) and another at Gilkicker Point on 11th and one each at Farlington Marshes (JCr) and Thatcher's Copse, Titchfield Haven (DH) on 15th. In addition there were five other records of arrivals of a total of nine at coastal locations in October which may have been part of the same continental influx.

Willow Tit   *Poecile montanus*
A local, declining and now scarce resident [Red].

There were 32 reports from 22 sites in 2005 compared with 15 reports from 12 sites in 2004. These reports showed an estimated minimum of 36 birds. The special attempt in 2003 to locate the species was not repeated; nevertheless, these are highly concerning statistics. Most records were of ones and twos. However five were seen in Butter Wood on Mar 27th. There were reports of two pairs breeding successfully at Ewhurst Park and one at Frame Wood, NF.

The BBS found just four in the 75 squares surveyed. The future status of this bird in Hampshire still appears threatened.

Marsh Tit   *Poecile palustris*
A common but declining resident [Red].

Over 300 records were received, exceeding last year's excellent figure of about 250; this indicates that the heightened awareness of observers to the plight of this species continues. Most records were from mainly well-wooded areas on the eastern side of the county and the New Forest. Usually seen singly or in pairs, the maximum winter count was eight at Selborne on Jan 22nd. There were eight records of birds visiting feeders in gardens during the winter periods, compared with five in 2004.

There were ten records of males singing or holding territory at eight sites during the breeding season. Additionally, there were reports of 23 assumed breeding pairs at 16 sites; in three cases young were known to have fledged, with a maximum of just two juveniles from any pair.

The BBS located just 18 birds, too small a number to provide convincing statistics. The number of squares occupied was 12% which compares favourably with the first ten years of the survey when the squares occupied averaged 9%.

Nuthatch   *Sitta europaea*
A numerous resident.

During the breeding season there were 15 reports of birds holding territory, including seven at Deadwater Valley and four at Longmoor Inclosure. The BBS showed a total of 40% of squares occupied compared with an average of 42% in the previous ten years. The mean of 1.0 per square is the same as the average for the previous ten years. It appears that the recent growth in the county population has now stopped.

Treecreeper   *Certhia familiaris*
A numerous resident.

During the breeding season 13 singing males and pairs were reported from Itchen Valley CP, while another 16 reports of birds holding territory were received. The BBS showed 23% of squares occupied compared with an average of 21% in the previous ten years. The mean per square was 0.48 compared with an average of 0.3 in the previous ten years.

Red-backed Shrike   *Lanius collurio*
A very scarce passage migrant; formerly a moderately common but localised summer visitor which last bred in 1984 (?,43 since 1984, 2).

Red-backed Shrike, Farlington   © Jason Crook

Only two were recorded. An adult male was at Mapledurwell on May 27th and was seen to have a very heavily inflamed vent area. It is the first confirmed record for the Basingstoke area this century (DK *et al*) and the first in spring since 2002. A juvenile was at Farlington Marshes from Sep 14th-16th (JCr *et al*).

Annual totals of bird-months for 1996-2005 follow:

	1996	1997	1998	1999	2000	2001	2002	2003	2004	2005
Red-backed Shrike	5	1	3	0	2	5	4	1	4	2

Great Grey Shrike
Lanius excubitor

A very scarce winter visitor and passage migrant (SPEC 3).

In the early year there were five in the county, around 17% of the small British wintering population. There were three, reported between Jan 2nd and Apr 2nd, in the New Forest. As in previous years, a bird in the west of the New Forest ranged widely and was reported on 17 dates, including the first and last dates above. Originally encountered about Hampton Ridge, it then moved to Milkham Bottom later in the winter. One returning to Vales Moor was seen up until Feb 2nd, it had previously been present in the first half of November 2004 and between early November 2003 and late February 2004. Another returning bird, at Holm Hill, was present from Jan 24th to Mar 11th, having formerly been present in the first half of December 2004.

Away from the New Forest, there were only two singles (*cf* four, 2004), as follows: Ludshott Common between Jan 17th and Feb 21st and on the north-eastern outskirts of Stockbridge on Jan 21st.

In the late year the population in the New Forest had dropped to just three; this mirrored the reduced numbers, perhaps only 20, noted throughout Britain. The first autumn return was at Stone Quarry Bottom, Hampton Ridge on Oct 8th; this bird relocated to Milkham Bottom on 18th, exactly the same date as first seen at this site in late 2004, then ranged widely to at least Dec 26th. Further wintering birds were at Holm Hill, returning from Nov 5th to the end of the year and Beaulieu Road Station, intermittently, from Nov 12th to Dec 23rd. Away from the New Forest, there were four records possibly involving two individuals: the first at Testwood Lakes and Broadlands Estate on Oct 16th, possibly the same south over the M27 at North Stoneham on 28th, and one near Mottisfont Abbey on Dec 16th.

Neighbouring counties also had a quiet year, with two in Sussex in the early year and three in the late year. The only record in Dorset was the returning wintering bird at Wareham Forest; whilst in Surrey one on Thursley Common/Ockley Common, on just three dates in the late year, was the only record.

Jay
Garrulus glandarius

A numerous resident and passage migrant.

Counts of birds holding territory included six at Longmoor Inclosure, five at Titchfield Haven and four at Lower Test Marshes NR. The larger counts were recorded during October. These included an influx of 30 at Itchen Valley CP on Oct 2nd, 22 flying west in the Keyhaven area on Oct 8th and 22 flying east at Normandy on Oct 15th. The BBS shows a small fall in both numbers and percentage of squares occupied following a significant year-on-year increase. In 2005 43% of squares were occupied with a mean of 0.73 per square compared with an average of 42% and 0.8 in the previous ten years.

Magpie
Pica pica

A numerous resident.

Winter roosts included 100 at Fleet Pond on Nov 26th and Dec 28th. The only reports of breeding territories in double-figures were 18 at Lower Test NR and 14 at Brownwich/Chilling.

The BBS shows 77% of squares were occupied with a mean of 3.1 per square. This compares with 79% of squares occupied at a mean of 2.9 per square in the previous ten-year data averages.

Jackdaw   *Corvus monedula*

A numerous resident.

The largest numbers were reported from roosts. There were 3500 at Greywell on the east side of the River Whitewater on Dec 27th and 3200 at Itchen Valley CP on Dec 9th.

The BBS shows 79% of squares occupied with a mean of 8.5 in 2005, compared with an average 76% and 8.9 in the previous ten years.

Diurnal migration was reported from just five observers during the year. In the spring a total of 13 flew east or north at Sandy Point between Mar 6th and Apr 14th, and at Lymington/Hurst eight moved north-east on Mar 20th. In the autumn reports were submitted from three areas. In the Langstone Harbour area 11 flew east and 11 west on three dates between Oct 11th and 20th over Farlington Marshes, 15 east over Milton Reclamation on Oct 16th, and a total of 56 west and 10 east on four dates between Oct 23rd and Nov 5th over Bedhampton and Broadmarsh. A total of 143 bird-days was reported from Sandy Point between Sep 3rd and Nov 23rd, including 37 on Oct 26th and 34 west on Nov 9th. In the Lymmington/Hurst area 134 flew east in three hours over Normandy on Oct 16th, 60 south-west on Oct 22nd and six south on Nov 12th.

Rook   *Corvus frugilegus*

A numerous resident and probable winter visitor.

The only roosts exceeding 500 reported were 3000 at Isle of Wight Hill, Porton on Oct 5th and 1500 at Bishops Green Farm on Dec 16th.

The largest rookeries were reported from Chattis Hill (200 nests), Dever Spring (90) and Wherwell (72). Unfortunately, there appears to be no systematic reporting of rookeries from year to year.

BBS data show that 63% of squares held this species, the highest ever percentage, and that there were 17 per square in 2005 compared with an average of 10 in the previous ten years.

Carrion Crow   *Corvus corone*

A numerous resident.

High counts in the early year were 110 at Oxleys Coppice, Fareham on Mar 25th/26th, with 110 also present at Foley Manor, Liphook on Apr 21st.

In the late year 170 were present at Oxleys Coppice, Fareham on Sep 4th. In October 104 were feeding at low tide at Chalkdock, Langstone Harbour on 3rd; 150 were at Titchfield Haven on 29th and 320 at Weston Shore on Oct 6th with 200 there on Nov 5th. Three-figure counts in December were 116 at Southsea Common on 7th and 100 at Darby Green on 11th.

A pair again held territory on the Langstone Harbour islands. Other breeding territories were five at Titchfield Haven and 11 at both Lower Test NR and Longmoor Inclosure.

BBS data indicate that 93% of squares held this species with a mean of 8.3 per square compared to an average 7.8 in the previous ten years.

Raven
Corvus corax

A very scarce but increasing resident, and a frequent visitor from neighbouring counties.

The number of records received was around 300, a similar level to 2004. However, there has been an increase in the number of sightings covering the breeding season from March-July, with 90 during this period. Breeding was confirmed at four sites, including the 2004 pylon location, with three tree-nesting pairs known to have raised broods of five, two and one. In addition, pairs were recorded at a further seven sites on a regular basis, with a confirmed juvenile seen at one and a presumed family party of three at another, and at ten sites on one date only. Records came from all parts of the county, apart from the extreme east, where it remains scarce.

Away from possible breeding sites, the most frequent reports came from the Lymington/Hurst area, where the Keyhaven rubbish tip is the main attraction. One or two were noted on eight dates up to Apr 2nd and on nine dates from Aug 29th, with four on Sep 12th and three on 17th. Other records of more than two included three south-west over Itchen Valley CP on Aug 21st, four moving south-east over Mogshade Hill on Sep 27th and four over Cockley Plain on Oct 18th.

Starling
Sturnus vulgaris

An abundant but declining resident, passage migrant and winter visitor [SPEC 3, Red].

Flocks of 100 or more were reported in all months of the year from a total of 28 sites. Peak monthly counts for Farlington Marshes are summarised below:

	J	F	M	A	M	J	J	A	S	O	N	D
Farlington Marshes	400	250	2000	200	600	600	10,000	2700	2500	300		600

Monthly maxima at other sites with counts exceeding 400, typically roost counts, were as follows:

Portsmouth Harbour area: Southsea Common 500, Aug 10th.
Beaulieu estuary: Park Shore 1000, Dec 13th.
Lymington/Hurst area: Pennington Marsh 600, Nov 8th and Avon Water 3000, Dec 27th.
Test valley: Fishlake Meadows 410, Nov 21st; Moorcroft farm 1000, Nov 20th; Testbourne 450, Oct 14th.
South Downs: Long Down 500, Mar 5th; Hambledon 1000, Mar 8th.
Alresford Pond: 5000, Feb 4th; 500, Nov 5th.
North-east: East Tisted 400, Feb 5th; Headley 400, Feb 25th.

A movement of 250 north over Keyhaven on Mar 20th was the only possible passage noted in the early year but corresponds with the early year peak at Farlington Marshes and high March numbers noted in the site list above. In the late year, apart from 197 east at Needs Ore on Oct 14th, all movements thereafter were westerly: 296 over Brownwich on Oct 23rd; a total of 680 over Sandy Point on three dates between Nov 9th and 20th; and a total of 3981 over Broadmarsh/Bedhampton and Farlington during the morning, on 13 dates between Oct 23rd and Nov 21st with a peak of 1339 on Nov 9th.

Ten-year BBS data provide the following mean detection counts per square:

	1996	1997	1998	1999	2000	2001	2002	2003	2004	2005
Starling	20.1	15.6	9.0	11.6	11.3	12.1	9.1	10.4	7.7	9.5

These county data parallel the survey results for the rest of England: a small increase compared to 2004 but a significant decline in the ten-year period.

House Sparrow

A numerous resident [Red].

Passer domesticus

There were reports of 20 or more from just 14 site, a small improvement on 2004, as were the two three-figure counts, both in August: 210 at Longstock on 9th and 100 at Crondall on 13th. The Longstock count was the highest since 225 roosted at Portsmouth in July 2001. All counts of 50 or more were as follows:

Chichester Harbour area: Lifeboat station 60, Sep 22nd.
Fareham; 55, Aug 27th.
Chattis Hill: 80, Jan 12th; 61, Apr 26th.
Romsey: 80, Dec 19th.
Hurstbourne Priors: 56, Feb 14th.
Itchen valley: Lakeside 50, Sep 22nd.

Ten-year BBS data provide the following mean detection counts per square:

	1996	1997	1998	1999	2000	2001	2002	2003	2004	2005
House Sparrow	7.8	7.4	6.9	6.9	7.6	8.8	6.4	6.3	6.5	7.4

The best year since 2001, with no evidence over the last ten years of the survey of any significant decline in the county breeding population. The position for the species in southern England appears to be clinal, still faring badly in south-eastern counties but steadily improving to the west.

Tree Sparrow

Passer montanus

A very scarce resident, passage migrant and winter visitor [Red, NBAP, HBAP].

At the regular wintering site in the north-east, one was seen on Jan 22nd, Feb 7th, 27th and Apr. 14th. There was one report in summer between May 9th and Aug 17th in the New Forest area but no evidence of breeding. Subsequently, the only records were in autumn, with two at Farlington Marshes on Sep 13th (JCr, KC), one at Hook-with-Warsash on Sep

18th (PCM), two east over Sinah Common on Oct 16th (TAL) and one at Hurst Beach on Oct 23rd which came in from the NNW and landed in nearby fields (MPM).

Chaffinch   *Fringilla coelebs*

An abundant resident, passage migrant and winter visitor.

In the early year flocks in excess of 50 were recorded up to Apr 10th and totalled 3811 at 23 sites, the maximum being 500 at Denny Wood on Jan 8th.

During the breeding season 24 territories were found at Lower Test Marshes (55 in 2004) and 272 at Longmoor Inclosure. BBS data indicate little change of status, with 97% of squares occupied at a mean count of 10.6 per square.

Autumn diurnal movement was noted between Sep 18th and Nov 23rd and totalled 2737 at 11 coastal and two inland sites, with a day maximum of 594 east at Barton on Sea GC on Oct 16th and 300 west at Broadmarsh on Nov 4th. Almost daily coverage at Sandy Point produced totals of 24 in September from 21st, 393 in October (max. 218 east or north on 15th/16th) and 186 in November up to 23rd (max. 83 W on 9th). Flocks in excess of 50 on the ground were noted from Oct 14th and totalled 2077 at 19 sites, the maximum being 290 at Sherfield-on-Loddon on Dec 29th.

Brambling   *Fringilla montifringilla*

A moderately common winter visitor and passage migrant [WCA 1].

Bramblings   © Nigel Jones

Just over 200 records were received, with 76 for early in the year, when numbers were generally low. Higher numbers were recorded during spring and late in the year; 75 records referred to double-figure flocks.

Prior to late February, the only sizable flock recorded was 42 at Cadmans Pool on Jan 23rd. Numbers built up from late February probably indicating the start of spring passage.

Maxima included 300 at Rushmoor Arena on Feb 27th, 350 at Willows Green Inclosure on Mar 13th and 300 at Barrow Moor on Apr 1st. Unusually, there were three May records: two at Mark Ash Wood on 2nd, with one at the same site on 3rd, and two east at Milford on Sea on 4th. Only two later birds were recorded between 1971 and 2004.

The first return was one at Sandy Point on Oct 6th. Autumn passage was generally light (county total of 430) but was locally good in south-east coastal Hampshire. At Sandy Point a total of 134 bird-days was logged, 77 in October (maximum 53 east on 31st) and 57 in November (maximum 26 east on 1st). At Broadmarsh/Bedhampton a total of 97 was recorded between Oct 14th and Nov 17th including 30 west on Nov 4th and 50 on Nov 5th comprised of 39 west or north-west and 11 south-east. Also on Nov 5th there was a peak day maximum of 80 east at Sinah Common. Notable late year flocks were 100 at Cheesefoot Head on Dec 17th, 100 at Romsey on Dec 18th and 125 at Winchfield Moor on Dec 29th.

The approximate monthly totals are tabulated below:

J	F	M	A	M	J	J	A	S	O	N	D
146	426	609	417	5	0	0	0	0	167	740	704

Serin
Serinus serinus

A rare visitor (1,26,2).

Two or three were recorded at Sandy Point. On Apr 25th, one called and circled as if going to land, but flew off west. This coincided with a fall of other migrants including three Ring Ouzels. The next day, one was heard calling as it flew east over the beach. This may have been the same individual recorded on 25th, as there was no significant arrival of migrants that day. Finally one, of indeterminate sex, was watched feeding in the reserve for over 30 minutes on July 7th (all records ACJ).

2004 addition: one was heard singing and seen for two minutes at West End on Feb 12th (GCB).

Greenfinch
Carduelis chloris

A numerous resident, passage migrant and winter visitor.

There were only seven records referring to grounded three-figure flocks. The largest flocks early in the year were 70 in a Petersfield garden on Jan 5th and 130 in a pre-roost gathering at Ashe on Jan 29th.

During the breeding season 33 territories were found at Lower Test Marshes and five at Longmoor Inclosure. BBS data indicate a slight decline with 82% of squares occupied at a mean count of 5.6 per square.

Autumn passage was noted at 11 coastal sites between Oct 5th and Nov 4th, totalling 4980 - far heavier than recent years. At Sandy Point 2838 were noted on five dates between Oct 14th and 23rd, with an estimated peak of 1000 north-east on both Oct 14th and 15th (ACJ). (The previous record one day autumn movement was 636 west at Weston Shore on Oct 12th 1960.) Elsewhere, the highest total was 500 north-west at Hook-with-Warsash on Oct 23rd. In reality the passage was even heavier than the records indicate as on many dates, at Sandy Point and elsewhere, no effort was made to count the species during heavy movements of other species. The largest flocks during the second half of the year were 200 at Old Winchester Hill on Oct 10th and 400 at Bur Bushes on 14th. No flock exceeded 100 during November or December.

Goldfinch   *Carduelis carduelis*

Present throughout the year. A numerous breeder and passage migrant with much reduced numbers in winter.

The largest flocks early in the year were 40 at Rotherlands, Petersfield on Jan 5th and Feb 2nd and 42 at Lakeside, Eastleigh on Jan 12th. Spring passage was typically light and at Hurst Beach totalled 58 NNE between Mar 26th and May 6th.

During the breeding season four territories were found at Long Valley and two at Longmoor Inclosure and Lower Test Marshes. BBS data indicate a decrease, with 60% of squares occupied at a mean count of 2.5 per square.

Between late July and October flocks on the ground totalled 2693 at 36 sites. Flocks in excess of 100 were recorded at seven sites, with 380 at Farlington Marshes on Oct 4th the largest. An exceptionally heavy coastal passage, mainly easterly, totalled 21,491 at 15 sites. The heaviest passage occurred at Sandy Point where a total of 12,854 was recorded between Sep 21st and Dec 18th including 12,203 during October and peaks of 2555 on 14th, 2214 on 15th and 1387 on 16th (ACJ). Elsewhere, there were 2275 past Needs Ore on Oct 14th (DJU) and 1719 past Barton on Sea on Oct 16th (SGK). (The previous record one day autumn movement was 1300 at Needs Ore on Oct 19th 2001.) The only sizable grounded flocks late in the year were 90 at Lakeside, Eastleigh on Nov 10th and 100 at Bisterne on Dec 26th.

Siskin   *Carduelis spinus*

Siskin   © *Nigel Jones*

Present throughout the year. A moderately common breeder (largely confined to the New Forest), common migrant and winter visitor.

During January and February very low numbers saw 256 at 11 sites, the only sizable flock being 100 at Fleet Pond on Feb 26th. Numbers were even lower during March and April with a maximum of 25 at Longmoor Inclosure on Apr 3rd.

Reports during the breeding season came from 16 sites: four in the New Forest; six in the north-east and six from widely scattered areas elsewhere. Five territories were located in one area in the north-east and juveniles were seen at five other locations.

Autumn passage was very heavy and was noted from early September. An early major easterly movement occurred on Sep 12th with three-figure counts coming from Farlington Marshes (220) and Sandy Point (266). At both sites these were the first Siskins detected during the month and coincided with large numbers of migrant Meadow Pipits. The total autumn passage at Sandy Point was 2244 up to Dec 10th, with a day peak of 374 on Oct 7th and four other three-figure counts. Elsewhere, coastal passage totalled 2271 at 17 sites, with a peak of 296 east at Needs Ore on Oct 14th. Little passage was noted inland.

Three-figure feeding flocks late in the year were recorded at nine sites with maxima of 250 at Somerley Lake on Oct 7th, 300 at Testwood Lakes on Nov13th, 450 at Fleet Pond and 320 at Yateley GP both on Dec 28th.

The approximate monthly totals are tabulated below:

J	F	M	A	M	J	J	A	S	O	N	D
96	171	38	63	12	10	28	3	1586	4206	1291	2202

Linnet
Carduelis cannabina

Present throughout the year. A numerous and passage migrant, but numbers are usually much reduced in winter [SPEC 2, Red, NBAP, HBAP].

Spring passage at Hurst Beach totalled 222 north or east between Mar 23rd and May 5th, with a day peak of 73 north-east on Apr 2nd. At Sandy Point 106 moved north on Apr 1st and at Pennington Marshes 145 flew north on Apr 2nd. Elsewhere, between mid March and late April, the largest flock was 210 at Hoe Cross on Apr 2nd. The highest count in May was 48 at Butser Hill on 27th.

Incomplete survey work on the north-east heaths produced a total of 87 pairs/singing males at ten sites, including the following: Blackbushe Airfield/Yateley Common, 18; Bramshill Plantation, 16; Hazeley Heath, 10; Longmoor Inclosure, 26. Elsewhere four territories were at Silchester Common. BBS data indicate a decline, with 46% of squares occupied at a mean count of 2.2 per square.

Post-breeding and passage flocks were noted between July and October, with three-figure flocks at 14 sites. Peak counts during October were: 700 at Newlands Farm, Fareham on 8th; 300 at Oklahoma Farm, Grateley on 4th, Nether Wallop on 5th and Green Lane, Hambledon on 25th. Passage during October at Sandy Point totalled 1832 east, with a day peak of 459 on 15th (a day of heavy passage for other finch species). Elsewhere, coastal passage totalled 2693 at 11 sites, daily maxima being 645 east at Needs Ore on Oct 14th.

Late year numbers were high. Three-figure flocks were recorded at eight sites, including 600 at Longwood Warren on Nov 27th and 220 at Newlands Farm, Fareham on Dec 10th.

Twite
Carduelis flavirostris

A very scarce winter visitor and passage migrant.

The only record was of one at Farlington Marshes on three dates between Nov 14th and 21st (JCr). The county totals of overwintering birds in the ten years to 2004/05 are as follows:

	95/96	96/97	97/98	98/99	99/00	00/01	01/02	02/03	03/04	04/05
Overwintering Twite	4	7	14	10	2	3	1	6	6	0

Lesser Redpoll
Carduelis cabaret

Present throughout the year. A common passage and winter visitor, recently a scarce breeder but now possibly absent from the county [Amber].

Numbers were low early in the year, but somewhat higher late in the year, with 56 records referring to flocks of ten or more. The only significant flocks during the early year were 50 at Wellington CP on Jan 16th and 40 at Ludshott Common on Feb 6th. Spring passage saw little rise in numbers, with the largest flocks being 36 at Yateley Common on Mar 31st and 20 at Liss Forest on Apr 14th. There was one report of birds coming to garden feeders between late January and early March and again in the late year, mostly singles, compared with up to ten in 2003 and 2004.

The only record during the breeding season was one at Holmsley Ridge on May 29th, otherwise the last were singles at Bishops Dyke and Dur Hill Down on Apr 30th.

The first returns were one west at Titchfield Haven and two at Hook-with-Warsash on Sep 17th. Autumn passage was light and totalled 331 at 14 coastal and five inland sites between Sep 17th and Nov 25th. Day peaks were 67 east at Needs Ore on Oct 14th and 41 east at Sandy Point on Oct 16th (a period of heavy finch passage at coastal sites). Large flocks late in the year were: 35 at Broomy Lodge on Oct 13th, 60 at Itchen Valley CP on Nov 13th and 150 at Roke Manor, Romsey on Dec 29th. A total of 36 was trapped at Southampton Common during the last three months of the year, which compares with a total of 36 during the previous 14 years.

The approximate monthly totals are tabulated below:

J	F	M	A	M	J	J	A	S	O	N	D
120	65	113	122	1	0	0	0	22	585	249	399

Common Redpoll   *Carduelis flammea*
A very scarce winter visitor.

An influx into eastern and southern England started in autumn 2005.

In Hampshire, the precursors to the main influx were a female at Moorgreen Farm on Nov 27th (GH-D); one at Moorcourt Farm, Broadlands on Nov 30th (RP); a group of at least four with a large flock of Lesser Redpolls at Roke Manor, Romsey from Dec 29th (SKW, RCh) which had been present for several days prior to this.

Crossbill   *Loxia curvirostra*

A scarce resident, whose numbers are periodically augmented by irruptions in late summer or autumn [WCA].

Numbers were low during the first half of the year, with breeding proved at three sites. An influx occurred during June and peaked in July, followed by a further influx during October and November. Most observed movements were small.

Between January and June most records were confined to the New Forest. The only flocks to exceed 12 were: 19 at Woolmer Forest on Jan 21st; 20 at Holm Hill on May 14th; 13 at Roydon Woods on May 15th and 21 at Frame Wood on June 12th. Breeding was proved at one north-west and two New Forest sites.

Counts of 30 or more from July are listed below:

North-east
Bramshill Police College: 40, July 8th.
Miles Hill: 34 (5 S), Nov 5th.
New Forest
Milkham Bottom: 45, Oct 22nd.
Acres Down: 40, Nov 5th.
Slufters Inclosure: 50, Nov 28th; 45, Dec 14th.
Elsewhere
Itchen Valley CP: 65 W, July 9th.
Sydmonton Common: 30, Aug 10th.
West Walk: 30, Nov 20th.

Ringwood Forest: 85, Dec 24th.

A winter survey in the New Forest found two at one site on Feb 19th/20th, 58 at seven sites on Nov 19th/20th and 63 again at seven sites on Dec 17th/18th.

A table follows summarising all the records received:

	J	F	M	A	M	J	J	A	S	O	N	D
In woodland	35	48	48	26	40	83	173	93	27	108	294	322
Flying over	0	0	0	0	0	5	72	15	6	25	35	0
Total	35	48	48	26	40	88	245	108	33	133	329	322
Number of sites	5	12	9	9	6	13	19	17	5	19	33	22

Bullfinch   *Pyrrhula pyrrhula*

A numerous resident [SPEC 2, Red, NBAP, HBAP].

Bullfinch   © Nigel Jones

The only double-figure flock early in the year was 13 at Binswood on Mar 19th. At Farlington Marshes, two on Mar 18th was the only early year record, then recorded on five dates from Nov 20th, with six on Dec 13th. At nearby Broadmarsh 14 moved, mainly west, on four dates between Nov 5th and 16th, with nine on 9th. At Sandy Point a total of 13 bird-days was recorded in November (max. 4 on 9th).

Survey work on the north-east heaths found 27 territories at five sites. At Itchen Valley CP 14 territories were found. BBS data indicate a slight increase to 37% of occupied squares at a mean count of 1.5 per square.

Double-figure flocks during the second half of the year were: ten at Lepe on Nov 1st; 12 at Titchfield Haven on Nov 30th and 11 on Dec 9th; 16 at Dibden Bay on Nov 30th and ten moving through gorse at dusk at Broxhead Common on Dec 29th.

Hawfinch

Coccothraustes coccothraustes

A resident, moderately common in the New Forest but thinly distributed and elusive elsewhere [Amber, HBAP].

A total of 142 records were submitted compared to 87 in 2004, of which 44 were away from the New Forest (just 14 in 2004).

In the New Forest 60 were recorded at 18 sites in the first quarter, the maxima being 19 at Blackwater Arboretum on Feb 13th and ten at Pitts Wood Inclosure on Mar 5th. No other site held more than five. Between April and June records suggestive of breeding came from 13 sites, with a count of at least five adults at one site. Five were at Holly Hatch Inclosure on July 9th, five at Bur Bushes on Aug 28th and six at Mark Ash Wood on Sep 18th.

From Oct 15th a total of 35 were recorded at nine sites, with maxima of seven at Bolder Wood on Nov 14th and 14 at Blackwater Arboretum on Dec 24th.

Away from the New Forest, the regular site at Romsey held birds at both ends of the year; reports were received for five dates between Jan 16th and Mar 2nd and ten dates from Nov 1st, with maxima of two on Feb 1st and Mar 2nd and six on Nov 28th. Elsewhere there were 28 records. Many of these were from an influx of presumed continental birds in to the county during October. All records are listed below:

Blackdam, Basingstoke: 1, Feb 25th.
Potbridge: 1 N, May 22nd.
Chawton Wood Park: 2, June 12th; 1, Dec 24th.
Itchen Valley CP: 2, Oct 7th.
Testwood Lakes: 4 NW, Oct 16th; 7, Dec 29th-31st.
Abbottswood, Romsey: 3, Oct 20th.
Hook-with-Warsash: 6 NW, Oct 23rd.
Woolmer Pond: 1 SSE, Oct 25th.

Casbrook Common: 1, Nov 4th; 2, Nov 19th and Dec 3rd.
Broadlands Lake: 1, Nov 6th.
Swaythling: 2 SW over observer's garden, Nov 7th.
Upham: 1, Nov 17th.
Nursling Mill: 1 N, Nov 19th.
Mockbeggar Lake: 1, Nov 23rd, Nov 26th and Dec 3rd and 27th.
Bushfield Camp: 4, Nov 24th.
Old Winchester Hill: 3 E, Nov 29th.
Lakeside, Eastleigh: 1, Dec 10th.
Moorcourt Farm, Broadlands: 1, Dec 10th.
Bishopstoke: 1, Dec 18th.
Moorgreen Farm, West End: 1, Dec 23rd.
Owlsbury: 1, Dec 28th.

Lapland Bunting   *Calcarius lapponicus*

A rare autumn passage migrant and winter visitor (1,72,2).

Two were reported. A confiding male was well-watched and photographed at Brownwich from Oct 29th-Nov 2nd (MLE *et al*). Another was noted at Titchfield Haven on Nov 30th (BSD).

Lapland Bunting, Brownwich   © *Richard Ford*

Snow Bunting   *Plectrophenax nivalis*

A very scarce autumn passage migrant and winter visitor .

There were no early year records, the fourth year over the past decade this has occurred and the tenth in the last two decades. By contrast the last quarter proved to be particularly good for sightings with a total of 40 submitted records relating to a probable 19 individuals, and all at coastal sites.

The first of autumn was a female/immature at Needs Ore on Oct 14th. The next was a first-winter male reported at Butts Point, Keyhaven Marsh on Oct 30th; this may have been the confiding male on Normandy sea wall on Nov 4th which, when flushed by a jogger, flew out across the saltmarsh where it was taken by a Peregrine. A juvenile was also at Hurst Castle on Oct 31st. Easterly movements, noted in mid November, were one over Sandy Point on 13th, and then three, including a male, over Farlington Marshes on 14th. A male was also in the Normandy area from Nov 17th to 21st. Next arrivals were probably different groups of two each on Nov 29th, on the links at Hook-with-Warsash and at Farlington Marshes. A party of three, comprising a male and two females, was at Hurst Beach on Dec 2nd, with a first-winter female flew west at Browndown on the same date. The Hurst Beach group, probably, accounted for another 17 bird-days in December until 24th, with a male reported again on 4th, two female types from 8th to 12th and a single female/immature thereafter. A bird was at Haslar Creek, Gosport on Dec 3rd. A male and female type were seen at Needs Ore on three dates from Dec 15th to 27th and the dates are consistent with the male and female first seen at Hurst. Lastly, one (presumably female/immature) keeping to the south-east corner of Sinah Golf Course, was first noted on Dec 17th and remained to the end of the year.

2004 correction: there were three, not one, at Needs Ore on Nov 23rd.

Yellowhammer   *Emberiza citrinella*

A numerous resident [Red].

Over 130 records were received, mainly relating to breeding birds and wintering flocks (*cf* 100 in 2004). The increased recording of this species was partly due to increased coverage during the Corn Bunting Survey. Double-figure counts were made at 42 sites during the year (29 sites in the early and 17 in the late year) with three three-figure counts.

In the first quarter, the largest wintering flocks noted were: 30 at Ashley Warren and 50 near Colemore on Jan 30th; 20 near Corhampton Down on Feb 1st; 100 in stubble at Lasham Airfield on Feb 20th; at least 200 at Toyd Down on 24th; 92 near Barton Stacey on Mar 8th and a flock of 25 in a count of 31 at Lomer Farm, Warnford on 13th. On Apr 18th two flocks of 75 and 20 were in a count of 107 on spring-sown arable crops at The Grange, Northington during a BBS transect count and 27 were at Hare Warren Farm. There were also 27 at Harbridge, Avon valley on Apr 23rd.

Coastal movements noted were restricted to three in March and five in October/November: one east at Gilkicker Point on Mar 25th; two north over Sandy Point on Mar 28th, one east on Oct 3rd and one west on Oct 11th; two west over Broadmarsh on Nov 9th and one west over the Lymington/Hurst area on Nov 13th.

On the north-east heaths, territories/singing males were noted as follows: Bramshill Plantation (5); Bramshott Common (1); Hazeley Heath (2); Heath Warren (1); Longmoor Inclosure (16); Shortheath Common (1) and Woolmer Forest (1). County BBS data indicate 45% of squares occupied (compared to 43 % across the UK and the previous ten-year county mean of 49%). Individual totals recorded during BBS for 1996-2005 follow:

	1996	1997	1998	1999	2000	2001	2002	2003	2004	2005
Yellowhammer	105	107	113	132	133	57	134	99	128	245

The large increase in 2005 is solely due to an unusually high count of 107 in one square described above and was presumably a late wintering concentration.

Late year gatherings of 20 or more included: 23 at Drayton Camp on Aug 31st; 24 at Testbourne on Dec 6th; 65 near Selborne on Oct 22nd, 25 there on Nov 19th and 28 on Dec 12th; 60 at Over Wallop on Dec 13th; 30 at Linkenholt on Dec 18th and 40 at Hare Warren Farm on Dec 27th.

Reed Bunting   *Emberiza schoeniclus*

A common resident, passage migrant and winter visitor [Red, NBAP, HBAP].

There were double-figure counts of birds on the ground from 20 sites during the year; in addition, seven further sites recorded double-figure movements in autumn. The highest count of the year was a roost of 150 in December.

In the early year, notable counts were as follows: 35 at Long Valley on Jan 15th; a maximum of 34 at Hazeley Heath on Feb 27th; 33 in a hedge at Chilbolton on Feb 28th; and 26 at Rotten Green, north of Fleet, on Jan 14th.

Breeding season counts of singing males or birds holding territories included: 29 at Lower Test Marshes (26 in 2004 and 38 in 2003); 11 in the Lymington/Hurst area, reported to be under-recorded (15 in 2004 and 22 in 2003); 22 at Titchfield Haven (a welcome increase on the all-time low of 17 in 2004); 13 in the Cheesefoot Head/Gander Down area, and nine in the Bourley Heath/Long Valley area (5 in 2004 and 12 in 2003). The previously reported decline in breeding numbers was not evident at all sites during 2004 as indicated with these counts.

BBS data indicate 8% of squares occupied (*cf* previous ten-year mean of 6%) at a mean of 0.2 birds per square (*cf* previous ten year mean of 0.1).

Individual totals recorded during BBS for 1996-2005 follow:

	1996	1997	1998	1999	2000	2001	2002	2003	2004	2005
Reed Bunting	10	13	3	8	3	0	3	9	10	15

Easterly movements were noted at several coastal sites in mid October, commencing with 21 at Needs Ore on 14th, then 10 at Sandy Point on 15th. On Oct 16th, counts came from four sites: 21, Sandy Point; 20, Farlington Marshes; 11, Gilkicker Point and 10, Barton on Sea GC, with 11 grounded birds at Hook Links. On Nov 5th, 19 moved south over Hill Head.

In the last quarter notable counts were reported from two roosts: Sparsholt College (150 on Dec 20th) and Fishlake Meadows, Romsey (64 on Nov 21st). Other significant counts were at Long Valley (a maximum of 45 between Dec 11th and 19th) and Milford on Sea (40 on Dec 29th).

Corn Bunting   *Emberiza calandra*

A declining and scarce breeding species, moderately common winter visitor (SPEC 2, Red, NBAP, HBAP).

Reports in both winter periods reached a new low, with records received from five sites in the early year and just three in the late year. Many HOS members took part in a county wide survey of this species (see later in this Report). This contributed to the majority of the 103 records received (*cf* 50, 2004). Despite the increased breeding survey effort, a disturbingly low total of 62 singing males was encountered at 29 sites in the county. Because of the increased recording effort, no direct comparisons can be made but these results, nevertheless, were better than recent years (*cf* 31, 2004; 26, 2003 and 56, 2002).

In the first quarter, wintering flocks of ten or more were found at just four sites (*cf* three, 2004 and six, 2003): 17 were at Old Winchester Hill on Jan 19th; 18 at Gander Down on Feb 16th; 20 at Chidden on 20th and 70 on Toyd Down on Feb 24th, with a flock of 45 there on Mar 28th, together with ten singing males.

During the breeding season an estimated minimum total of 62 singing males was reported from 29 sites as follows (2004 figures in brackets): Toyd Down 10 (5); Hoe Cross 2 (3); Windmill Hill 3 (-); Over Wallop 2 (-); Jacks Bush 7 (-); Middle Wallop 3 (-); Old Winchester Hill 2 (-); and Fawley Down, Morestead 2 (1). Further reports of single singing males came from 12 sites.

In neighbouring Sussex, 74 singing males were located on territory in suitable breeding habitat.

For the BBS the number of squares occupied and total numbers seen 1996-2005 are as follows:

	1996	1997	1998	1999	2000	2001	2002	2003	2004	2005
Occupied squares	3	4	2	2	2	1	2	1	1	1
Number seen	5	9	4	2	2	1	3	3	3	1

There were only four records submitted for the final quarter involving just 14 individuals.

APPENDIX ONE-ESCAPES

Fulvous Whistling-Duck   *Dendrocygna bicolor*
Escape (widespread in Africa, Americas and Asia)

One occurred at Petersfield Heath Pond from Aug 12th –18th.

White-faced Whistling-Duck   *Dendrocygna viduata*
Escape (South America and South Africa)

An unringed individual was at Farlington Marshes on the evening of May 31st (photo).

Black Swan   *Cygnus atratus*
Escape, although it is possible that breeding attempts have been made in the wild. (Australia)

All records below relate to singles unless indicated otherwise:

Breamore-Searchfield: Jan 11th and Mar 6th, then not seen again until Sep 25th, Nov 27th and Dec 18th.
Lymington/Hurst area: Jan 4th at Keyhaven, Jan 7th at Mount Lake and Jan 8th at Sturt Pond.
Fleet Pond: the individual present in 2004 was back from Jan 22nd-28th. This had previously mated with a Mute Swan. Also noted on Feb 13th and 28th, Mar 13th and 24th; 2, Apr 7th and Aug 10th. The final sighting was on Sep 16th.
Headley Mill Pond: Feb 6th, and 2, Dec 28th.
Tundry Pond: Sep 5th and 14th.
Wellington CP: Feb 15th-20th, Mar 13th, June 1st, Sep 17th and Nov 26th.
Petersfield Heath Pond: Jan 1st-3rd; then on five dates from Apr 16th-May 9th.
Portsmouth Harbour: Mar 12th.
Walpole Park Gosport: on 12 dates from Jan 1st-Mar 12th.
Romsey Water Meadows: May 15th and Aug 17th.
Saddlers Mill Romsey: Jan 1st-Mar 13th, and on Oct 16th-Dec31st.
Testbourne Lake: Nov 8th.

Lesser White-fronted Goose   *Anser erythropus*
All records are presumed to relate to feral or escaped birds. (A rare vagrant to Britain from Siberia).

There were five sightings of this species during 2005 probably relating to two individuals. One was present at Hurstbourne Priors on Jan 29th and again on Feb 14th. One was nearby later in the year at Testbourne Lake on Aug 10th. The long-staying individual noted on the river Avon at Sabines Farm was seen on Sep 24th, and in the same area at Bisterne on Dec 18th.

Bar-headed Goose   *Anser indicus*
A small feral population has become established in the north-east of the county. (Central Eurasia)

Records, mostly of singles, were received from Ibsley Water on July 2nd and 16th; Beaulieu Lake on Sep 17th; Lymington on Aug 30th, Sep 18th and 25th; Tanners Lane on Aug 14th and New Woolmer Pond on Apr 11th and 12th. Up to two birds were present at Needs Ore on May 1st-20th.

Speckled Teal
Anas flavirostris

Escape (South America).

Two birds were present in the Lymington/Hurst area all through the year although one temporarily disappeared during April. At another regular site, Bramshill Police College, two were observed on Feb 13th, Mar 13th, Apr 10th, May 8th, Nov 6th and Dec 4th; these were joined by a third on Sep 18th.

Ringed Teal
Callonetta leucophrys

Escape (South America).

One male was reported from The Vyne Watermeadow on Apr 19th, and again on Oct 16th and Dec 4th.

Muscovy Duck
Cairina moschata

Originating from domestic stock (Central and South America).

Four were present on January 4th at Eyeworth Pond. One was reported on Mar 13th on Fishlake Meadows.

Wood Duck
Aix sponsa

Escape (North America).

One was present at Buriton on Mar 12th; a second report came from The Vyne Watermeadow on Mar 22nd.

Chiloe Wigeon
Anas sibilatrix

Escape (South America).

There were numerous sightings this year. A female was at Highwood Reservoir on Aug 16th and on Lakeside at Eastleigh on Sep 8th. A male was at Mansbridge on Jan 9th and 30th and Feb 23rd-24th. There were further sightings (sex unspecified) at the balancing pool at Keyhaven on Sep 19th and Nov 20th; Baffins Pond on Feb 6th, 8th and 22nd; Lower Test Marshes on Oct 9th and 19th and Andover Rivers on Feb 14th and Mar 7th.

Red-crested Pochard
Netta rufina

Escape (SW Eurasia).

Three male and two female hybrids of this species crossed with Mallard were on Kings Pond, Alton from Aug1st-8th.

Rosy-billed Pochard
Netta peposaca

Escape (Southern South America).

Individuals were observed at Baffins Pond on Feb 8th and Mar 31st; Andover Rivers on Jan 1st and Feb 28th; and Anton Lakes on Feb 13th and Mar 13th.

Saker
Falco cherrug

Presumed escaped falconer's bird (Normally resident in Eastern Europe)

An immature male soared low over Itchen Valley CP on May 22nd.

Harris Hawk   *Parabuteo unicinctus*

(Southern USA, Central and South America)

One with jesses was at Lepe on Mar 6th.

Cockatiel   *Nymphicus hollandicus*

Escape (Australia)

There were two sightings of this species on Mar 28th: one at Selborne Common and another at Stokes Bay.

Budgerigar   *Melopsittacus undulatus*

Escape (Australia)

Only one record this year of one at Farlington Marshes from Aug 8th-Sep1st, in bushes east of the sea wall.

Regent Parrot   *Polytelis anthopeplus*

Escape (Australia)

One flew through Farlington Marshes on Sep 8th.

[African] Grey Parrot   *Psittacus erithacus*

Escape (West and Central Africa)

One was present all day at Farlington Marshes on Sep 8th and left in the evening, flying south!

Yellow-crowned Bishop   *Euplectes afer*

Escape (Africa)

A male was at Chichester Harbour Lifeboat Station on Aug 19th.

Zebra Finch   *Taeniopygia guttata*

Escape (Australia)

One was at Petersfield on Jul 22nd.

APPENDIX TWO

PENDING RECORDS / RECORDS NOT ACCEPTED

List of Records still under consideration by the *British Birds Rarities Committee*.

2004

Baltic Gull *L.f. fuscus*: Keyhaven, Aug 23rd. **Pallid Swift** *Apus pallidus*: 2, Bedhampton, Oct 12th.

2005

Cattle Egret *Bubuculus ibis*: Stoney Cross NF, July 21st. **Red-rumped Swallow** *Cecropis daurica*: Titchfield Haven, May 21st.

List of Records for which descriptions are required for consideration by the HOS Records Panel or the *British Birds Rarities Committee*.

2005

Great White Egret *Ardea alba*: Farlington Marshes, Apr 17th; Denmead, June 11th; Lepe, July 23rd; 2, Hill Head, Oct 10th. **White Stork** *Ciconia ciconia*: 2, Milford on Sea, June 11th. **Black Kite** *Milvus migrans*: Old Winchester Hill, June 4th; St Catherine's Hill, June 14th; Southwood, Farnborough, Sep 10th. **Iceland Gull** *Larus glaucoides*: first-winter, Hurst Beach, Apr 19th. **Gull-billed Tern** *Gelochelidon nilotica*: Hill Head, May 23rd. **Caspian Tern** *Hydroprogne caspia*: Tanners Lane, Aug 9th; Hill Head, Aug 11th. **Hoopoe** *Upupa epops*: Langstone, Mar 29th; Shatterford, NF, Apr 21st. **Swift** *Apus apus*: Chandler's Ford, Oct 30th. **Yellow-browed Warbler** *Phylloscopus inornatus*: Bedhampton, Oct 30th. **Arctic Redpoll** *Carduelis hornemanni*: Blackwater Arboretum, Oct 30th; **Common Redpoll** *Carduelis flammea*: 3, Blackwater Arboretum, Oct 30th. **Rosy Starling** *Sturnus roseus*: adult, Hedge End, June 30th.

Several of the above were reported via various information services; anyone who can supply information about any of the records should contact the Recorder. Unless they can be substantiated, they will not form part of the Hampshire avifaunal record.

List of Records not accepted by the *HOS Records Panel* or the *British Birds Rarities Committee*.

In the vast majority of cases, the records below were not accepted because the relevant committee was not convinced that the identification was fully established; only in a very few cases was it believed that a mistake had been made.

2003

Black Kite *Milvus migrans*: Cheesefoot Head area, Jan 19th-Mar 5th. [*Editor:* An explanation was sought by the Recorder from the chairman of BBRC for the rejection of this well-watched bird. The reply indicated that for such an unusual date the committee felt the identification would have needed to be established in every detail. Earlier, Dick Forsman, the acknowledged foremost European expert on raptor identification, had commented on published photographs taken by SKW: "*It looks like an adult nominate Black Kite in most respects but for a few minor details*".]

2004

White-rumped Sandpiper *Calidris fuscicollis*: Keyhaven Marsh, Aug 4th. **Caspian Gull** *Larus (a.) cachinnans*: first-summer, Needs Ore, June 29th. **White-winged Black Tern** *Chlidonias leucopterus*: Titchfield Haven, Sep 26th. **Little Swift** *Apus affinis*: Southampton University, Feb 6th.

2005

Great White Egret *Ardea alba*: Hayling Oysterbeds, May 10th. **White Stork** *Ciconia ciconia*: Frame Wood, May 9th. **Least Sandpiper** *Calidris melanotus*: Keyhaven Marsh, Sep 9th. **Ring-billed Gull** *Larus delawarensis*: Keyhaven, Nov 8th. **Caspian Tern** *Hydroprogne caspia*: Chilling, June 27th. **Nightjar** *Caprimulgus europaeus*: New Forest, Apr 18th. **Richard's Pipit** *Anthus richardi*: call only, Needs Ore, Sep 17th. **Red-throated Pipit** *Anthus cervinus*: call only, Sandy Point, Oct 13th. **Marsh Warbler** *Acrocephalus palustris*: singing in scrub, Anton Lakes, July 30th. **Sardinian Warbler** *Sylvia melanocephala*: Farlington Marshes, June 17th. **Yellow-browed Warbler** *Phylloscopus inornatus*: call only, New Milton, Oct 16th.

Arrival and Departure Dates of Summer Visitors

	Earliest Ever	Average 1971-2005	Earliest 2005	Latest Ever	Average 1971-2005	Latest 2005	Wintering Records
Garganey	02/03/2003	20-Mar	01-Apr	29/11/1953	28-Sep	25-Oct	6 (Dec-Mar)
Quail	09/04/1991	17-May	23-May	19/11/1958	25-Aug	10-Oct	
Honey Buzzard	21/04/1996	n/a	06-May	30/10/1976	n/a	08-Oct	
Montagu's Harrier	08/04/1979	01-May	02-May	02/11/1960	02-Sep	29-Sep	
Osprey	06/03/1954	09-Apr	19-Mar	11/12/1999	13-Oct	07-Nov	
Hobby	16/03/2002	13-Apr	02-Apr	06/11/2001	10-Oct	16-Oct	2 (Dec)
Stone Curlew	25/02/1998	29-Mar	14-Mar	06/11/1966	05-Oct	early Nov	
Little Ringed Plover	05/03/1997	19-Mar	16-Mar	**29/10/2006**	19-Sep	**29-Oct**	1 (Feb)
Wood Sandpiper	23/03/2003	05-May	28-May	26/10/1975	23-Sep	25-Sep	7 (Dec/Jan)
Pomarine Skua	09/03/1995	25-Apr	23-Apr	17/11/1963	16-Oct	19-Oct	9 (Dec/Jan)
Arctic Skua	20/03/1979	09-Apr	04-Apr	25/11/2000	26-Oct	24-Nov	1 (Jan)
Little Tern	24/03/1957	11-Apr	02-Apr	22/10/1972	01-Oct	01-Oct	
Black Tern	11/04/1979	23-Apr	23-Apr	15/11/1967	07-Oct	23-Sep	3 (Dec)
Common Tern	17/03/2003	06-Apr	23-Mar	30/11/1973	25-Oct	08-Nov	
Roseate Tern	21/4/96&03	05-May	02-May	10/10/1999	10-Sep	05-Sep	
Arctic Tern	29/03/1958	22-Apr	22-Apr	**19/11/2006**	08-Oct	**19-Nov**	5 (Dec-Feb)
Turtle Dove	25/03/1970	17-Apr	25-Apr	11/10/1986	05-Oct	01-Oct	
Cuckoo	15/03/1989	02-Apr	28-Mar	13/10/1974	13-Sep	06-Aug	1 (Nov)
Nightjar	26/04/1995	06-May	01-May	17/11/1974	07-Sep	01-Sep	
Swift	08/04/97&01	17-Apr	15-Apr	19/11/2004	05-Oct	30-Oct	2 (Dec/Jan)
Sand Martin	27/02/1990	17-Mar	18-Mar	22/12/1977	16-Oct	16-Oct	6(Jan)
Swallow	27/02/1994	22-Mar	13-Mar	22/12/1982	27-Nov	08-Dec	2 (Jan/Feb)
House Martin	10/02/2004	28-Mar	27-Mar	25/10/1979	19-Nov	08-Nov	
Tree Pipit	16/03/92&03	30-Mar	16-Mar	20/11/1976	03-Oct	24-Oct	
Yellow Wagtail	10/03/1968	01-Apr	18-Mar	09/10/1985	20-Oct	08-Oct	5 (Dec-Feb)
Nightingale	03/04/1975	15-Apr	04-Apr	24/11/1989	21-Aug	21-Aug	
Redstart	17/03/1968	04-Apr	24-Mar	21/12/2003	14-Oct	13-Oct	1 (Mar)
Whinchat	21/03/1968	17-Apr	24-Apr	31/12/1994	30-Oct	23-Oct	5 (Jan/Feb)
Wheatear	06/02/1989	12-Mar	14-Mar	04/12/2003	10-Nov	22-Dec	3 (Jan)
Ring Ouzel	03/03/1996	30-Mar	02-Apr	20/10/2001	29-Oct	12-Nov	4 (Dec-Feb)
Grasshopper Warbler	01/04/1997	13-Apr	12-Apr	09/11/1963	17-Sep	09-Oct	
Sedge Warbler	17/03/1963	08-Apr	18-Mar	02/12/1984	11-Oct	06-Oct	1(Dec)
Reed Warbler	01/04/1994	15-Apr	02-Apr	29/11/1987	20-Oct	22-Oct	
Garden Warbler	17/03/1974	12-Apr	11-Apr		02-Oct	04-Oct	1 (Dec-Feb)

Arrival and Departure Dates of Summer Visitors (continued)

	Latest Ever	Average 1971-2005	Latest 2005	Earliest Ever	Average 1971-2005	Earliest 2005	Summering Records
Lesser Whitethroat	01/04/1989	17-Apr	16-Apr	31/10/1982	03-Oct	08-Oct	6 (Nov-Mar)
Whitethroat	07/03/1997	10-Apr	03-Apr	17/11/1995	06-Oct	11-Oct	4 (Dec/Jan)
Wood Warbler	09/04/1988	21-Apr	09-Apr	29/09/1964	24-Aug	09-Aug	
Willow Warbler	16/03/2001	25-Mar	18-Mar	01/12/1990	09-Oct	20-Oct	
Spotted Flycatcher	08/04/2000	28-Apr	27-Apr	29/10/1961	02-Oct	01-Oct	
Pied Flycatcher	26/03/96&05	14-Apr	26-Mar	22/10/1977	27-Sep	20-Sep	

NB: Sandwich Tern: with more individuals overwintering on a regular basis, has been removed from the table as it is no longer possible to accurately determine 'first' and 'last' dates

Departure and Arrival Dates of Winter Visitors

	Latest Ever	Average 1971-2005	Latest 2005	Earliest Ever	Average 1971-2005	Earliest 2005	Summering Records
Bewick's Swan	24/03/1976	14-Mar	11-Mar	15/10/2004	01-Nov	05-Nov	9 (Apr/May,
White-fronted Goose	20/05/1984	27-Mar	26-Jan	05/10/1952	14-Nov	n/r	5 (Jun-Sep)
Scaup	19/05/1977	08-Apr	14-Mar	09/09/2000	26-Oct	18-Sep	3 (Jun-Aug)
Long-tailed Duck	27/05/2000	05-May	11-Apr	23/09/1961	04-Nov	18-Sep	2 (Jun-Aug)
Velvet Scoter	26/06/1997	02-May	21-Apr	29/09/1991	01-Nov	12-Nov	
Black-throated Diver	02/06/1987	04-May	02-May	14/09/2001	06-Nov	28-Sep	1 (May-Aug)
Great Northern Diver	05/06/2004	11-May	15-May	16/09/2003	09-Nov	05-Nov	1 (Aug)
Red-necked Grebe	09/05/1968	31-Mar	25-Mar	24/08/1994	16-Oct	15-Oct	
Slavonian Grebe	22/05/1959	08-Apr	08-May	25/09/1988	28-Oct	08-Oct	2 (Aug)
Hen Harrier	10/06/1986	09-May	14-May	29/08/1979	27-Sep	29-Sep	
Merlin	24/05/2003	23-Apr	15-May	**30/07/2005**	26-Aug	**30-Jul**	1 (July)
Purple Sandpiper	31/05/1961	23-Apr	04-May	07/07/1969	08-Oct	26-Sep	2 (July)
Jack Snipe	09/05/1977	16-Apr	09-Apr	05/09/1990	29-Sep	24-Sep	
Water Pipit	06/05/1998	16-Apr	09-Apr	26/09/1993	15-Oct	07-Oct	
Fieldfare	23/05/1980	30-Apr	12-Apr	03/09/2004	30-Sep	29-Sep	3 (June)
Redwing	12/05/1981	23-Apr	17-Apr	11/09/1999	28-Sep	01-Oct	2 (June)
Great Grey Shrike	08/05/1983	05-Apr	02-Apr	04/10/1972	18-Oct	08-Oct	
Brambling	13/05/1983	19-Apr	04-May	22/09/1996	06-Oct	06-Oct	1 (July)

GUIDELINES FOR THE SUBMISSION OF RECORDS

All observers birding in Hampshire are urged to submit their sightings to the Recorder on Hampshire Ornithological Society record forms. The form is available from the Recorder or on the society web site at http://www.hos.org.uk/ recordingforms/recform.doc. Records can be entered on line and e-mailed to the Recorder at johnclark@cygnetcourt.demon.co.uk. Records can also be submitted in an excel spreadsheet or word file by e-mail or on disk by post. It would be of great assistance if observers would comply with the following points when submitting records.

- Records should be listed either by species (in the order used in the *Hampshire Bird Report*) or by date.

- Please submit your sightings to the Recorder quarterly and at the latest by Jan 31st of the following year.

- Information on the records required for species occurring **annually** is tabulated below.

A =All records; details of age, plumage, time, direction of movement etc should be included as appropriate, especially for birds seen in places where not usually recorded or out of season.
B =All breeding records, with type of evidence obtained: confirmed, probable or possible.
CB =Counts of breeding pairs/singing males/territories in clearly defined areas.
F =Flocks, roosts and falls: minimum number required is given in parentheses.
F&L=First and last dates of summer and winter visitors.
M =Observations of birds moving on migration: give each day's count separately, with time of observation and direction the birds were moving.
MM =Dated monthly maxima from localities you regularly watch, counts may be below the threshold in F when submitting a complete year's data.
R =All records from localities where not normally recorded.
S =All summer records.
W =All winter records.
N =Brief notes of diagnostic identification features observed should be written on the record form.

Mute Swan	B, F (20), MM		F&L, S, elsewhere A
Bewick's Swan	A, N away from Avon valley	Smew	A
White-fronted Goose	A	Red-breasted Merganser	main coastal sites F (20), MM, F&L,
Greylag Goose	B, F (20), MM, R		S, elsewhere A
Snow Goose	A	Goosander	A
Bar-headed Goose	A	Ruddy Duck	A
Greater Canada Goose	B, F(100), MM	Red-legged Partridge	B, F (50), R
Lesser Canada Goose	A, N	Grey Partridge	A
Barnacle Goose	A	Quail	A
Brent Goose (Dark-bellied)	F (100), MM, F&L, M, R, inland A	Pheasant	CB, R
Brent Goose (Pale-bellied)	A	Golden Pheasant	A
Egyptian Goose	A	Lady Amherst's Pheasant	A, N
Ruddy Shelduck	A	Red-throated Diver	A, inland & summer N
Shelduck	coast: B, F (50), MM, inland: A	Black-throated Diver	A, inland & summer N
Mandarin	B, F (10), MM, R	Great Northern Diver	A, inland & summer N
Wigeon	F coast & Avon valley(100),	Little Grebe	B, F (5), MM, R
	elsewhere (25), MM, F&L, R	Great Crested Grebe	B, F (10), MM, R
Gadwall	B, F (25), MM, F&L, R	Red-necked Grebe	A, inland & summer N
Teal	B, F coast & Avon valley (100),	Slavonian Grebe	A, inland & summer N
	elsewhere (25), MM, F&L, R	Black-necked Grebe	A
Mallard	CB, F (100), MM	Fulmar	A, inland N
Pintail	main coastal sites F (20), MM, F&L,	Manx Shearwater	A, N
	S, elsewhere A	Gannet	A, inland N
Garganey	A	Cormorant	F (20), MM, R
Shoveler	B, F (10), MM, F&L, R	Shag	A, inland N
Red-crested Pochard	A	Bittern	A
Pochard	B, F (10), MM, F&L, R	Little Egret	B, F (10), MM, M, R
Tufted Duck	B, F (25), MM, F&L, R	Grey Heron	B, F (10), MM, M
Scaup	A, inland & summer N	Spoonbill	A, inland N
Eider	A, inland N	White Stork	A, N
Long-tailed Duck	A, inland & summer N	Honey Buzzard	A, away from New Forest N
Common Scoter	A	Red Kite	A
Velvet Scoter	A, inland N	Marsh Harrier	A
Goldeneye	main coastal sites F (10), MM,	Hen Harrier	A

Montagu's Harrier	A, N	Razorbill	A
Goshawk	A, away from New Forest N	Little Auk	A, N
Sparrowhawk	B, M	Puffin	A, N
Buzzard	B, F (10), M, R	Feral Pigeon	CB, F (100)
Osprey	A	Stock Dove	CB, F (25), M
Kestrel	B, M	Woodpigeon	CB, F (500), M
Merlin	A	Collared Dove	CB, F (50), M
Hobby	B, F&L, M, R	Turtle Dove	A
Peregrine	A	Cuckoo	A
Water Rail	A	Ring-necked Parakeet	A
Spotted Crake	A, N	Barn Owl	A
Moorhen	CB, F (20), MM	Little Owl	A
Coot	CB, F (20), MM	Tawny Owl	B, R
Oystercatcher	B, F (100), MM, M, inland A	Long-eared Owl	A
Avocet	A	Short-eared Owl	A
Stone Curlew	A	Nightjar	A
Little Ringed Plover	A	Swift	CB, F (100), MM, F&L, M
Ringed Plover	B, F (50), MM, M, inland A	Kingfisher	A especially B
Dotterel	A, N	Hoopoe	A, N
Golden Plover	F (20), F&L, S, R	Wryneck	A, N
Grey Plover	F (50), F&L, S, R, inland A	Green Woodpecker	CB, M, R
Lapwing	B, F (100), MM, M, R	Great Spotted Woodpecker	CB, M, R
Knot	A	Lesser Sptted Woodpecker	A
Sanderling	A	Woodlark	A
Little Stint	A	Skylark	CB, F (50), M
Temminck's Stint	A, N	Shore Lark	A, N
Curlew Sandpiper	A	Sand Martin	B, F (100), F&L, M
Purple Sandpiper	A	Swallow	CB, F (100), F&L, M
Dunlin	F (100), MM, M, S, inland A	House Martin	CB, F (100), F&L, M
Ruff	A	Tree Pipit	A
Jack Snipe	A	Meadow Pipit	B, F (25), M
Common Snipe	B, F (5), MM, F&L, R	Rock Pipit	A
Woodcock	CB, R, W	Water Pipit	A
Black-tailed Godwit	F (50), F&L, S, R, inland A	Yellow Wagtail *flavissima*	A
Bar-tailed Godwit	F (50), F&L, M, S, R, inland A	Yellow Wagtail other races	A, N
Whimbrel	F (10), F&L, M, W, R, inland A	Grey Wagtail	B, F (5), M
Curlew	B, F (100), MM, M, inland A	Pied Wagtail	CB, F (50), M
Spotted Redshank	A	White Wagtail	A, autumn N
Redshank	B, F (100), MM, M, inland A	Dipper	A, N
Greenshank	A	Waxwing	A, N except in invasion years
Green Sandpiper	A	Wren	CB, F (25)
Wood Sandpiper	A	Dunnock	CB, F (10)
Common Sandpiper	A	Robin	CB, F (25)
Turnstone	F (20), F&L, M, S, R, inland A	Nightingale	A
Grey Phalarope	A, N	Black Redstart	A
Pomarine Skua	A, other than coast in spring N	Redstart	A
Arctic Skua	A, inland N	Whinchat	A
Great Skua	A, inland N	Stonechat	A
Mediterranean Gull	A - include age/plumage	Wheatear	A
Little Gull	A - include age/plumage	Ring Ouzel	A
Black-headed Gull	B, F (500), MM, S	Blackbird	CB, F (25), M
Common Gull	B, F (50), MM, S	Fieldfare	F (25), M, F&L
Lesser Black-backed Gull	B, F (50), MM, S	Song Thrush	CB, F (10), M
Herring Gull	B, F (50), MM, S	Redwing	F (25), M, F&L
Yellow-legged Gull	A - include age/plumage	Mistle Thrush	CB, F (25), M
Iceland Gull	A, N	Cetti's Warbler	A
Glaucous Gull	A, N	Grasshopper Warbler	A
Great Black-backed Gull	B, F (20), MM, S, inland A	Sedge Warbler	CB, F (5), F&L, R
Kittiwake	A - include age/plumage	Reed Warbler	CB, F (5), F&L, R
Little Tern	B, F (25), F&L, M, R	Blackcap	CB, F (10), W
Black Tern	A	Garden Warbler	CB, F (5), F&L, R
Sandwich Tern	B, F (50), F&L, M, R	Lesser Whitethroat	A
Common Tern	B, F (50), F&L, M, R, inland A	Whitethroat	CB, F (10), F&L
Roseate Tern	A, inland N - include age / plumage	Dartford Warbler	A except New Forest CB
Arctic Tern	A - include age/plumage	Yellow-browed Warbler	A, N
Guillemot	A	Wood Warbler	A

Chiffchaff *collybita*	CB, F (10), W	Carrion Crow	CB, F (50), M
Chiffchaff other races	A, N	Hooded Crow	A, N
Willow Warbler	CB, F (10), F&L	Raven	A
Goldcrest	CB, F (10)	Starling	CB, F (100), M
Firecrest	A	House Sparrow	CB, F (20), M
Spotted Flycatcher	A	Tree Sparrow	A
Pied Flycatcher	A	Chaffinch	CB, F (50), M
Bearded Tit	A	Brambling	A
Long-tailed Tit	CB, F (20), M	Serin	A, N
Blue Tit	CB, F (50), M	Greenfinch	CB, F (50), M
Great Tit	CB, F (20), M	Goldfinch	CB, F (20), M
Coal Tit	CB, F (20), M	Siskin	B, F (10), M, S
Willow Tit	A	Linnet	CB, F (50), M
Marsh Tit	A	Twite	A, N
Nuthatch	CB, F (10)	Lesser Redpoll	A
Treecreeper	CB, F (10)	Crossbill	A
Golden Oriole	A, N	Bullfinch	CB, F (10), M
Red-backed Shrike	A, N	Hawfinch	A
Great Grey Shrike	A, N away from New Forest	Lapland Bunting	A, N
Jay	CB, F (10), M	Snow Bunting	A, inland N
Magpie	CB, F (50), M	Yellowhammer	CB, F (10), M,
Jackdaw	CB, F (500), M	Reed Bunting	A
Rook	CB, F (500), M	Corn Bunting	A

- Records of the following scarcer species must be supported by the completion of *Unusual Record Forms* which are available from the Recorder or on the society web site at:

- http://www.hos.org.uk/recordingforms/UnusualRecordForm.doc

- The submission of digital photographs or sound recordings in electronic format is also welcomed.

Whooper Swan, Tundra Bean, Taiga Bean and Pink-footed Geese, Black Brant, American Wigeon, Green-winged Teal, Ring-necked and Ferruginous Ducks, Surf Scoter, Sooty, Cory's and Balearic Shearwaters, Storm-petrel, Wilson's and Leach's Storm-petrels, Night Heron, Great White Egret, Purple Heron, Black Kite, Rough-legged Buzzard, Red-footed Falcon, Corncrake, Common Crane, Kentish Plover, American Golden Plover, White-rumped Sandpiper, Pectoral Sandpiper, Buff-breasted Sandpiper, Red-necked Phalarope, Long-tailed Skua, Sabine's, Ring-billed and Caspian Gulls, Iceland Gull of race *kumleini*, White-winged Black Tern, Black Guillemot, Alpine Swift, Bee-eater, Short-toed Lark, Red-rumped Swallow, Richard's, Tawny and Red-throated Pipits, Bluethroat, Aquatic, Marsh, Icterine, Melodious, Barred, Subalpine, Greenish, Dusky, Radde's and Pallas's Warblers, Red-breasted Flycatcher, Woodchat Shrike, Rosy Starling, Common (Mealy) Redpoll, Arctic Redpoll, Common Rosefinch, Cirl, Ortolan, Rustic and Little Buntings

All exceptionally early or late migrants.

- Records of the above will not be published unless they have been accepted by the *HOS Records Panel.*

- Heard only records of the above species. Criteria for the acceptance of these by the panel are set out below.

1. The observer must have good experience of the species concerned and demonstrate that the call of the bird claimed was heard clearly.

2. The call must be described in detail including transcription on paper (length, quality, tone etc of call), comparison with other calls heard either at the same or around the same time (or failing that, qualification as to why it was different to commoner species) and frequency/ timing of call(s).

3. The observer should provide an exact account of the record including, for example, how many times the bird called and over what period of time it was heard.

Records of rarer species are dealt with by the *British Birds Rarities Committee*. Space does not permit a full listing of these species which can be found on the BBRC website. BBRC record forms are available from the Recorder.

John Clark, County Recorder   *Tel:*   01252 623397
4 Cygnet Court   *e-mail:*   *johnclark@cygnetcourt.demon.co.uk*
Old Cove Road
Fleet, Hampshire
GU13 8RL

MAP OF HAMPSHIRE
AND SITE GAZETTEER

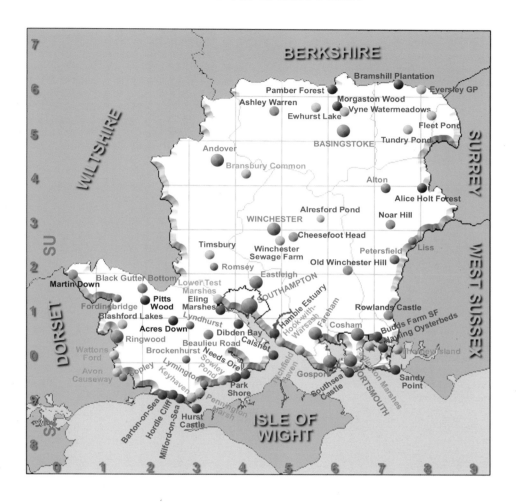

The map shows a representative selection of the main sites in the county, but is not intended to be exhaustive.

The reader is referred to the Society's website for a comprehensive listing of sites in Hampshire, at `http://www.hos.org.uk/recordingforms/gazab.html`.

HAMPSHIRE BIRD RINGING REPORT 2005

Duncan Bell

Introduction

This annual report summarises data gathered from fieldwork and research undertaken by bird ringers in Hampshire during 2005. It records the number of birds ringed in the county and publishes details of bird migration and movements in the form of recoveries and controls. The report aims to increase the awareness of birdwatchers and the general public to the ringing activities and project work that occur in the county. All records presented in this report are a reflection of the time and effort invested in the field by individuals and groups involved with ringing studies in Hampshire. Records from research workers and occasional / holiday ringers that operate in the county, are not necessarily included in this report.

Summary of Bird Ringing Activities

Project Work

The BTO Swallow Roost Project was launched in 2002 as part of a major EURING project for Swallows. The EURING (European Union for Bird Ringing) project aims to collect data on breeding and migration as well as studies on the African wintering grounds. Over 25 countries have participated in the study resulting in over half a million Swallows ringed. In Britain and Ireland over 38,000 juvenile and 4,505 adult Swallows have been ringed at roosts since 2002. Swallows ringed at Farlington Marshes, Titchfield Haven and at Winchester have contributed to the study. The project initially was to run for three years but due to its success has been extended by two years, ending in 2006.

The Re-trapping Adults for Survival (RAS scheme) set up in 1998 encourages all fully-trained ringers to get involved in collecting high-quality information on survival rates. Long term monitoring of bird populations is needed to conserve them efficiently. RAS aims to provide information on adult survival for a range of species in a variety of habitats, particularly for birds of conservation concern. In 2005 RAS studies were undertaken on House Sparrows at Manor Farm Country Park and Sand Martins at the Kimbridge colony.

The Constant Effort Sites (CES) Scheme is the first national standardised ringing programme within the BTO Ringing Scheme and has been running since 1983. Ringers set their nets in the same pattern, for the same period at regular intervals throughout the breeding season. The scheme provides valuable key information on 1) changes in population size, 2) changes in breeding success and 3) adult survival rates for 28 species of common songbird. In 2005 CES sites were operational at Winchester College and Winchester SF.

Ringing Totals

The total number of birds ringed in the county was 13,619 of 106 species, this is a 19% decrease compared with 2004. Totals for all species are summarised below (abbreviations used are F for full grown and P for pulli [nestling]):

Species	F	P	Total
Mute Swan	63	0	63
Canada Goose	6	0	6
Teal	11	0	11
Mallard	7	0	7
Little Egret	0	11	11
Grey Heron	0	2	2
Goshawk	0	3	3
Peregrine	1	0	1
Sparrowhawk	15	51	66
Buzzard	1	38	39
Kestrel	5	34	39
Hobby	1	0	1
Water Rail	1	0	1
Moorhen	1	0	1
Oystercatcher	23	0	23
Avocet	1	0	1
Little Ringed Plover	0	4	4
Ringed Plover	2	0	2
Grey Plover	4	0	4
Knot	23	0	23
Little Stint	1	0	1
Lapwing	0	39	39
Dunlin	115	0	115
Common Snipe	2	0	2
Black-tailed Godwit	26	0	26
Curlew	1	0	1
Redshank	225	6	231
Greenshank	19	0	19
Common Sandpiper	4	0	4
Black-headed Gull	1	279	280
Sandwich Tern	0	22	22
Common Tern	0	41	41
Stock Dove	0	9	9
Wood Pigeon	34	0	34
Collared Dove	4	2	6
Barn Owl	1	85	86
Little Owl	0	3	3
Tawny Owl	6	3	9
Nightjar	2	2	4
Swift	0	3	3
Kingfisher	18	0	18
Green Woodpecker	5	0	5
Great S. Woodpecker	18	0	18
Lesser S. Woodpecker	1	0	1
Woodlark	0	7	7
Skylark	0	1	1
Sand Martin	684	13	697
Swallow	257	677	934
House Martin	1	5	6
Meadow Pipit	42	0	42
Grey Wagtail	22	0	22
Pied Wagtail	133	13	146
Waxwing	33	0	33

Species	F	P	Total
Wren	288	1	289
Dunnock	275	5	280
Robin	338	35	373
Black Redstart	2	0	2
Nightingale	1	0	1
Redstart	3	0	3
Whinchat	5	0	5
Stonechat	51	6	57
Blackbird	357	18	375
Fieldfare	3	0	3
Song Thrush	78	6	84
Redwing	92	0	92
Mistle Thrush	1	5	6
Cetti's Warbler	70	0	70
Grasshopper Warbler	349	0	349
Aquatic Warbler	3	0	3
Sedge Warbler	1168	0	1168
Reed Warbler	813	1	814
Blackcap	1026	0	1026
Garden Warbler	63	0	63
Lesser Whitethroat	19	0	19
Whitethroat	73	0	73
Dartford Warbler	1	0	1
Chiffchaff	767	2	769
Willow Warbler	237	0	237
Goldcrest	329	0	329
Firecrest	17	0	17
Spotted Flycatcher	3	11	14
Pied Flycatcher	3	0	3
Bearded Tit	11	0	11
Long-tailed Tit	288	0	288
Blue Tit	739	158	897
Great Tit	418	129	547
Coal Tit	88	0	88
Marsh Tit	6	0	6
Nuthatch	50	22	72
Treecreeper	25	0	25
Jay	11	0	11
Magpie	2	0	2
Jackdaw	2	2	4
Carrion Crow	1	0	1
Starling	31	10	41
House Sparrow	139	4	143
Chaffinch	220	0	220
Brambling	220	0	220
Greenfinch	474	1	475
Goldfinch	545	2	547
Siskin	5	0	5
Lesser Redpoll	39	0	39
Linnet	7	0	7
Bullfinch	124	0	124
Yellowhammer	0	3	3
Reed Bunting	145	0	145
GRAND TOTALS	**11845**	**1774**	**13619**

Table 1   Summary of 2005 ringing totals

Nestlings

The year's total of 1,774 nestlings is a decrease on 2004 by 8%. There are now only a few ringers in Hampshire specialising in ringing nestlings of the commoner species.

Of particular note, Brian Dudley ringed 406 birds mainly in the New Forest, including 3 Goshawks, 46 Sparrowhawks, 35 Buzzards, 4 Little Ringed Plovers, 23 Lapwings, 6 Redshanks, 3 Woodlarks, 183 Swallows, 8 Pied Wagtails, 11 Spotted Flycatchers and 15 Nuthatches.

Graham Giddens, continued his study of gulls and terns at Pylewell Lake, Lymington where 279 Black-headed Gulls, 22 Sandwich Tern and 41 Common Terns were ringed. Elsewhere on the coast 11 Little Egrets were ringed at confidential sites.

The EURING Swallow Project again prompted great effort and resulted in 677 Swallow nestlings being ringed in the county.

Large nest box schemes operate throughout the county, this year 85 Barn Owl nestlings were ringed but surprisingly only 3 Tawny Owls and 3 Little Owls.

Woodlark, Romsey 21/05/05   ©Nigel Jones

Full grown

In the early year finch numbers were disappointing and made worse by an almost total absence of Siskins. Nigel Jones in Romsey still managed to attract good numbers to his baited site with highlights of 140 Chaffinches, 219 Bramblings, 30 Greenfinches, 109 Goldfinches but only two Siskins. The year's total of 474 Greenfinches is low in comparison to past years but 545 Goldfinches continues to show that the species is an increasing visitor to garden feeding stations, where they are attracted to niger seed. On Southampton Common 36 Lesser Redpolls were caught.

Up to 352 Waxwings were present near the Ordnance Survey Building in Maybush, Southampton during February and March, allowing Tim Walker to catch an incredible 33 birds, which is an unprecedented ringing total for the county.

A baited site in the north field at Farlington Marshes attracted good numbers of wintering Reed Buntings. During the summer the mixed scrub habitat at the site was used successfully to capture and study breeding species. The site can also be good for migrants as demonstrated by a Nightingale caught and ringed on May 1st.

Unfortunately the CES site at Lower Test Marshes ceased to operate in 2005 leaving CES studies confined to those worked by Wilf Simcox at Winchester College and by Tim Walker at Winchester SF.

Aquatic Warbler, Farlington Marshes   *Duncan Bell*

Reedbeds and associated sites at Farlington Marshes, Titchfield Haven and Lymington were worked from mid July into the autumn. At Farlington Marshes the traditional reedbed trapping site produced totals of 18 Cetti's, 19 Grasshopper, 351 Sedge and 298 Reed Warblers and nine Bearded Tits. Also of particular interest was the capture of three Aquatic Warblers in August including one adult.

At Titchfield Haven netting was carried out on three mornings each week, birds trapped included 38 Cetti's Warblers, 329 Grasshopper Warblers, 767 Sedge Warblers, 409 Reed Warblers, 545 Blackcaps, 43 Garden Warblers, three Lesser Whitethroats, 30 Whitethroats, 323 Chiffchaffs, 123 Willow Warblers, and two Firecrests. Some of the data collected from handling the large numbers of Grasshopper Warblers at the site are now being used in the preparation of a paper on the species.

Swallow roosts were monitored as part of the EURING Swallow Project, the initial three year study was completed at the end of 2004 but has been extended by the BTO for a further two years. Sand Martins were caught at colonies near Kimbridge and at Wickham; a total of 697 were ringed during the year.

Wader ringing sessions at Farlington Marshes, Hayling Island and at various sites on Southampton Water, provided good opportunities to study species, including: 23 Oystercatchers, one Avocet, four Grey Plovers, 23 Knot, one Little Stint, 115 Dunlins, 26 Black-tailed Godwits, 231 Redshanks and 19 Greenshanks. Target species were colour-ringed and a sample of Greenshanks was fitted with radio transmitters.

Pete Carr was active at Eastleigh SF throughout the year for the first time; this produced some good species totals and a few surprises, including 41 Meadow Pipits, 18 Grey Wagtails, 129 Pied Wagtails, 92 Wrens, 89 Dunnocks, two Black Redstarts, 33 Robins, five Cetti's Warblers, 71 Blackcaps, five Lesser Whitethroats, 246 Chiffchaffs, 31 Willow Warblers and 109 Goldcrests. Pied Wagtails were colour ringed at the site for a third winter.

At the Hawthorns Urban Wildlife Centre on Southampton Common nets worked throughout the year were particularly productive for migrant species. Andy Welch trapped one Redstart, 53 Blackbirds, 23 Song Thrushes, 80 Redwings, 155 Blackcaps, 62 Chiffchaffs, 11 Willow Warblers, 103 Goldcrests, 11 Firecrests, one Spotted Flycatcher

189

and one Pied Flycatcher.

Colour Ringing
Conventional metal rings usually require birds to be recaptured to allow re-identification. With colour rings and other marking techniques it is possible to correctly identify individual birds in the field without recapture.

The Farlington Ringing Group in The Solent is currently studying waders and waterbirds. Each bird when captured is marked with a combination of metal ring and four plastic colour-rings to make it possible to identify them as individuals in the field. A huge number of sightings have been reported to date and a large database has been compiled. The selection of sightings and movements reproduced in the recoveries section of this report is testimony to the value of colour ringing. This unique dataset is of considerable conservation value. It has been possible to demonstrate how long individual birds stay in The Solent and how they use different areas, both within, and between winters, showing how separate feeding and roosting areas are of importance to The Solent. Also being undertaken is the analysis of data to allow estimates of adult survival rates and the collection of sightings to record migration routes and dates.

What to do with your records of ringed and colour marked birds
Birds found with metal rings should be reported to the Ringing Unit of the BTO or for foreign marked birds, reported direct to the address of the scheme shown on the ring. All finders will receive a report of the original ringing details and confirmation of the finding details direct from the appropriate Ringing Unit. Reported information should include ring number, date, time and place of sighting, species involved and condition of the bird when found.

When reporting colour-marked birds, great care is required in reading colour rings or other colour marks in the field. The information gained can be essential to the success of a project and the organisers of these schemes rely on your accurate records. Waders, for example, are fitted with four colour rings which can be on the upper leg (tibia) as well as the lower leg (tarsus), therefore care is required when checking both legs. Records of colour marked birds should be reported to the organiser or group listed below. Sightings will be fully acknowledged:

Brent Goose, Wigeon, Oystercatcher, Grey Plover, Dunlin, Curlew, Redshank and Turnstone individuals have been marked in Southampton Water as part of a environmental impact assessment into the use of the estuary.

Black-tailed Godwit, Greenshank and Grey Plover individuals have been marked in Langstone and Chichester Harbours as part of a long-term population study of how the species use The Solent and also to understand migration routes in more detail.

Any sightings of the above species should be sent to The Farlington Ringing Group c/o Pete Potts, Solent Court Cottage, Hook Lane, Warsash, Southampton, SO31 9HF.

Nightingale with colour rings: Trevor Codlin, 37 Roebuck Avenue, Funtley, PO15 6TN

Little Egret and Pied Wagtail Two Little Egret nestlings were colour-ringed at a colony in Gosport and Pied Wagtail were colour-ringed at Eastleigh SF. Sightings of these species should be referred to: Pete Carr, 6 Grenadier Close, Locks Heath, SO3 6UE.

Stonechats continue to be colour-marked in the New Forest, all sightings should be sent to: Andy Welch, 90 Woodmill Lane, Bitterne Park, Southampton SO18 2PD

Recoveries

A selected list of ringing recoveries reported during 2005 follow and as in previous years these include sightings of colour marked birds. Recoveries are arranged in species order with the ringing information on the first line and recovery data below (NB a few recoveries are multiple records with two or more lines of recovery information). The symbols and conventions used are listed below.

Age when ringed
This is given according to the EURING code. The figures do not represent years but are interpreted as follows:

1	nestling or chick not yet able to fly
2	fully grown, year of hatching quite unknown
3	definitely hatched during current calendar year
4	hatched before current calendar year, exact year unknown
5	definitely hatched during last calendar year
6	hatched before last calendar year of ringing, exact year unknown
7	definitely hatched two calendar years before ringing
8	hatched more than two calendar years before ringing

Sex

M = male   F = female

Condition on recovery

X	found dead
XF	found freshly dead or dying
XL	found dead (not recent)
+	shot or intentionally killed by man
S	sick or injured - not known to have been released
VV	alive and probably healthy, ring or colour marks read in field
R	caught and released by ringer
N	caught and released but not by a ringer + breeding
/?/	condition on finding wholly unknown

Date of recovery
Where this is unknown, the date of the reporting letter is given.

Mute Swan *Cygnus olor*

W19196	5M	24.07.05.	Christchurch, (Dorset)
Orange F5D	VV	01.10.05.	Walpole Lake, Gosport 45 km E
	VV	30.12.05.	Walpole Lake, Gosport

Wigeon *Anas penelope*

FA72685 6F		31.01.00.	Dibden, Hythe, Southampton Water
	XF shot	15.10.05.	Bolshakovo, Olonetskiy district, 61 00N 32 46E, (Karelia), **Russia** 2395 km ENE
FP00624 6M		06.01.01.	Dibden, Hythe, Southampton Water
	XF shot	14.10.05.	Saranpaul, Berezovskiy district, 64 14N 60 55E, (Khanty-Mansi), **Russia** 3956km ENE

These are the 4th and 5th Russian recoveries from 340 birds ringed at Dibden Bay during 2000 and 2001. Two birds shot in France are the only other foreign recoveries.

Cormorant *Phalacrocorax carbo*

SVS	1	25.06.98.	Skrakhallen, Alvkarleby 60 39N 17 51E, (Uppsala), **Sweden**
9263873	XF	10.11.04.	Between Aldershot & Fleet, 1547 km SW

This bird is the 15th Swedish Cormorant to be recovered in Britain and Ireland. .

Sparrowhawk *Accipiter nisus*

EG07707	1F(4/4)	04.07.05.	Coppice of Linwood, New Forest
	XL	20.12.05.	Overleigh, Street, (Somerset) 79 km WNW
EG07715	1F(1/1)	06.07.05.	North Oakley Inclosure, New Forest
	XF	29.08.05.	Bridport, (Dorset) 80 km WSW

On leaving the nest at about four weeks old, young stay in the immediate vicinity for another four weeks, when still being fed by their parents. At the end of this period they become independent and disperse in a random direction, remaining on the move for no more than two to four weeks, they then remain sedentary for the rest of their lives. Males move less distances than females and only 23% of all fledglings disperse more than 20 km from their natal site.

Buzzard *Buteo buteo*

| GH47171 | 3M | 04.08.05. | Walditch, (Dorset) |
| | XF | 23.09.05. | Marchwood, Southampton 91 km ENE |

Kestrel *Falco tinnunculus*

ER90767	4F B	17.06.98.	Chilling, Warsash, Southampton
	XF	23.02.05.	Southsea, Portsmouth 16 km ESE
EG76225	1(5/5)	06.06.03.	The Hawk Conservancy, Weyhill, near Andover
	XF	25.02.05.	Porton, Salisbury, (Wiltshire) 15 km SW

Oystercatcher *Haematopus ostralegus*

BLB	1	10.05.98.	Kallo, 51 15N 04 17E, (Oost-Vlaanderen) **Belgium**
L80303	R	23.02.01.	Dibden, Hythe, Southampton Water 402 km W
FP36797	8	04.01.01.	near Weston, Southampton Water
	VV	03.06.04.	Wieringen, 52 53N 04 55E, (Noord-Holland) **The Netherlands**
	VV	02.08.05.	Strekdam, Thv Straandpaal 1, Den Hender, 52 57N 04 43E, (Noord-Holland), **The Netherlands** 477 km NE

Avocet *Recurvirostra avosetta*

One of two young hatched and raised in the east Solent during the summer of 2002 was reported breeding in the Netherlands in 2005 (still awaiting full recovery details).

Little Ringed Plover *Charadrius dubius*

| NS64756 | 1(3/3) | 03.06.01. | Ellingham, Ringwood |
| | R | 07.08.03. | Laguna de Petrola, (Albacete), 38 49N 01 33 W **Spain** 1339 km S |

Very few Little Ringed Plovers are ringed in Hampshire, so it is especially pleasing to have one ringed as a chick recovered abroad. Autumn departure starts in late July, by the end of that month many have crossed the channel. By late August, quite a number have already reached Mediterranean coasts where the Camargue is an important staging post before reaching their likely wintering destination in the western Sahel. By mid September few remain in Britain.

Dunlin *Calidris alpina*

BX16441	3	13.12.92.	Farlington Marshes
	R	18.05.04.	Dieksanderkoog-Vorland 53 58N 08 53E Schleswig-Holstein, **Germany** 757km ENE
NT46996	3	28.12.01.	Sinah, Hayling Island
	R	22.07.05.	Butterwick, (Lincolnshire) 255km NNE
NT46127	6	11.02.05.	near Weston, Southampton Water
	XF 31.03.05.		Harlingen, Noorder Zeedijk, 53 12N 05 26E Friesland **The Netherlands** 533km ENE

Nominate race *C.a.alpina* from northern Fennoscandia and European Russia are the most abundant population in Britain and Ireland outside the breeding season, and they form the vast flocks in winter estuaries. NT46996 has returned early for an individual of this race. Recoveries in Germany and the Netherlands record birds on spring migration at traditional feeding/fattening sites.

Black-tailed Godwit *Limosa limosa*

ER61994	4	20.10.95.	Farlington Marshes, Portsmouth
	VV	15.11.95.	Cley-Next-The Sea, (Norfolk) 277 km NE
	VV	22.07.05.	Alresford Creek, Colne Estuary, (Essex) 178 km NE
ER62000	3	20.10.95.	Farlington Marshes, Portsmouth
	VV	14.08.03.	Alresford Creek, Colne Estuary, (Essex)
	VV	17.07.04.	Siglufjordur, 66 07N 18 54W, **Iceland**
	VV	05.08.05.	Colne Estuary, Brightlingsea, (Essex) 178 km NE
ES74707	4	16.11.98.	Farlington Marshes
	VV	04.10.05.	Mistley, River Stour, (Essex) 192 km NE
ES74755	3	25.11.98.	Eling Marsh, Totton, Southampton
	VV	03.09.03.	River Stour, (Essex)
	VV	21.09.04.	Heybridge Basin, Blackwater Estuary, (Essex) 176 km NE
	VV	01.08.05.	Heybridge Basin, Blackwater Estuary, (Essex)
ES74713	4	16.11.98.	Farlington Marshes
	VV	19.04.00.	Elmley RSPB Reserve, (Kent) 141 km ENE
	VV	12.07.03.	Breydon Water, (Norfolk) 272 km NE
	VV	28.02.04.	Breydon Water, (Norfolk) 272 km NE
	VV	21.07.04.	Breydon Water, (Norfolk) 272 km NE
	VV	15.08. 04	Breydon Water, (Norfolk) 272 km NE
	VV	07.11.04.	Breydon Water, (Norfolk) 272 km NE
	VV	04.02.05.	Breydon Water, (Norfolk) 272 km NE
ER61994	4	20.10.95.	Farlington Marshes, Portsmouth
	VV	15.11.95.	Cley-Next-The Sea, (Norfolk) 277 km NE
	VV	22.07.05.	Alresford Creek, Colne Estuary, (Essex) 178 km NE
ER62000	3	20.10.95.	Farlington Marshes, Portsmouth
	VV	14.08.03.	Alresford Creek, Colne Estuary, (Essex)
	VV	17.07.04.	Siglufjordur, 66 07N 18 54W, **Iceland**
	VV	05.08.05.	Colne Estuary, Brightlingsea, (Essex) 178 km NE
ES74707	4	16.11.98.	Farlington Marshes
	VV	04.10.05.	Mistley, River Stour, (Essex) 192 km NE
ES74755	3	25.11.98.	Eling Marsh, Totton, Southampton
	VV	03.09.03.	River Stour, (Essex)
	VV	21.09.04.	Heybridge Basin, Blackwater Estuary, (Essex) 176 km NE
	VV	01.08.05.	Heybridge Basin, Blackwater Estuary, (Essex)
ES74713	4	16.11.98.	Farlington Marshes
	VV	19.04.00.	Elmley RSPB Reserve, (Kent) 141 km ENE
	VV	12.07.03.	Breydon Water, (Norfolk) 272 km NE
	VV	28.02.04.	Breydon Water, (Norfolk) 272 km NE
	VV	21.07.04.	Breydon Water, (Norfolk) 272 km NE
	VV	15.08. 04	Breydon Water, (Norfolk) 272 km NE
	VV	07.11.04.	Breydon Water, (Norfolk) 272 km NE
	VV	04.02.05.	Breydon Water, (Norfolk) 272 km NE

Turnstone *Arenaria interpres*

NLA	5	08.09.02.	Noordzeestrand Paal, 53 17N 04 59E, (Vlieland) **The Netherlands**
K907973	VV	23.09.02 to 07.05.03	Sturt Pond, Hurst Spit, near Lymington 533 km SW seen on 22 dates and recorded in each month
	VV	24.08.03 to 05.09.03	Sturt Pond, Hurst Spit, near Lymington 533 km SW seen on 5 dates

Red-necked Phalarope *Phalaropus fulicarius*

A juvenile watched and photographed at Cherque Farm Lake, Lee-on -the-Solent on Oct 7th/8th was colour-ringed with sky blue above the knee on the left leg and mint green above the knee on the right leg. Correspondence with Malcolm Smith, Warden in Shetland, revealed that this combination had been used in conjunction with a metal ring below the knee and the only information he could give was that the bird was ringed as a chick in Shetland in late July 2005 (report *per* Peter Raby). A scarce breeder and passage migrant, none of the 294 individuals ringed in Britain and Ireland between 1909 and 1997 were subsequently reported (Wernham 2002).

Mediterranean Gull *Larus melanocephalus*

FRP	1	03.07.03.	Oye Plage, Les Huttes d'Oye, 50 59N 02 03E, (Pas-de-Calais), **France**
FX 14327	VV	03.08.03.	Dannes, Plage, 50 35N 01 34E, (Pas-de-Calais), **France**
UO3 left	VV	10.08.03.	Keyhaven Lagoon, Keyhaven

Black-headed Gull *Larus ridibundus*

EN99276	6	02.02.94.	Hook Tip. Warsash
	XF	02.02.05.	Gosport 9 km ESE
EN99814	4	22.12.94.	Hook Tip. Warsash
	XL	20.08.03.	Malbork, Pomorskie, 54 02N 19 02E, (Szcezecin), **Poland** 1421 km ENE

Great Black-backed Gull *Larus marinus*

NOS 347829	1	21.06.03.	Midtre Katland, Farsund, 58 03N 06 50E (Vest-Agder), **Norway**
J42C	VV	23.04.05	Selsey Bill, (Sussex)
	VV	11.05.05.	Sandy Point, Hayling Island

Ringing on the east coast of England has shown that many wintering birds are Norwegian migrants. From July onwards birds from Arctic Norway and the Murmansk region migrate westwards and southwards, where they are joined by more southerly Norwegian breeders before crossing to Britain. The vast majority of this population stays to winter on the east coast with only a few reaching southern or western Britain. Return migration usually starts in late February, by March the birds have left the east coast. Eastern migration on the Hampshire coast in early April, could involve immature Norwegian birds.

Sandwich Tern *Sterna sandvicensis*

DD43204	1(2/2)	23.06.05.	Pylewell Lake, Lisle Court, Lymington
	V	21.11.05.	Cap Skiring, 12 22N 16 46W, **Senegal** 4487 km SSW

Ringing studies in Britain and Ireland has shown that the wintering range of first-years is largely centered on the western African coast from Senegal to Ghana where most remain during the following summer. See HBR 2003 for details of individuals recovered in Liberia and Ghana. The bird reported above was apparently caught alive and subsequently released. It was one of 22 ringed by Graham Giddens at Pylewell Lake this year.

Barn Owl *Tyto alba*

GF78133	1(4/4)	27.05.03.	Rockley Manor Farm, Up Somborne
	XF	29.01.05.	A380, Kingskerswell, (Devon) 167 km WSW
GH68867	1(2/2)	26.06.03.	Solent Court Farm, Warsash
	XL	02.09.05.	Southwick, Fareham 13 km ENE

This is a long distance recovery for a Barn Owl, only 4.6 % of all recoveries reported to the British and Irish ringing scheme were found over 100 km . The cause of death in 82% of cases reported involved suspected collision with road or rail traffic.

Tawny Owl *Strix aluco*

GK33457	3	28.08.05.	Crampmoor, Romsey
	XF	26.09.05.	Winchester 13 km ENE

Woodlark *Lullula arborea*

VT90231	1	25.04.02.	Avon Heath Country Park, (Dorset)
	N = M	18.04.04.	Somerley, near Ringwood 4 km NNE

A nestling remaining in its natal area to breed.

Sand Martin *Riparia riparia*

R275453	4M	24.07.02.	Kimbridge, near Romsey
	X	22.01.05.	Las Cruceras, El Barraco, 40 26N 04 35W, (Avila), **Spain** 1199km SSW
R282645	3	01.09.03.	Farlington Marshes, Portsmouth
	R	23.07.05.	Baston Gravel Pits, Baston, (Lincolnshire) 215 km NNE
T251596	3J	26.07.04.	Waterhay, Ashton Keynes, (Wiltshire)
	R	17.06.05.	Wickham, 96 km SE
T251766	3J	07.08.04.	Waterhay, Ashton Keynes, (Wiltshire)
	R=F	22.06.05.	Kimbridge, near Romsey 74 km SSE
T241329	3J	11.08.04.	Kimbridge, near Romsey
	R	22.06.05.	Le Braye Sandpit, St Ouen, Jersey, Channel Islands 208 km SSW
T244917	3J	15.06.05.	near Nettleton, (Lincolnshire)
	R	04.07.05.	Wickham, 295 km S
T167462	4	17.06.05.	Wickham
	R	10.07.05.	Laguna de Sarinena, Sarinena, 41 46 N 00 10W, Huesca, **Spain** 1018 km S
T673504	3J	24.06.05.	Woodington, East Wellow
	R	07.07.05.	Dukes Brake, (Wiltshire) 80 km NNW
T284778	3	09.07.05.	Seaside Dyke, Errol, (Tayside)
	R	03.08.05.	Kimbridge, near Romsey 606 km SSE

Many other recoveries were reported during the year involving inter-colony movements between Kimbridge, Fair Oak and Wickham

Swallow *Hirundo rustica*

R488798	1(5/5)	08.06.03.	Headbourne Worthy, near Winchester
	R	03.10.03.	Talavera, Pina de Ebro, (Zaragoza) **Spain** 1068 km S
P625461	1(4/4)	09.08.03.	Chilling, Warsash, Southampton
	X	04.07.05.	Littlehampton, (Sussex) 52 km E
R152418	3J	04.08.04.	Titchfield Haven, Hill Head, Fareham
	X	21.06.05.	Wootton Rivers, near Marlborough, (Wiltshire) 70 km NNW
T167291	1(4/4)	08.08.05.	Ashton, Bishop's Waltham
	R	08.09.05.	Icklesham, (Sussex) 134 km E

Fieldfare *Turdus pilaris*

A first-winter bird ringed at Sparsholt College, Winchester on 10.12.03. was re-trapped there on 13.11.05.

Cetti's Warbler *Cettia cetti*

R152502	2F	03.09.04.	Titchfield Haven, Hill Head, Fareham
	R	30.01.05.	South Milton Ley, (Devon) 193 km WSW

This is the third year in succession that we have had recoveries involving juvenile female Cetti's Warblers moving west in their first autumn. The previous recoveries involved movements of 19 km and 109 km.

Sedge Warbler *Acrocephalus schoenobaenus*

FRP 2132185	3 R=M	21.08.01. 31.07.04.	Camp du Hode, Sandouville, 49 29N 00 19E, (Seine-Maritime), **France** Titchfield Haven, Hill Head, Fareham 186 km NW
FRP 4715574	3 R	18.08.03. 22.08.04.	Baie de Seine, Saint-Vigor-d'Ymonville, 49 29N 00 21E (Seine-Maritime), **France** Titchfield Haven, Hill Head, Fareham 187 km NW
ESA A92193	3 R=F	03.10.03. 24.07.04.	Irun, 43 20N 01 47W, (Guipuzcoa), **Spain** Titchfield Haven, Hill Head, Fareham 833 km N
R153190	3 R	24.07.04. 18.06.05.	Titchfield Haven, Hill Head, Fareham Stanborough Reed Marsh, (Hertfordshire) 128 km NE
FRP 4850119	3 R=M R	28.04.04. 01.05.05. 12.06.05.	Mars-Ouest, St Philbert-de-Grand-Lieu, 47 02N 01 38W Loire-Atlantique **France** Farlington Marshes, Portsmouth 425 km N Farlington Marshes, Portsmouth

R155562	3	21.08.04. Titchfield Haven, Hill Head, Fareham
	R	06.09.05. Lagoa de Santo Andre, Baixo Alentejo, 38 06N 08 48W Baixo Alentejo, **Portugal** 1533km SSW
T387550	4	21.05.05. Blackgrange, (Central Region)
	R	09.08.05. Titchfield Haven, Hill Head, Fareham 614 km SSE
R943398	4M	21.07.05. near Youghal, 51 56N 07 54W, (Cork), **Eire**
	R	03.08.05. Farlington Marshes, Portsmouth 492 km ESE
T284905	3	23.07.05. Seaside Dyke, Errol, (Tayside)
	R	07.08.05. Farlington Marshes, Portsmouth 633 km SSE
T452473	3	06.08.05. Llangorse Lake, (Powys)
	R	07.08.05. Titchfield Haven, Hill Head, Fareham 187 km SE
T821074	3	06.08.05. Farlington Marshes, Portsmouth
	XF	11.08.05. Outreau, 50 42N 01 37E, (Pas-de-Calais), **France** 210 km E
R154782	3	06.08.05. Titchfield Haven, Hill Head, Fareham
	R	18.08.05. Ingooigem, 50 49N 03 26E (West-Vlaanderen), Belgium 329 km E
T821167	3	07.08.05. Farlington Marshes, Portsmouth
	R	20.08.05. Icklesham, (Sussex) 120 km E
T813286	3	09.08.05. Uskmouth, Newport, (Gwent)
	R	17.08.05. Titchfield Haven, Hill Head, Fareham 144 km SE
T291525	3J	09.08.05. Westwick, near Bishop Monkton, (North Yorkshire)
	R	21.08.05. Farlington Marshes, Portsmouth 363 km S
T732760	3	17.08.05. Walberswick, (Suffolk)
	R	23.08.05. Titchfield Haven, Hill Head, Fareham 258 km SW
R154065	3	27.08.05. Titchfield Haven, Hill Head, Fareham
	R	10.09.05. Rye Meads, Hoddesdon, (Hertfordshire) 137 km NE

Sedge Warblers undertake pre-migratory movements from late July, when birds search for sites with high densities of Plum-reed Aphids. They move both east and west along the south coast and may need to cross into France to find sufficient food to accumulate fat reserves. R154782 is interesting and suggests that some birds continue their search east into Belgium rather than turning south to follow the French coast. This puts into questions the continental origin of many birds trapped in Britain that wear Belgian rings. Also of interest are the two recoveries in Iberia, which are unusual in autumn as Sedge Warblers, in contrast to Reed Warblers, are believed to undertake long-haul flights directly from western France to Africa.

Reed Warbler *Acrocephalus scirpaceus*

P118576	3J	11.08.01.	Titchfield Haven, Hill Head, Fareham
	R	30.07.05.	Salburua-Betono Victoria, 42 54N 02 44W, (Alava), **Spain** 887 km S
R080257	3	02.08.02.	Lymington River, Lymington
	R	01.05.05.	Thatcham Marsh, Thatcham, (Berkshire) 74 km NNE
P876916	3J	30.08.03.	Titchfield Haven, Hill Head, Fareham
	R	27.06.05.	Hams Hall, Whitacre Heath, (Warwickshire) 191 km N
R334823	3	06.09.03.	Shoreham-by-Sea, (Sussex)
	R =4F B	01.08.05.	Farlington Marshes, Portsmouth 53 km E
T117269	3	31.07.04.	Pylewell Lake, Lisle Court, Lymington
	R	26.08.05.	Embalse de Castrejon, Polan, 39 48N 04 14W (Toledo), **Spain** 1235 km SW
T536256	3J	30.07.05.	Woolston Eyes, Warrington, (Cheshire)
	R	16.08.05.	Titchfield Haven, Hill Head, Fareham 298 km SSE
T732506	3	13.08.05.	Walberswick, (Suffolk)
	R	04.09.05.	Titchfield Haven, Hill Head, Fareham 258 km SW
R154998	3	17.08.05.	Titchfield Haven, Hill Head, Fareham
	R	19.08.05.	Hundred Acre, Frampton, (Gloucestershire) 130 km NW

A bird ringed at Lymington reedbeds on 25.7.98 was re-trapped there on 13.07.05. when at least eight years old. The Reed Warbler longevity record holder from BTO-ringing was 12 years, 11 months and 21 days old.

Blackcap *Sylvia atricapilla*

P154265	2M	19.09.01.	Southampton Common
	R	07.06.03.	Thetford, (Norfolk) 221 km NE
R971699	3J	07.07.04.	St Cross, Winchester

	R=M	21.06.05.	Berkhamstead Common, (Hertfordshire) 97 km NNE
R587283	3F	20.09.05.	Titchfield Haven, Hill Head, Fareham
	R	09.10.05.	Finca Castillejos, Guadalajara, 40 37N 02 59W (Guadalajara), **Spain** 1142 km S

Birds ringed on the coast between July and September are normally undertaking south to south-west movements. At this time their weights suggest they are capable of making non-stop flights to Iberia. Very few birds are re-trapped on the Hampshire coast, as even light-weight birds move on quickly.

Garden Warbler *Sylvia borin*

R152949	3	26.07.05.	Titchfield Haven, Hill Head, Fareham
	XF	29.08.05.	Saughall, Chester, (Cheshire) 289 km NNW

The first movements of Garden Warblers are recorded at Titchfield during the last week of July with peak passage occurring in the third and fourth weeks of August, The recovery of R152949 almost 300 km to the north is unusual but not unprecedented. This is a case of post-juvenile dispersal rather than reverse migration and had the bird survived would most likely have moved south again by late August

Chiffchaff *Phylloscopus collybita*

AJA876	3	16.09.04.	Titchfield Haven, Hill Head, Fareham
	R	05.10.05.	Villeton, 44 21N 00 16E, (Lot-et-Garonne), **France** 727 km S
AVA104	3J	31.07.05.	Slapton Ley, (Devon)
	R	17.09.05.	Titchfield Haven, Hill Head, Fareham 179 km ENE
BEX939	3JM	25.08.05.	Wintersett Reservoir, Wakefield, (West Yorkshire)
	R	04.10.05.	Titchfield Haven, Hill Head, Fareham 311 km S
5Z9854	3J	03.09.05.	Woolston Eyes, Warrington, (Cheshire)
	R	17.09.05.	Titchfield Haven, Hill Head, Fareham 298 km SSE
BET614	3	09.10.05.	Titchfield Haven, Hill Head, Fareham
	R	16.10.05.	Brook Farm, Reculver, (Kent) 180 km ENE

Goldcrest *Regulus regulus*

AEX684	3M	18.10.05.	Landguard Point, Felixstowe, (Suffolk)
	R	08.11.05.	Fareham 211 km SW

A large fall of continental migrants occurred on the east coast from Oct 14th to 20th, headings involving the English east coast are usually in a northeast to southwest direction. Many birds involved in these movements come from Norway with some from Sweden and even northern Finland. Birds on this heading often penetrate into Dorset and Devon.

Bearded Tit *Panurus biarmicus*

R971591	2M	29.09.04.	Thorney Island, (Sussex)
	R	31.07.05.	Farlington Marshes, Portsmouth 8 km E
T241588	2F	29.09.04.	Thorney Island, (Sussex)
	R	31.07.05.	Farlington Marshes, Portsmouth 8 km E

Long-tailed Tit *Aegithalos caudatus*

8L2826	4	23.01.00.	Ringwood
	R	29.12.05.	Ringwood

This individual is at least six years old, the longevity record from BTO-ringing is 8 years, 8 months and 5 days.

Blue Tit *Parus caeruleus*

T241469	3	05.12.04.	Timsbury
	R	21.03.05.	St Cross, Winchester 14 km ENE

Chaffinch *Fringilla coelebs*

R431225	6M 22.02.03.	Crampmoor, Romsey
	XL 17.03.05.	Haaksbergen, Enschedesestraat, 52 10N 06 45E (Overijssel), **The Netherlands** 582 km ENE
SUM	4M 29.09.03.	Rybachiy, Zelenogradskiy, 55 05N 20 44E, (Kaliningrad), **Russia**
O964969	XL 17.02.04.	Plaitford Green 1559 km WSW

Fennoscandian Chaffinches seldom cross the North Sea directly into Britain, instead they migrate on a narrow front through Denmark, north-west Germany, the Netherlands, Belgium and north-east France. They cross the channel at the narrowest point and spring migration is the reverse of autumn. Adult males often remain close to their breeding areas in winter, with any subsequent movements of these birds being in response to adverse weather, it is unlikely that R4315225 entered Britain in 2004/05 and O964969 does not fit the usual pattern of migration for Chaffinches wintering in Britain and Ireland.

Brambling *Fringilla montifringilla*

| T044437 | 5M 11.03.04. | Crampmoor, Romsey |
| | XF 12.10.05. | Murchison Oil Rig, Murchison Oil Field, 61 23N 01 44E, at sea, off Shetland 1171 km NNE |

Bramblings migrate at night directly across the North Sea, this is usually in a broad front with migrants being seen the length of the east coast from late September until early November. Large numbers have been ringed in the Northern Isles, where relatively few winter. T044437 arrived on the oil rig, located to the north-east of Shetland, with 150 other birds but died the next day.

Siskin *Carduelis spinus*

R831176	6F 03.02.04.	Crampmoor, Romsey
	R 24.04.05.	Inshriach House, Aviemore, (Highland Region) 700 km NNW
R831205	6M 05.02.04.	Crampmoor, Romsey
	R 01.06.05.	Inverarnie, (Highland Region) 731 km NNW

Reed Bunting *Emberiza schoeniclus*

R727614	5F 04.03.04.	Farlington Marshes, Portsmouth
	R 01.06.05.	Icklesham, (Sussex) 120 km E
T066560	3M 02.10.04.	Icklesham, (Sussex)
	R 29.12.05.	Ringwood 174 km W

Acknowledgements

Thanks go to all ringing groups and licensed ringers who submitted records for use in this paper and are listed at the beginning of The Report. On behalf of all Hampshire ringers I would like to thank the public bodies and individuals that gave access to reserves, farms, woodlands, gardens, etc, without their continued support and understanding ringing in Hampshire would not be possible. Thanks also to J.M. Clark, J. Crook, M.P. Moody, P.Raby and S.R. Ruscoe who submitted species finder records or sightings of colour-ringed birds.

References

In preparing this report reference has been made to the following:

Annual Report on Bird Ringing in Britain. December 2005 *Ringing and Migration* 23: Pt 2.

Wernham, C.V. *et al.* 2002. *The Migration Atlas: movements of the birds of Britain.* BTO Berkhamsted.

Duncan A. Bell, 38 Holly Grove, Fareham, Hampshire, PO16 7UP

E-mail duncan.bell5@ntlworld.com

BTO TAWNY OWL SURVEY 2005

Brian Sharkey

Background

The Tawny Owl *Strix aluco* is a nocturnal bird and monitoring its population is, therefore, more difficult than for many other species. Both the Common Bird Census and its successor, the Breeding Bird Survey, were not suitable for monitoring the species; consequently, in 1989 a specific Tawny Owl survey was carried out using a night-time point count methodology. The 1989 survey established a baseline for the species - the same methods being employed when the survey was repeated in 2005..

Aim

The aim of the Tawny Owl Survey was to provide reliable means of monitoring Tawny Owl populations more accurately than through general population monitoring surveys such as BBS. The results from the 2005 survey were used:

a)   To compare with the results from the 1989 survey.

b)   To compare Tawny Owl numbers in different areas and habitats.

Methodology

In 1989 a special survey methodology was developed that would be easily repeatable in future years, providing a reliable assessment of any changes in Tawny Owl national population levels. The survey involved night-time 10-minute counts of calling birds at a suitable watch point near the centre of selected tetrads (2km-squares) in randomly-selected 10-km squares. Counts were undertaken in the early autumn, between mid-August and mid-October, when Tawny Owls were considered to be most vociferous. The same methodology was employed for the 2005 Survey of the same locations so that comparisons could be made.

The 2005 Survey

Within Hampshire the survey covered the 10km-squares: SU20, SU23, SU50, SU53 and SU83. A total of 113 locations were allocated to observers – including 78 identified as key squares within the county by the BTO for the BTO Breeding Atlas 1988-91 (Gibbons *et al* 1993). In addition four locations in Hampshire were allocated in SU56 (which lies partly in Berkshire) and surveyed by the author; these locations were not surveyed in 1989.

Of the 113 locations allocated, 110 were actually surveyed. A summary of the results can be found below:

10 km-square	2005 Survey		1989 Survey		Difference 2005 - 1989
	Points surveyed[t]	Number of pairs [tt]	Points surveyed[t]	Number of pairs [tt]	
SU20	21	17	24	11	+6
SU23	22	25	20	14	+11
SU50	23	11	20	5	+6
SU53	25	26	24	11	+15
SU83	15	11	15	20	-9

Table 1 – Total number of birds recorded in 2005 and 1989

t = Number of points surveyed in 10km-square
tt = Estimate of total of number of pairs at all survey points within 10km-square

10 km-square	No of points surveyed in both years	No of birds recorded at same survey points in both years			% change
		2005	1989	Comparison (numbers)	
SU20	21	17	10	7	70
SU23	20	24	14	10	71
SU50	20	9	5	4	80
SU53	24	25	11	14	127
SU83	15	11	20	-9	- 45

Table 2 – Total number of birds recorded at points counted in both years.

Discussion

For this discussion, I have used only the data from those points that were counted in both years:

SU20 – New Forest. There was a 70% increase in the number of Tawny Owls recorded at the 21 points surveyed.

SU23 – Winterslows area (partly in Wiltshire). Fewer birds were recorded at four of the 20 points surveyed, but ten were noted at eight points where none had been detected in 1989 – an overall net change of the order of +71%

SU50 – Fareham and Lee-on-the-Solent area. There was an increase of four in the overall number recorded at the points in this square. In the Titchfield area (point in SU50H), two were recorded in 2005 whereas four were recorded in 1989. There was an increase of 80% in the number of birds counted at the points in this square.

SU53 – Alresford area. The largest (127%) increase recorded in 2005 was in this 10km-square: 25 Tawny Owls recorded in 2005 as against 11 recorded in 1989. The increase in numbers in this 10km-square came from six recorded at the Northington area point (in

SU53T), with none recorded in 1989, and a further eight recorded at seven points in 2005 that were not recorded in the previous survey.

SU83 – Hindhead area. This was the only square of the five main squares in which the number recorded in 2005 (11) was lower than those recorded in 1989 (20). No Tawny Owls were recorded on Weavers Down, Bramshott Common, Shottermill and Hindhead Common points in 2005, but eight were recorded in total at these sites in 1989. There were also reduced numbers recorded at Headley Down, Brookham Plantation and Ludshott Common points in 2005. Churt was the only point with an increase, with two recorded in 2005, whereas only one was noted during the previous survey. This represents a decline of 45%.

Conclusion

In this small sample of Hampshire count points, there were nearly 31% more owls detected in the 2005 survey than in 1989. To the extent that the methodology is valid for such a relatively small sample of points, it would therefore suggest that the population may well have increased. The reduced number recorded in SU83 suggests the possibility of an east-west difference, but the sample is probably too small for this to be significant.

It will be interesting to compare the regional and national results when available, to assess whether the overall Hampshire results are typical.

Acknowledgements

My thanks go to the volunteers who participated in the survey and to Dawn Balmer at the BTO for providing the 1989 data for the county. I would also like to thank John Eyre, Glynne Evans and Alan Cox for a constructive review of the draft paper.

References:

Baillie, S.R., Marchant, J.H., Crick, H.Q.P., Noble, D.G., Balmer, D.E., Beaven, L.P., Coombes, R.H., Downie, I.S., Freeman, S.N., Joys, A.C., Leech, D.I., Raven, M.J., Robinson, R.A. and Thewlis, R.M. 2005, Breeding Birds in the Wider Countryside: their conservation status 2004. *BTO Research Report No. 385. BTO, Thetford.* (http://www.bto.org/birdtrends2004)

Gibbons, D.W., Reid, J.B. & Chapman, R.A. 1993, *The New Atlas of Breeding Birds in Britain and Ireland: 1988-1991.* London: Poyser.

Brian Sharkey, 99 Stephens Road, Tadley, Hampshire RG26 3RT

THE HOBBY IN THE TWILIGHT ZONE

Ian Pibworth

Introduction

I regularly watch Testwood Lakes Nature Reserve and the following account is taken from my field notes written up at the time in 2000. As you will read, a first dramatic encounter was the spark that ignited my interest. It resulted in a summer of twilight observations and a series of records that provide a fascinating insight into the hunting behaviour of one of our most attractive raptors - and the evasion tactics of its prey! Simon King (SSK), the local HWT warden, later joined me for the nightly vigil - a great help as two pairs of eyes were definitely better than one in the fading light.

First Encounter

On the evening of May 2nd 2000, I was standing by the gates of Testwood Lakes at 2115 hrs, 46 minutes after sunset, waiting for the inland feeding flock of Whimbrel to return. My attention was diverted to a group of Noctule bats *Nyctalus noctula* hunting insects along the top of the hedge that borders Little Testwood Lake. I was getting good close views, thanks to the street lights behind me. As I was watching one particular bat, about five metres away, seemingly from out of nowhere a falcon swept past my shoulder to intercept the bat sideways on. It missed. At the same time the bat cried out, three or four loud squeaks in quick succession. Whether this was because it had been injured by a talon, or whether it was the bat's warning call, I do not know, but all the bats disappeared.

A few minutes later the falcon swept by again and, looking down on it against the backdrop of the lakes surface, I could see it was the unmistakable silhouette of a Hobby *Falco subbuteo*. I waited transfixed for five more minutes. I was about to give up and started to breath normally again when I spotted a single Noctule coming towards me. It was directly in front of me at eye level and so close that I felt I could reach out and touch it. That's when the Hobby struck. It had come from behind and below and, this time, the bat did not utter a sound. The only noise was a loud slap as the contact was made. The Hobby then rose up, re-adjusting the bat in its talons, whilst pecking at it several times. It then flew leisurely across the lake, where I lost sight of it against the tree line. Noctules did not reappear along this particular hedge for several weeks.

I was astounded at how easy it had been for the Hobby and concluded that bats didn't

202

stand a chance, for if the falcon attacks from behind, it silhouettes its prey against the sky whilst staying out of the range of the bat's sonar - especially Noctules, which tend to fly in straight lines. I feared for the bat population at Testwood Lakes - I was determined to try and discover whether this had been a one-off event, or was it to be repeated?

Next Encounters

Some time later, however, my theory was proved wrong. At a different part of the site, again some time after sunset, I noticed a Hobby above a tree line and right on *its* tail was a Noctule! The bat remained close up behind while the Hobby cruised back and forth along the tree tops. I do not know if the falcon even knew the bat was there, but the bat knew exactly where its predator was! Then, when the Hobby made yet another U-turn, the Noctule didn't follow this time, but flew off at speed in the opposite direction to the Hobby. It flew for about 100 metres before diving sharply down and out of sight. It was now 54 minutes after sunset.

This Hobby was seen again on numerous occasions, hunting well after sunset throughout the summer. As it was seen in silhouette, it was difficult to age or sex; however, it was seen close up on Aug 7th and its size suggested that it was probably a male. All its flight and tail feathers were intact but they did show signs of wear, especially to the tips of the primaries. A first-summer female Hobby, recorded on and off during the summer, was definitely ruled out as this bird was missing a tail feather. Consequently, a male of a locally breeding pair seems logical; the female would need to spend most of her summer at the nest site, getting into condition for egg laying, incubating and then brooding her young.

The Hobby's purpose for being out so late (bird prey had long gone to roost), was to hunt bats and hawk insects, especially small moths that clustered in clouds around the top branches of Oaks *Quercus rober*. On one occasion, it was seen following a hunting Barn Owl *Tyto alba*, possibly with the intention of robbing it of its prey; unfortunately, this was not witnessed as both birds were soon lost in the darkness. When hawking moths, its flight was casual, up and down the sky line above the tree tops, easily manoeuvring itself through the maze of electricity pylon cables, making observation relatively easy. It usually stayed longer at the location as well. When seen chasing bats, its flight was usually straight and fast, with the end result not usually visible as both bird and prey would disappear below the tree line. When the Hobby was seen only briefly hunting in this mode, it was thought it possibly indicated success.

The full record of my sightings of hunting Hobby after sunset, together with an indication of its intended prey, is summarised in table 1. From May 19th to June 11th and on Aug 3rd, the bird was looked for but not seen. On other missing dates in the table neither of us was available to watch; it does not mean that the Hobby did not put in an appearance. The abbreviation TaS represents "time after sunset" and records marked † indicate the presence of at least two birds during the observation period:

Date	SUNSET	TIME FIRST SEEN		TIME LAST SEEN		LOCATION	HUNTING MODES
	BST(hrs)	BST	TaS(m)	BST	TaS(m)		
02 May	2029	2115	46	2123	54	Gate	*Mostly bats*
03 May	2030	2120	40	2120	40	Gate	*Mostly bats*
07 May	2037	2120	43	2120	43	Gate	*Mostly bats*
14 May	2048	2130	42	2130	42	Gate	*Mostly bats*
17 May	2052	2130	38	2130	38	Gate	*Mostly bats*
18 May	2053	2140	47	2140	47	Gate	*Mostly bats*
12 Jun	2120	2140	20	2140	20	Oaks	*Moths/Other insects*
15 Jun	2121	2210	49	2215	54	Oaks	*Moths/Other insects*
18 Jun	2122	2200	38	2205	43	Oaks	*Mostly bats*

Table 1. Times of hunting and intended prey of Hobby (continued overleaf)

Date	SUNSET BST(hrs)	TIME FIRST SEEN BST	TIME FIRST SEEN TaS(m)	TIME LAST SEEN BST	TIME LAST SEEN TaS(m)	LOCATION	HUNTING MODES
21 Jun	2123	2215	52	2215	52	Oaks	Mostly bats
25 Jun	2124	2205	41	2210	46	Oaks	Moths/Other insects
26 Jun	2124	2210	46	2220	56	Oaks	Moths/Other insects
28 Jun	2124	2200	46	2200	46	Oaks	Mostly bats
29 Jun	2123	2155	32	2206	43	Oaks	Moths/Other insects
02 Jul	2122	2200	38	2200	38	Oaks	?
03 Jul	2122	2155	33	2205	43	Oaks	Moths/Other insects
05 Jul	2121	2200	39	2206	45	Oaks	Moths/Other insects
06 Jul	2121	2155	34	2200	39	Oaks	Mostly bats
07 Jul	2121	2155	34	2205	44	Oaks	Moths/Other insects
09 Jul	2120	2155	35	2205	45	Oaks	Mostly bats
10 Jul	2119	2155	36	2203	44	Oaks	Mostly bats
12 Jul	2117	2154	37	2159	42	Oaks	Mostly bats
13 Jul	2116	2156	40	2158	42	Oaks	Mostly bats
16 Jul	2113	2148	35	2200	47 †	Oaks	Moths/Other insects
17 Jul	2112	2148	36	2205	43	Oaks	Moths/Other insects
19 Jul	2110	2145	35	2200	50	Oaks	Moths/Other insects
20 Jul	2109	2143	34	2155	46	Oaks	Moths/Other insects
23 Jul	2105	2139	34	2150	45 †	Oaks	Moths/Other insects
24 Jul	2103	2138	35	2138	35	Oaks	Mostly bats
26 Jul	2101	2132	31	2143	42 †	Oaks	Moths/Other insects
27 Jul	2100	2130	30	2133	33 †	Oaks	Food pass
30 Jul	2057	2125	28	2131	35	Oaks	Mostly bats
31 Jul	2054	2125	31	2134	40	Oaks	Moths/Other insects
02 Aug	2051	2140	49	2140	49	Field	Chasing Barn Owl
04 Aug	2047	2120	33	2124	37	Oaks	Mostly bats
06 Aug	2043	2110	27	2125	42	Oaks	Moths/Other insects
07 Aug	2041	2110	29	2125	44	Oaks	Moths/Other insects

Table 1 (continued). Times of hunting and intended prey of Hobby

Summary of Observations

The following are the key statistics of the encounters listed in Table 1:

The earliest, first seen after sunset:	20 minutes
The latest, first seen after sunset:	52 minutes

Disregarding these two extremes, the average time first seen after sunset was 37 minutes.

The earliest, last seen after sunset:	20 minutes
The latest, last seen after sunset:	56 minutes

Disregarding these two extremes, the average time last seen after sunset was 43 minutes.

The average time seen hunting:	6 minutes
The latest time recorded:	June 26th at 2220 hrs

On most occasions, after hunting, the Hobby would simply disappear into the darkness; occasionally, the bird was seen to leave to the west. Most of the time, the Hobby, after its first appearance, would simply disappear into the darkness; however, on a few occasions, the bird was seen to leave the site to the west. This, together with day-time sightings of a Hobby carrying prey in the same direction, probably indicated that this, indeed, was the male of a local breeding pair.

Discussion

This set of records proves that this Hobby was a regular twilight hunter at this site; you could almost set your watch on when it would appear. The species is known to be an opportunistic feeder, exploiting local food availability, including bats. Hobbies are also known to be active in the evening, even hunting exceptionally by moonlight (Cramp S. 1980). Do the males of all breeding pairs hunt at twilight? The answer to this is probably yes, as I have also witnessed this behaviour at another site, as has SSK at different Hobby breeding sites.

It also raises some questions: there were no sightings between May 19th and June 11th. I do not think that I would have failed to see the Hobby over this period if it had been present, even though it had moved location by about 100 metres. My assumption is: through this phase of the breeding cycle, getting into condition and the egg laying stage, the male was watching the female very closely, afraid to leave, in case another male Hobby interceded. However, once the clutch was complete, he resumed his twilight hunting. In the early sightings the male was always silent. Does the little cluster of four sightings of a second larger Hobby in late July, including a food pass, indicate that it was the breeding female and she now felt safe to leave her half-grown youngsters for a short time? When the two were seen together, they were heard calling and on at least two other nights in this period; the second bird was not seen, but calls were heard so possibly the second bird was, indeed, nearby. Why did all sightings abruptly end on August 7th - was this the fledging date?

In early May, bats were very abundant, which may account for the short time the Hobby was seen in this period and, again, might indicate hunting success.

What twilight prey is preferred, moths or bats? I would have to conclude - bats, because the Hobby always arrived just as bats came out to feed, even though moths had been flying for up to thirty minutes earlier. Secondly, the moths are quite small and a Hobby must expend a lot of energy catching them in twilight conditions. Thus, were moths just a snack whilst waiting for the main course?

I did not submit these records at the time as I was hoping that the birds would return in following years; this would have enabled SSK and me to collect a more comprehensive and complete set of records - sadly, this was not to be.

Conclusion

Hobbies were seen repeatedly hunting between 20 and 56 minutes after sunset over a summer period. On several occasions, the quarry species was seen to be Noctule bats and the timing of the Hobby's hunting coincided with the evening flight period of the Noctules, rather than alternative insect prey. A witnessed successful interception involved a rear attack from beneath the flight line of the Noctule. This provided maximum visibility for the Hobby in the low-light conditions and minimal defence for the Noctule with its essentially forward-looking sonar sensing capability. These tactics are consistent with recognised Hobby hunting strategy (Cramp S. 1980). The periods when this behaviour was observed is possibly linked with one or two phases of the breeding cycle.

Acknowledgements

I would like to thank Simon King (SSK) for assistance with these observations and for discussing some of my ideas above.

References:

Cramp, S. and Simmons, K.E.L. (editors) (1980) *The Birds of the Western Palearctic*, Vol. II.

Ian Pibworth, 28 Rosoman Road, Sholing, Southampton, Hants. SO19 7PN

A BREEDING SEASON SURVEY OF CORN BUNTINGS IN HAMPSHIRE, 2005.

Keith Betton

Introduction

The UK Corn Bunting population declined by 89% between 1970 and 2004 and BTO Breeding Bird Survey results show a 32% decline across the country just between 1994 and 2005 (Eaton *et al* 2006). The species has been included in both the list of national Biodiversity Action Plan priority species, and the RED list of Birds of Conservation Concern. In Hampshire's Biodiversity Action Plan, Corn Bunting was included in the Species Action Plan for seed-eating farmland birds. (Hampshire Biodiversity Partnership, 2001). One of the HOS actions in this Action Plan was to carry out a survey of the species' distribution to bring our knowledge up-to-date.

Objectives

Between May and July 2005 an attempt was made to carry out a basic two-hour survey in every tetrad that had held territorial Corn Buntings since 1986. Volunteers agreed to visit a total of 234 tetrads over the three months. The objective was to establish the presence, or absence, of the species rather than make a full assessment of its abundance.

Methods

All 194 tetrads that held the species during the 1986-1991 Atlas were selected for the survey. A further 40 tetrads which had produced additional breeding season records between 1992 and 2004 were also visited. Some 41 tetrads that held birds in the period 1971-1985 (but not since) were not surveyed. Apart from reporting on Corn Buntings, observers were also asked to note numbers of Yellowhammer, Lapwing, Grey Partridge and Tree Sparrow. Each tetrad survey card included a 1/25,000 map of the area to be visited and observers were asked to mark their Corn Bunting sightings on this. They were asked to visit each tetrad for two hours. If no Corn Buntings were found in that time they were encouraged to make a second visit of two hours a few weeks later. In theory they were allowed to stop surveying once a Corn Bunting was found, but the cards indicate that most visits continued for at least two hours. Observers were free to visit at any time during the day and this information was requested. The area around Hambledon was selected for a more intensive search beyond the recommended two-hour counts.

A national BTO survey in 1993 collected a lot of information about habitat types and song-post preferences but we decided against seeking this level of detail as we felt its value was not great and the challenge of collecting it might have reduced the number of survey volunteers. The same concern about demands on observers' time was behind the decision to limit the amount of time required in each tetrad.

Workforce

HOS members really took up the challenge of looking for Corn Buntings with enthusiasm – even visiting areas where they had not been seen for ten years. Some 57 volunteers took part, representing at least 500 hours in the field. A request on *Hoslist* also brought in supplementary records of additional territories that had not been found in the survey.

Results

The apparent absence of Corn Buntings from so many areas in recent years was reconfirmed by the survey. The results show that Corn Buntings occupied 36 tetrads (compared to 194 tetrads during the 1986-91 Atlas). A schematic map showing the historical and current distribution of these is shown:

```
          0  0  1                              0  0  0
       2  8 13  3  3                     0  3  1  3  1
    8  2  1  1  4 10  1                3  0  0  1  2  5  0
    8 10  7  4  5  0                   4  7  4  5  0  1
2 4 2  3  4 13  7  2              1 1  0  0  4  7  4  0
7 3 0  0  0  2 14 12              5 0  0  0  4  3  6  7
    0  0  0 10  4  7                   0  0  1  5  3  3
    3  3  0  0  0  1                   0  2  0  1  1  1
```

1986-1991 194 tetrads occupied   *1993-2004 99 tetrads occupied.*

```
              0  0  0
        0  1  0  0  0
     0  0  1  0  0  0  0
     5  4  0  0  0  0
  1  3  0  0  0  3  3  0
  4  0  0  0  0  1  8  1
     0  0  0  0  0  0
     0  0  0  0  0  0
```

2005 36 tetrads occupied.

Figure 1 The county distribution of occupied tetrads 1986-2005 on the Ordnance Survey 10km square grid.

The species has disappeared from many areas. Figure 2 overlays the three survey results above to illustrate this:

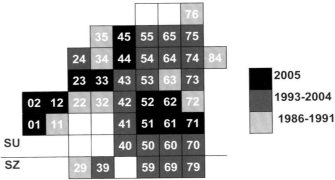

Figure 2 Ordnance Survey (OS) designation of 10km squares in Hampshire and reducing numbers of squares occupied (shades of grey) 1986-2005.

In some regular strongholds (such as SU61) they were found to be significantly scarcer than just five years previously. A total of 74 Corn Bunting territories were found in the 36 tetrads as shown at 10km grid resolution in Figure 3:

				0	0	0	
		0	1	0	0	0	
	0	0	2	0	0	0	0
	18	10	0	0	0	0	
1	5	0	0	0	6	5	0
11	0	0	0	0	1	11	3
	0	0	0	0	0	0	
	0	0	0	0	0	0	

Figure 3   County distribution of territories 2005 on the OS 10km square grid.

For reference in any follow up visits in future years the distribution of territories per tetrad was as follows:

SU 10-km square	01	01	01	01	02	12	12	12	23	23	23	23	23	33	33	33	33	33
Tetrad	P	T	U	Z	V	A	B	F	P	T	U	Y	Z	B	H	I	J	N
Total	3	1	1	6	1	1	2	2	1	10	3	3	1	2	3	1	1	2
SU 10-km square	33	44	45	51	52	52	52	61	61	61	61	61	61	61	62	62	62	71
Tetrad	Z	S	T	Z	I	N	P	D	H	I	M	N	P	T	A	F	K	C
Total	1	2	1	1	4	1	1	1	2	2	1	3	1	1	1	1	3	3

Table 1. The 2005 breeding survey - location of territories by tetrad.

Six-figure grid references were collected for all the Corn Buntings which were later correlated with habitat data held by Hampshire County Council.

With the exception of sightings at Ladle Hill in SU45 and near Tufton in SU44 there were no reports from the north of the county. The remaining concentrations were around Martin and Toyd Downs, Over Wallop, Stockbridge, Chilbolton, Longwood Warren, Cheesefoot Head, Hambledon, Chidden and Corhampton. There were none in coastal locations and the immediate hinterland.

Almost 700 Yellowhammers were found. In some areas (such as SU74) they were widely reported, while in others (such as SU61) they were largely absent. Around 260 Lapwings and 50 Grey Partridge were also reported, although no Tree Sparrows were seen. Other observations submitted included a Quail and several Red Kites.

The Hambledon Experience

The proposed intensive survey in SU61 was impossible to achieve due to lack of observers in that area, however John Shillitoe has kindly provided details of his annual Corn Bunting census in the area immediately around Hambledon and the accompanying maps show the distribution of territorial males between 2001 and 2005. During this period the local population declined from 16 to four territories.

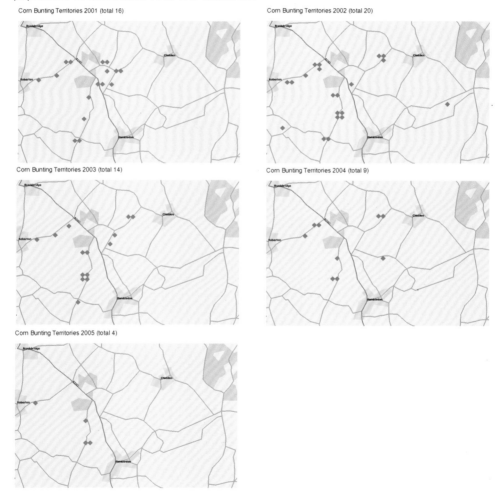

Corn Bunting Territories 2001 (total 16)

Corn Bunting Territories 2002 (total 20)

Corn Bunting Territories 2003 (total 14)

Corn Bunting Territories 2004 (total 9)

Corn Bunting Territories 2005 (total 4)

The area that was surveyed is very approximately triangular in shape, to the south of Old Winchester Hill, east of Soberton and west of Hambledon. It is undulating, with relatively few areas which are steeply sloping. It is open in aspect, with many field boundaries being fences and relatively few hedgerows. There are small areas of mainly deciduous woodland. Land use is a mixture of arable and cattle grazed pasture, with one or two fields grazed by sheep and horses. The proportion of land under pasture has increased over the last five years. The main crops are wheat, barley and oilseed rape. There have also been fields planted with maize and linseed. Apart from the increase in pasture there have been no obvious changes in land use.

Conclusion and Actions

The decline in numbers in recent years shows no signs of abating. However, the results from this survey have been used by the Rural Development Service (Defra) in presentations to farmers in the areas critically important for the species, Several of these farmers have subsequently applied for entry to the Environmental Stewardship Higher Level Scheme (HLS), with plans that include conservation measures for the species. The data gathered by HOS can now be used to prioritise assistance grants against the plans submitted. It is to be hoped that measures can be implemented to halt the decline. Otherwise, it would appear that extinction looms for this species in the county. It is very important that the survey work initiated in 2005 forms the basis of annual monitoring; particularly of those tetrads occupied by Corn Buntings in 2005 and of neighbouring areas

Acknowledgements

Glynne Evans, John Eyre and Peter Thompson advised on possible fieldwork techniques, and for promoting the use of the survey results. David Harper offered much useful advice based on his work in Sussex. However the big thanks must go to the 57 HOS members that agreed to take part in the survey and are responsible for delivering the results.

References

Donald, P. F. & Evans, A. D. 1995. Habitat selection and population size of Corn Buntings breeding in Britain in 1993. *Bird Study* 42:190-204.

Donald, P. F., Wilson, J. D. & Shepherd, M. 1994. The decline of the Corn Bunting. *Brit Birds* 87: 106-132.

Donald, P. F. 1997. The Corn Bunting in Britain: a review of current status, patterns of decline and possible causes. *The ecology and conservation of Corn Buntings, Miliaria calandra. P. F. Donald & N. J. Aebischer (Editors)*, 11-26. Peterborough, Joint Nature Conservation Committee. (UK Nature Conservation, No 13.)

Eaton, M. A., Ausden, M., Burton, N., Grice, P.V., Hearn, R. D., Hewson, C. M., Hilton, G. M., Noble, D. G., Ratcliffe N., Rehfisch, M. M. 2006. *The Status of the UK's Birds 2005.* RSPB, BTO, WWT, CCW EN, EHS and SNH, Sandy, Bedfordshire.

Hampshire Biodiversity Partnership, 2001
http://www.hampshirebiodiversity.org.uk/pdf/PublishedPlans/Seed-eatingFarmlandBirds.pdf

Manns, L. 1996. The Corn Bunting Survey 1993-94. *Sussex Bird Report 1994,* 133-137.

Mason, C. F. & Macdonald, S. M. (2006). Recent marked decline in Corn Bunting numbers in northeast Essex. *Brit Birds* 99: 213-214.

Keith Betton, 8 Dukes Close, Folly Hill, Farnham, Surrey GU9 0DR

INVASION OF WAXWINGS *Bombycilla garrulus* INTO HAMPSHIRE, DECEMBER 2004 – MAY 2005

John Clark

Abstract

An unprecedented invasion of Waxwings into Hampshire took place during the winter of 2004/05. Sightings were spread over the period from Dec 5th 2004 until May 8th 2005. Analysis of the available records on a weekly basis indicates that peak numbers were present in late January/early February, when a minimum of 1332 was present in the county. Summation of the maximum count in each area gives a total in excess of 3600, but as flocks deserted areas once food supplies were exhausted and moved elsewhere, this clearly exaggerates the total involved in the invasion. A figure of around 2000 seems realistic. The Hampshire picture is put into context with national events, and also compared with previous invasions to the county.

First sightings

The invasion started in the extreme north of Scotland, with a single on Fair Isle on Oct 7th, followed by flocks of 30-42 in Shetland, Lewis and Highland during Oct 15th-18th. Small numbers were also recorded on the east coast south to Norfolk at this time. By the end of the month, an estimated 7300 had been reported in Birding World (Gantlett 2004), mostly in Scotland, including a flock of 1168 counted from digital images taken at Forres (Birdguides). Birds continued to arrive in north-east Scotland in early November, but by the end of that month substantial three figure flocks also appeared in eastern England south to Norwich, north-west England and in Ireland south to Dublin.

During December, three figures flocks were present at several sites in East Anglia but only small numbers were found in the midlands, Wales and central southern England, with virtually none in the south-east. The first sighting in Hampshire involved a flock of 12 roosting in a small tree on heathland at Woolmer Forest on Decr 5th. This group possibly remained in the area as 16 were seen there on Dec 13th and 31st. Other records during that month, listed in the 2004 Hampshire Bird Report, involved a further eight individuals, although interestingly a flock of 36 was located feeding on rowan berries at Frimley Green, Surrey, just a mile from Hampshire, on Dec 24th.

Main arrivals in Hampshire

In the first week of the new year, the roosting group at Woolmer remained until at least Jan 3rd, three were reported at Hedge End on 2nd, and a flock of 37 was at Dibden Purlieu on 7th. The second week brought increased numbers. A group of 15 was discovered feeding on rowan berries at Hedge End retail park on Jan 8th; these proved to be a magnet for Hampshire birders on a daily basis up to Jan 23rd with a maximum of 49 recorded on 13th. This period also saw the first signs of the invasion in the north-east of the county, with 50 in Fleet on Jan 8th rising to 72 two days later. An estimated 186 were recorded in the county during that week. Numbers then increased steadily, with some established flocks growing and new gatherings in excess of 50 being discovered in Aldershot, Alton, Chandlers Ford, Cove, Hawley, Lymington, Nursling, Romsey, Rowledge, Southampton Common and nearby Bevois Valley, and Totton/Calmore. There was also a report of 25 at the Ordnance Survey site in Maybush, Southampton on Jan 28th. The total number jumped from an estimated 689 in the week of Jan 22nd-28th to 1332 in the following week.

This period also saw peak numbers in other parts of England. In January, there were flocks of 1200 at Newcastle-under-Lyme (Staffordshire) on 20th, 1115 near Cardiff on 22nd and 1200 at Bilston (West Midlands) on 31st. The movement southwards was demonstrated by peak numbers in other southern counties – 180 at Farnham (Surrey) on Jan 23rd, quite probably also recorded at nearby Hampshire sites, 460 at Canford Heath, Dorset on 24th, 207 at Melksham (Wiltshire) on 25th, 300+ at Southwater (West Sussex) on Feb 2nd and 95 at Newton Abbot (Devonshire) on 11th. The relocation of birds from Scotland was confirmed by sightings of birds colour-ringed in Aberdeen as far south as Devon. One individual noted by Jonathan Mist at Church Crookham on Jan 22nd had been ringed at Bridge of Don, Aberdeen on Nov 13th. It was still in that area on Nov 30th, but was sighted at Macclesfield, Cheshire on Dec 28th before ending up in Hampshire. Birding World estimated that 26,000 were present in Great Britain and Ireland in February (Gantlett 2005).

© Nigel Jones

In Hampshire, the total number present evidently fell from the second week of February, although the largest gathering, centred on the Ordnance Survey site and the nearby Oaklands School, held at least 314 on 10th and 352 at 0800 hrs on 13th, the latter counted from a video grab taken by Alan Lewis.

Subsequently, this flock broke up as food supplies diminished, evidently dispersing to other areas of Southampton with abundant supplies of rowan berries. It may well be that they continued to roost in the vicinity of the Ordnance Survey site, since 310 were present there early on Mar 9th.

Later in February and early March, new flocks were discovered at Woolmer Trading Estate, Bordon, and by the Basingstoke Canal near Aldershot, while numbers in excess of 50 continued to be recorded in Alton, the Chandlers Ford/Boyatt Wood area, the Greatham/Woolmer area, Hawley/Blackwater/Sandhurst and Romsey. An estimated 847 was still present in the week of Mar 5th-11th. March produced several records of small flocks flying over inhospitable habitat. These probably referred to birds relocating to new food supplies, although four flocks totalling 71 moving north over Basing on Mar 19th may well have been leaving the county. Birds may have been starting a return to the east, since numbers increased in south-eastern counties at this time, with large flocks appearing in Greenhithe (Kent) (150 on Mar 7th), central London (132 on Mar 11th) and Grays, Essex (160 from Apr 11th-13th).

© Alan Cox

Last Sightings

In early April, flocks in excess of 20 were still present in six areas, and, much to the delight of Basingstoke observers, after a series of brief glimpses of ones and twos and a short staying flock of up to 28 at the Tesco supermarket in nearby Chineham between Mar 8th and 16th, a flock finally became established in the town at the Houndmills Industrial Estate, with 20 on Apr 18th rising to a peak of 140 two days later and the last being 34 on 27th. The second half of April also produced notable flocks in Southampton and Church Crookham. In Southampton, a maximum of 41 was recorded at the Polygon School on Apr 18th, but subsequently most sightings were from Trinity Road car park, St Mary's, where there were 32 on Apr 27th and 28th and the last were eight on May 1st. In Church Crookham, there were 75 on Apr 18th, 65 still present on 28th and the last record of 18 on May 2nd. However, the latest to be recorded was one seen well by Marc Moody in his garden in Lymington on May 8th. Nationally, around 680 were still present in May, including flocks of 60-65 in Aberdeen, Edinburgh and Nottingham and the last two in Oxfordshire on May 16th.

Weekly maxima from sites where in excess of 50 were recorded are shown below:

	January			February				March					April				May	
	8/14	15/21	22/28	29/4	5/11	12/18	19/25	26/4	5/11	12/18	19/25	26/1	2/8	9/15	16/22	23/29	30/6	7/13
Fleet/Crookham	72	81	105	184	135			41	25			50	50	50	75	65	18	
Farnborough/Cove		81	60	44					12	10		8						
Hawley			180	180			90		40	50								
Aldershot			65	90	150	100	80											
Basingstoke Canal							80	80	110	140	78							
Rowledge				90														
Alton	8	25	40	174	45	40	40	130	18	50	50	50	40	60	22	20		
Bordon					11	53	76	74	70	19		16						
Woolmer/Liss		17	20	20	18	50	34	72	10		15	57	22	24				
Maybush area			25	92	314	352	203	115	310	150	130	63	30	54				
Southampton		36	70				24		58	40		50		17	7	41	32	14
Totton/Calmore				150		25			50	100								
Romsey				63				78			17		2	25				
Nursling				80								38						
Boyatt Wood		19		30	45		50	111										
Chandlers Ford		1		50	100	38		58	84	157	60	50	35					
Andover					51			15				11						
Basingstoke							20	1	28	10					56	140	100	
Lymington			29	85	60	7	4	4				52						1
Lyndhurst											54							
Estimated total	186	342	689	1332	1001	775	771	904	847	700	625	420	267	349	289	152	32	1

Distribution

Figure 1 shows the distribution of all records by tetrad. This confirms the majority of sightings were in the Southampton/Romsey/Chandlers Ford area, Lymington and the north-east and east of the county:

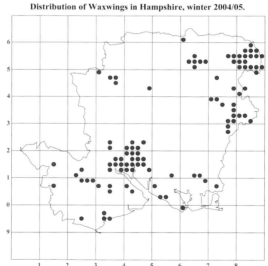

Distribution of Waxwings in Hampshire, winter 2004/05.

Most of the other registrations referred to small numbers seen on one date only. The lack of sightings from Winchester and the Gosport/Fareham/Portsmouth area is surprising.

Comparison with previous invasions

Between 1951 and 1992, Waxwings were recorded in 22 winters, including 202 in 1965/66, 134 in 1970/71 and 51 in 1988/89 (Clark and Eyre 1993). Since then, there were 215 in early 1996, one in early 1997, 47 in early 2001, 35 in early 2003 and singles in December 2003 and February 2004. Thus the 1996 arrival was the largest prior to the 2004/05 invasion, although, given the much poorer observer coverage in the mid 1960s, it is likely that the 1965/66 influx involved more birds.

Acknowledgements

My thanks to James Lidster, Christian Melgar, Graham Sparshatt, Jeffery Wheatley and Tim Walker for their help in compiling this paper.

References:

Clark, J.M. and Eyre, J.A. 1993, *Birds of Hampshire.* Basingstoke: HOS.
Gantlett, S. (Ed.) 2004, Bird news October 2004, *Birding World* 17,10.
Gantlett, S. (Ed.) 2005, Bird news February 2005, *Birding World* 18,2.

J. M. Clark, 4 Cygnet Court, Old Cove Road, Fleet, Hampshire GU51 2RL.

BIRDING BY BIKE: THE HAMPSHIRE "BIG YEAR" 2005

Simon Woolley

In December 2005, my wife and I arrived in Sri Lanka on Boxing Day - for a birding trip – so we very narrowly missed the appalling events of that day. On my return home, as a combined get-fit plan and fundraising scheme (for a Sri Lankan school affected by the tsunami disaster), I decided to make an attempt on the Hampshire non-motorized year list record. Fortunately for me, no-one had tried this before, so I was guaranteed the record as soon as I had heard that Robin singing before dawn in January! Nevertheless, I set myself a working target of 180 (which I thought would be pretty outstanding), and got pedalling early in the new year.

I was by no means a fit or experienced cyclist, but I was pleasantly surprised to find that I was covering some serious mileage and seeing plenty of birds – by the end of January, I'd banked 89 species (including Waxwing, Short-eared Owl, Woodcock, Ring-billed Gull and Great Grey Shrike), and cycled 272km, giving me a return of a bird every 3.1km – not bad! Of course, things quickly got rather harder....

February turned up another 25 species, including Jack Snipe and Razorbill, but the main highlight was the purchase of a sleek new road bike – more speed and narrower tyres. Lance Armstrong I wasn't, but I was now moving about much more speedily, and rather more sexily (in my lycra bib-shorts)!

Spring kicked off with a consolidation of my owl list, displaying Goshawks, a Spoonbill, Ravens and even a Red-crested Pochard – which after some soul-searching I decided was 'tickable'. The miles were now racking up fast and I was learning the times needed to travel to the various key sites – from Winchester, Hayling Island is 2½, Farlington 1¾ , Titchfield 1½ and Pennington about 2 hours – and then back rather more slowly!

April and May gave me a major species rush, of course, and Hampshire was liberally blessed with rarities for me to 'twitch' – Spotted Crake, Roseate Tern, Whiskered Tern and Montagu's Harrier spring to mind! But rare breeders like Red Kite, Honey Buzzard, Stone Curlew, Firecrest, Grasshopper Warbler and Willow Tit were just as valuable, and some big sea watches from Hurst gave me skuas, divers and other seabirds - including awkward species like Kittiwake, Arctic Tern and Fulmar. By the end of May, I was looking at 14.3km per species – and it would only get tougher. On the up-side, I reached my provisional target of 180 as early as May 25th! 200 was looking possible – but how high could I go by December?

Summer brought an inevitable birding lull, punctuated by the ghastly event of my wife, Julia Casson, getting hospitalised for six weeks with a broken back. She's absolutely fine now! We had planned a trip to Malaysia, but that didn't happen...probably for the good as far as my year list was concerned.

Autumn began well, with an Osprey at Lower Test Marshes, a Great White Egret at Blashford, Wilson's and Grey Phalaropes, Wryneck, and Wood, Baird's and Pectoral Sandpipers. It suddenly looked like Hampshire was having an *exceptional* birding year and I'd picked a good one to go for this record! Indeed, in late October, a ride to Titchfield Haven did the business – I relocated the Lesser Yellowlegs on the floods to bring up the double century of birds. But just two days later, I was at Titchfield once again for a Lapland Bunting, and just a week later, Gosport struck back with Laughing Gull – I got there in time!

The major highlight of the year came on November 13th (see Desert Wheatear article later in this Report) – after 4000km on the road, this felt like a good reward! But still autumn refused to die – I rode to Pennington in late November for Snow Bunting, saw it, but also found a Richard's Pipit, and refound the Gosport Laughing Gull. On that coldest of days, I was on fire.

Now over 210, my year list continued to tick over with some truly bitter winter rides – 0530 hrs setting off from Winchester for Gosport, in December, in temperatures of -3°C, anyone? But Guillemot, Red-necked Grebe, Great Northern Diver, Marsh Harrier, Bittern – who's complaining? At the very end of the year, the Normandy Scaup and the Romsey Common (Mealy) Redpolls brought my list to a very pleasing 217, collected over 4868km (or 3025 miles), at 22.4km per species. I'd pedalled for eight days and 17 hours (that's actual time on the bike), or 2.4% of the whole year, averaging 23.3 km/h.

In what was truly a stunning Hampshire birding year, 216 represented 79% of the approximately 273 species recorded or, more meaningfully, 88% of those which I feasibly *could* have seen, given more time, luck and even more energy! (My personal record stands at 217 – I myself count Black Brant as a good species!)

I found 165 species for myself, and outscored all other Hampshire year-listers in 2005, even those using cars. Such a total would have been quite impossible without help, support, information and even accommodation from a whole legion of Hampshire birders, to whom I owe a very great deal.

Sponsorship and sales of my book (still available from the author at £5 – contact me on skw@wincoll.ac.uk) raised over £2200 for St Thomas' College, Matara, in southern Sri Lanka – my thanks go to all those who contributed. I am also indebted to Debbie Thrower for putting together a short film package for Meridian Tonight, featuring the lycra-clad bird fiend pedalling about manically. This really helped to boost sales. It was an eventful day – I punctured, Debbie struggled badly with some of the hills and she also fell off her bike into rather a large pile of what might optimistically be called "mud"!

Notwithstanding Debbie's misfortune, I would recommend at least occasionally abandoning the car to anybody who wants to see more of Hampshire's birds, and/or who wants to lose a few pounds. You'll learn to see the county in a new light, and you'll never take those quick drives to the coast for granted again! I genuinely look forward to seeing my record bettered one day – but so far no challengers have come forward. If anyone *does* fancy "going for it", here are my top ten tips:

1. Get a fast bike – it makes a huge difference

2. Go for those potentially tricky winter species in the *early* part of the year – time flies at the end of the year!

3. Abandon any thought of carrying a 'scope

4. Develop a network of contacts across the county for "gen" and in the field assistance (including use of a 'scope)

5. Marry a wonderful woman like Julia who will occasionally drive out and meet you with a flask of soup

6. Plan cross-country routes that avoid major roads where possible – though you'll find that sites like Farlington Marshes are pretty tough in this respect

7. Forget having a social life

8. Expect punctures and crashes – I had eight of the former, but (only by luck) none of the latter. The going rate is one crash per 8000km, apparently!

9. Buy the best carry bag, clothing, helmet and lights you can afford

10. Never, ever give up – bike birding forever!

Simon Woolley 2 Culver Lodge, Culver Road, Winchester SO23 9JF

DESERT WHEATEAR, HAYLING SEAFRONT, NOVEMBER 13TH 2005

Andy Johnson and Simon Woolley

This joint account of the finding of the second-ever Desert Wheatear for Hampshire has been compiled from the separate notes made by the authors [Editor].

[SKW]: In 2005, I successfully attempted to set a new Hampshire record for a non-motorised year list. I travelled over 3000 miles, mostly by bike, and recorded 216 species. This proved to be 79% of all the species recorded in the county, and 88% of the theoretically (if not practically) 'twitchable' ones.

Following an incredible late autumn run of twitches for Baird's Sandpiper, Lesser Yellowlegs, Lapland Bunting and Laughing Gull, the undoubted highlight of the year came on Sunday Nov 13th.

About a month previously (and as God is my witness, this is true), when contemplating the likely autumn cycle rides, I casually mentioned to my wife, Julia Casson, that it was amazing that Hayling Island never seemed to get really rare passerines. "I reckon it'd be good for a Desert Wheatear in late October or November", I said.

Fast forward to Nov 13th, a freezing cold morning after a clear, starry night. Frost covered the cars and the water meadows as I pedalled off to the south, departing just after 0700 hrs, my face chilled by the cold air. My target for today was Black Redstart, which had become a bogey bird after two long-distance dips last winter, both at Hayling Island. The information was that one had been in residence for about a week, and was site-faithful. I decided to go for it, especially as there were some other possible year-ticks (e.g. Great Northern Diver, Velvet Scoter, Red-necked Grebe, Little Gull, Long-tailed Duck), which might be on offer.

[ACJ]: I had long contemplated finding a Desert Wheatear at Sandy or Black Point (amongst a myriad of other hopes, which, let's be honest, is what keeps us going); in particular, I had been checking the rocks at Hayling Island Sailing Club on an almost daily basis – my only previous one being in similar habitat at Rosslare Harbour in 1997.

Sunday 13th November started well enough with a Snow Bunting rippling and "pewing" over Sandy Point at 0720 hrs , followed by 1550 Woodpigeons, 148 Goldfinches, 20 Meadow Pipits, five Siskins, four Bramblings, two Redpolls and a few other bits and bobs. Up to the Black Point wader roost for a nine o'clock high tide produced the usual selection of waders in unremarkable numbers (and nothing on the Sailing Club rocks), whilst a look at the sea revealed the usual lack of any activity whatsoever.

[SKW]: The familiar ride to the furthest, most south-easterly corner of the county took me just over 2¼ hours for the 42 miles. My feet felt like blocks of ice. But it was with optimism that I started combing the seafront houses, though with no luck in the first twenty minutes.

[ACJ]: It was now around 1000 hrs. Knowing that Simon was cycling down from Winchester to try for Black Redstart, I decided to see if I could pin one down for him along the seafront. Before I'd got far, I met Simon coming the other way. He'd had no luck so we idly chatted about various birds, including the Sabine's Gull I had found two days previously - virtually where we were standing!

Simon then got a phone call from his back-up team (Julia) in the car. While he took the call, I noticed a small bird about 150m along the seafront drop off the wall onto the footpath and then flick up on to the shingle, showing an obvious buffy rump! Knowing full well what I'd just seen, I scrabbled to get the 'scope' up. Meanwhile, I could hear Simon on the phone:

"Where are you? ... OK, stay where you are and I'll come and find you," he said and I interjected:

"You'd best not go anywhere – I've just had a wheatear up the beach!"

Simon (still on the phone, and also knowing full well what I was implying):

"OK, stay where you are, Julia, and I'll come and find you – in a minute!"

[SKW]: I scanned, picking up first a Pied Wagtail, and then a brown, vaguely variegated passerine on the beach. I directed Andy to where I was looking, and to our mutual delight (we were both thinking the same thing, I am certain), he called it:

"DESERT WHEATEAR!"

There it was, a cracking first-winter male! Joyous celebration – and the first round of hurried texts fired off - joy unrivalled, delight unparalleled, adrenaline rush, simply awesome. I ran down to the road to fetch Julia, but returned briefly to see the bird again while she got organised.

[ACJ]: Now to get closer and keep tabs on the bird. An immediate problem became apparent: the bird was favouring the seafront path, hopping up onto the wall or beach either side. But it was a beautiful Sunday morning, and increasing numbers of happy walkers materialised seemingly from nowhere. Much gratitude to the gentleman who volunteered to go round the beach and not disturb the wheatear, now happily foraging at much closer range. However, faced with half a dozen walkers converging from both sides, the bird disappeared.

Julia arrived – no Desert Wheatear! - the ultimate flashpoint for married birders. Until now, the bird had reappeared on the wall after each danger had passed. This time it didn't. It was 1024 hrs, and the first person to respond to the text message arrived, Tim Lawman, followed by an increasing number of expectant birders. It became apparent that the wheatear was not where it had been, and a rather panicked search of the immediate surrounds began. If it's gone over/through the houses, it could be anywhere! The search widened, and I began to fear the worst.

[SKW]: What had looked like being the greatest morning of them all was swiftly becoming an anti-climax – we simply couldn't find the bird, and Julia was dipping on what would have been not only a British but also a Life tick. Within half an hour, perhaps 30 birders had arrived, but still we could not relocate it. Those dark questions began: "What age was it?", "Did you see the tail?", "Did you rule out aberrant Northern Wheatear?" Andy and I were having none of this, but it was still a brief two observer sighting at this stage.

[ACJ]: What relief to get a call from Tim at 1238 hrs! He and Pete Gammage had relocated the bird west of Beachlands car park – a full mile and a half along the seafront. Now with plenty of beach in which to avoid the still numerous walkers, the bird performed superbly for an ever-increasing crowd of admirers.

[SKW]: By now, Julia and I had adjourned to the Hayling Oysterbeds to avoid the crowds and to try and eke out a Long-tailed Duck for my year list – but the pager alert soon had us back at the beach! Julia connected at last, and I had much better and more relaxed views.

[ACJ]: The wheatear continued to work its way westwards, pursued at a suitably respectful distance by birders, ending up between the funfair and the Inn on the Beach by the time the light was failing. Unfortunately, there was no sign of the bird the next day, although several hopefuls were treated to a fly-by Hoopoe, which had made the trip westwards from Sandy Point even faster!

Desert Wheatear   *© Richard Ford*

Description.

Even at a quick glance, the bird was clearly an *Oenanthe* species – jaunty, essentially chat-like, and yet immediately obvious as not a Northern Wheatear *Oenanthe oenanthe*, on account of its less sleek, more petite and rounded appearance, not to mention some pretty obvious fieldmarks. The bird appeared noticeably smaller than *O. oenanthe* (maybe 90% the size?), with a more rounded head shape and, perhaps, a proportionally slightly shorter tail.

One of our immediate descriptions written in the field (SKW) described a small wheatear: buffy brown on mantle and crown; paler sandy coloured below, with markedly black-flecked cheeks and ear coverts; largely dark wings (with pale edges) connecting up to the dark on the sides of the head and, crucially, showing a wholly black tail, apparently

220

Flaring pale supercilium shown well here

Good view of the blackish marks joining wing to head markings at this angle

Note much paler underparts

Very variegated wing pattern evident in this flight shot

Solidly black tail

Dark legs

Darker, more intensely warm buff

Buffish fringes on inner tertials

Solidly black tail – possibly the *tiniest* amount of white visible at base in this photo?

flying up
vertically
catching flies

tail wagging

imm ♂
Desert Wheatear
Hayling Beach
13 Nov 05

dead dock

Desert Wheatear

© David Thelwell

squarely cut off against buff uppertail coverts.

Our later views were much better, and confirmed what had been seen earlier - in more detail: the bird's whitish supercilium was noted, narrow over the bill and in front of the eyes, but flaring softly behind the eyes; the slightly darker buff tone to the mantle and scapulars (reminiscent of Western Black-eared Wheatear *O. hispanica*, but not as intense), and more detail on the wing pattern.

The visible tertials and primaries were essentially very dark brown, with striking whitish (outer) and buffish (inner) webs on the former, and fine, slightly diffuse whitish-buff tips on the latter.

The wing coverts were distinctly darker (virtually black), with strong whitish fringes. SKW noted (from photographs above) an apparent slight moult contrast in the greater coverts – the outer feathers appeared to have a pale brownish wash on the feather edges. This suggested that bird was a first-winter, rather than an adult male, as first suspected. Based on the pale throat and generally warm buff upperparts most authorities ascribe this plumage to first-winter rather than to an adult male Desert Wheatear *O. deserti* .

The blackish feathering on the head was essentially restricted to the cheeks and the ear coverts, though with a distinct and quite marked blackish 'smear' between the lower cheek and the bend of the wing, thus essentially 'joining up' the head and wing markings. The cheeks and ear coverts were apparently very heavily pale-fringed, obscuring what would become a solid black area next spring. The throat appeared essentially pale, though suffused darker, and with a hint of a dark sub-moustachial line.

The tail was solidly dark (almost black) – no trace of white near its base could be seen in the field - and even in SKW's photographs, only the vaguest possible hint of whitish feather bases was visible. There was certainly no inverted-T-shape as present in most other *Oenanthe* species.

Both bill and legs were dark. No call was heard.

Andy Johnson 5/22 Bracklesham Road, Hayling Island PO11 9SJ

Simon Woolley 2 Culver Lodge, Culver Road, Winchester SO23 9JF

Editorial Note

Desert Wheatear *Oenanthe deserti* breeds widely but discontinuously from North Africa across the Middle East and Asia to North China. The Asian breeders are long distance migrants, wintering from India to the Arabian peninsular. It is the Asian form which, following the post-juvenile moult, is presumed to be responsible for the extraordinarily consistent very late autumn arrivals: a few in October from mid month, most in November and even some in December. Between 1958 and 1983, there were just ten British records, including one at Farlington Marshes from Nov 4th-19th 1961. Favourable weather conditions are clearly required for such extreme vagrancy and 27 of the 66 records between 1984 and 2004 have been in just two years - 1997 and 2003 - with five years blank, including 2004. None of these records was in Hampshire - Andy and Simon changed that in 2005. The delightful first-winter male described above was one of four and, additionally, three females found in Britain. The other six records were more typically on the east coast, from Kent to Shetland. Across the English Channel, two arrived on Guernsey in November, including one on 13th - the same day as the Hayling Island record.

FIRST FOR HAMPSHIRE – LAUGHING GULL

John A Norton & Peter N Raby

© Richard Ford

During late October Hurricane Wilma tracked up the eastern seaboard of the USA and by 27th had reached Newfoundland. The subsequent depression moved quickly across the Atlantic and was centred on the Azores by the 29th, it then moved up through the Western Approaches carrying the largest group of Nearctic gulls ever to arrive in Britain. The first noted was a Franklin's Gull *Larus pipixcan* in Cornwall on Nov 1st with another in Ireland the following day and Laughing Gulls *Larus atricilla,* on both the Isles of Scilly and Cornwall (Ahmad, 2005), then a first-winter arrived at Weymouth, Dorset on 3rd.

During the afternoon of Friday 4th November 2005 one of us (PNR) briefly noted a medium-sized dark gull at Stokes Bay, Gosport. It was seen from a passing car (PNR works as a driving instructor) and it was not possible to stop to check the bird further. PNR thought nothing more of it until looking at the 'Surfbirds' web site that evening and discovered that there had been an influx of Laughing Gulls into south-west England. Later that evening, whilst discussing our forthcoming WeBS count on the following day, PNR commented to JAN that the dark gull he had seen earlier at Stokes Bay could possibly have been a Laughing Gull.

On Nov 5th we started off the day checking out a number of known gull feeding areas in the town, with no sign of the Stokes Bay individual. As part of our WeBS high tide count we arrived at Walpole Park Boating Lake, Gosport at 1245 hrs. After counting Mute Swans and Canada Geese we checked and counted the gulls on the lake and disappointedly could not find the regular Ring-billed Gull or one of the many Mediterranean Gulls that also frequent the Walpole Park area. We then headed towards Haslar Creek to the south, walking around the Boating Lake and across some adjacent amenity grassland to do so. Before reaching the creek, a noticeably dark gull of roughly Black-headed Gull size landed on the grass about 10m in front of us. A quick check through binoculars revealed the most obvious features, especially those to eliminate Franklin's Gull: long slightly drooping dark bill, dark rakish looking legs, elongated tail end and white eye crescents, and we both realised straight away that it was a first-winter Laughing Gull.

The Laughing Gull stayed on the grass for three minutes or so until it was enticed into flight by someone feeding the Mute Swans on the Boating Lake. It flew around for the next ten minutes, giving excellent flight views. Eventually it settled on the water and allowed PNR to take some digiscoped photos. The news was put out to local birders and Birdguides and the first arrived within the next half an hour. At 1336 hrs the gull flew off and headed west over Gosport town. We continued on our WeBS count and returned to the lake at roughly 1430 hrs, to find that the gull had returned. It was present until 1540 hrs at least when we decided to leave.

Laughing Gull, Normandy Marsh   © *Marcus Ward*

Description

A general description of the bird was noted. No attempt was made to write a detailed description of all plumage characters; as it was felt that good photographs would be more useful in later analysis (e.g. comparison of other similar-plumaged Laughing Gulls that had arrived in the country).

Size/shape: In flight, appeared roughly the same size as a Black-headed Gull, *Larus ridibundus* but the wings were slightly longer and broader, giving the gull an elegant, almost tern-like flight. The bird showed good manoeuvrability in flight, also like a tern. On the ground, the body size of the bird was about the same size as Black-headed Gull. When swimming on the water the long wings gave the gull an attenuated look towards the tail and wing tips.

Laughing Gull, Normandy Marsh   © *Marcus Ward*

Plumage: In general the head, hind-neck, breast, flanks and mantle were a dark, metallic grey, with white on the fore-crown, forehead, chin and throat. The bird had obvious white eye crescents above and below the eye, set off

225

against the dark grey ear coverts and slightly lighter grey extending below and above the eye, and also a black, horizontal line extending a short distance forwards from the eye. The folded wing was dark grey with some pale brown juvenile feathers still showing on the lesser and greater coverts. The tertials were predominately muddy chocolate brown with whitish fringes and the primaries were black. In flight, the whole of the wings appeared rather dark, with blackish primaries and grey-brown inner wing, with contrasting black secondary bar edged with white along the trailing edge of the wing. The tail was white with a broad dark band at the tip.

Bare parts: The bill was all black and slightly longer and broader than Black-headed Gull, with a noticeable bulge towards the end, caused by an angled lower mandible; it had a nostril hole that one could see straight through. The legs were also black. Eye/iris: black.

We presumed this to be a juvenile Laughing Gull, moulting into first winter plumage. There was no doubt as to the gull's identification. The bird was still present when we first submitted this description to *HOSRP* and *BBRC*. It was last seen at Gosport on Nov 13th (see Systematic List for later sightings in Hampshire).

Reference

Ahmad M., 2005 Franklin's and Laughing Gulls in Britain & Ireland in November 2005, *Birding World*, 18, 461-464.

J. A. Norton, 215 Forton Road, Gosport PO12 3HB.

P. N. Raby, 141 Tukes Ave, Gosport PO13 0SB.

Editor's Comment:

Laughing Gull is an abundant species in North America with a summer population in north-eastern USA and eastern Canada which migrates to southern USA and south to Venezuela in autumn. Hurricane Wilma reversed this migration and drew an estimated 60 Laughing Gulls (50 records already accepted by *BBRC* in 2005 report) and seven Franklin's Gulls into Britain and Ireland with further sightings in continental Europe (see Ahmad, 2005). Of these at least two went further east along the English Channel to Kent and Sussex with a report of an adult at Lepe on Nov 5th (Birdguides) possibly being the Sussex adult, although this report has not been submitted as a record to *HOSRP*. What was presumed to be the Gosport bird was reported at Hill Head on Nov 7th. Then, as included in the systematic list, *This Report,* there was a further sighting of a first-winter by several observers at Normandy, Lymington on Nov 20th which was documented and submitted to *HOSRP/BBRC*. This was in the same state of moult as the Gosport bird and was judged to be the same individual (SKW) and accepted as such by *BBRC*. The last sighting report was of a first-winter at Haslar Creek on Dec11th. The *BBRC* has notified acceptance of John and Peter's discovery of the Gosport Laughing Gull which becomes the first record for Hampshire.

FIRST FOR HAMPSHIRE – RADDE'S WARBLER

Nicholas Montegriffo & Andy Pink

© Mike Duffy

At approximately 1045 hrs on the morning of Dec 7th 2005 I (NM) received a phone call from Andy Pink who was visiting an industrial area at Stokes Lane, Sherborne St John, Basingstoke - a few hundred metres from my home. Andy had found a "Phylloscopus" which he described as calling like a Blackcap *Sylvia atricapilla* or Lesser Whitethroat *S. curruca*, a hard "chack", but Andy had no binoculars, or field guide. Knowing Andy to be an extremely reliable and skilful observer, I immediately grabbed two pairs of bins and joined him on the site. Andy told me he thought the bird was either a Radde's Warbler *Phylloscopus schwarzi* or Dusky Warbler *P. fuscatus*, and we both thought the latter more likely given the late date (how wrong we were), but we first had to see it properly! I had found a Radde's Warbler on St Mary's, Isles of Scilly in November 2002, so had some prior experience of separating this species from Dusky Warbler.

The bird was confined to an area approximately the size of a football pitch, covered in nettles and obviously used as an informal dumping ground for old farm machinery and paving tiles, and very overgrown. Adjacent to this area was a small car park, surrounded by an embankment, the entire area bordered by a system of old hedges. It had clearly been a very cold night, with heavy frost still evident. We parked up and worked the area for nearly an hour, by which time Andy had to leave, so we returned to our vehicles. Fortunately, Andy then relocated the bird on top of the embankment surrounding the small car park, but the bird dropped out of site before I had a chance to see it. We stalked the bird around the back of the car park, and it eventually popped into view. One glance at the long supercilium and apricot-coloured undertail coverts was enough: it was clearly a Radde's Warbler!

For the next few minutes, the bird showed very well on the embankment down to five meters, and I cursed at not having brought my camera. The hard "chack" call was also heard. A small group of seven Lesser Redpoll *Cardeulis cabaret* was also present. Unfortunately, the site was on private land, so the news could not be put out to the general public until Andy had obtained permission from the land-owner; however it was possible for a few people to visit the site with Andy's permission. Meanwhile, Andy had to go back to work, so I went home to pick up my camera, and returned an hour later with my video camera, with which I was eventually able to get enough footage to take the video grabs shown below.

While at home, I looked through *Birds of Hampshire* and all the recent *Hampshire Bird Reports* to get some idea of how many Radde's Warblers had been recorded in Hampshire. It was only then that the significance of Andy's find became apparent: it was the first for the county, and one of the latest Radde's Warblers ever recorded in Britain.

Video grabs from footage taken on JVC Digital Video (GRD63EK) camera

By now, Mike Duffy, Peter Hutchins and Trevor Ellery had arrived, and Mike had managed to photograph the bird, which had settled down in an area of nettles mid-way along the western edge of the site, in which it appeared to be actively searching for food (very many small flying insects were gradually emerging). It continued to show on and off until dusk, but was generally very skulking, spending most of its time on the ground or in low vegetation, with occasional visits onto the embankment, and it no longer called. Everybody who saw the bird agreed that it was a Radde's Warbler. Unfortunately for those who turned up next day, there was no further sign of the bird.

Description.

General impression was of a large, brown Phylloscopus warbler.

Upperparts. Quite a warm brown colour.

Underparts. Pale grey-buff, with distinctive apricot-coloured undertail coverts and vent (best seen in photographs four and five).

Head. Very prominent, long supercilium, with dark border to upper edge of supercilium and dark eye-stripe. The supercilium had a distinctive upwards "kink" towards the rear (photograph six). It was also a warmer buff in front of the eye (photograph three). The ear coverts appeared slightly mottled (photograph three).

Tail. Quite broad, medium length.

Legs. Pale pink (photograph 2)

Eyes. Dark.

Bill. Pale.

Call. A hard "chack", like a Lesser Whitethroat or Blackcap.

Nicholas Montegriffo, 13 Cranesfield, Sherborne St John, Basingstoke RG24 9LN.

Editor's Comment:

As Nick has explained, it was not possible to release the information of this extraordinary late year find by Andy Pink until the site owner's had been contacted and agreed to site management arrangements for the mass invasion that would have surely followed. As it was the bird had moved on before arrangements could be made. Although over 250 have now been accepted by BBRC in the last 58 years – they have been almost exclusively confined to the peripheral shores of Great Britain and typically in late October or early November. Breeding in the remote Siberian taiga and wintering in south-east Asia, Radde's Warbler is one of the "must see" classic reverse migration "Sibes" for the listing community. By coincidence I was one of the fortunate few who had a close encounter with the individual that Nick located on St Mary's, Isles of Scilly – again there was much human agony as this individual's stopover was similarly brief. The record has been accepted by *BBRC* and becomes the first record for Hampshire.

HAMPSHIRE ORNITHOLOGICAL SOCIETY

Honorary Officers, 2006-2007

President: Chris Packham

MANAGEMENT COMMITTEE

Chairman:	John Eyre	3 Dunmow Hill, Fleet, GU51 3AN
Honorary Secretary:	Barrie Roberts	149 Rownhams Lane, North Baddesley, SO52 9LU
Honorary Treasurer:	Nigel Peace	4 Wincanton Close, Alton, GU34 2TQ
Membership Secretary:	Dawn Russell	5 Grange Cottages Broadoak Botley SO30 2EU
Membership Records Officer:	John Norton	215 Forton Road, Gosport PO12 3HB
County Recorder:	John Clark	4 Cygnet Court, Old Cove Road, Fleet, GU51 2RL

Chairman of Scientific Committee:	Glynne Evans
Chairman of Membership Committee & Meetings Co-ordinator:	Andrew Walmsley
Conservation Liaison Officer:	Colin Young
Bird Report Editor:	Alan Cox
Production:	Mike Wall
Newsletter Editor:	David Thelwell
Publicity Officer:	Keith Betton
Sales Officer:	Margaret Boswell
Co-opted Members:	Brian Sharkey, Mike Wall, Matt Coumbe
Ordinary Members:	Barrie Roberts, John Shillitoe

MEMBERS OF SUB-COMMITTEES

Scientific:
Glynne Evans (Chairman), Brian Sharkey (Secretary), Duncan Bell, John Clark, Alan Cox, John Eyre, John Shillitoe, Colin Young, Keith Wills

Membership:
Andrew Walmsley (Chairman), Margaret Boswell, Judith Chawner, Matt Coumbe, David Gumn, John Norton, Dawn Russell and David Thelwell

Membership enquiries should be addressed to the Membership Secretary, subscription renewals and membership enquiries to the Membership Records Officer and other general correspondence to the Secretary.

POLICY, MEMBERSHIP AND ORGANISATION OF THE HAMPSHIRE ORNITHOLOGICAL SOCIETY

The Society was founded in March 1979, following the winding-up of the Ornithological Section of the Hampshire Field Club. The objects of the Society are, within the County of Hampshire:

a. To advance the education of the public in all aspects of ornithology.

b. To promote research into ornithology and to publish the useful results of such study and in particular to publish reports, newsletters and other papers of ornithological interest or as may be deemed by the Management Committee suitable or desirable for promoting the Society's objects.

c. To support and encourage the preservation and conservation of wild birds and places of ornithological interest.

Membership is available for the following subscriptions (effective after September 30th, 1994):

Ordinary Membership (including students))...................}	
Family Membership)...}	
(2 or more members at the same address).................}	£9.00
Corporate Membership (Schools, NHSs etc.)................}	
Junior Membership (Members under 18)	£3.00

Application should be made to the Membership Secretary. Subscriptions are renewable on January 1st each year, but the subscription of any new member joining after September 30th shall cover the succeeding calendar year, save that members so joining shall not be entitled to receive the Bird Report published in the year they are joining.

All members receive the annual Bird Report and quarterly newsletters, which give details of all indoor and outdoor meetings arranged on behalf of the Society.

The Newsletters contain Society news and views, articles on various aspects of Hampshire ornithology, recent reports, ringing news, and details of the organisation, progress and results of surveys organised by the Scientific sub-committee. These include wildfowl and wader counts, breeding censuses and migration watches. New surveys are started every year, and the newsletter serves as a medium to contact potential volunteers. All members are invited to take part in surveys and contribute articles to the newsletter.

Regular field meetings are arranged for all parts of the county throughout the year. Meetings are designed to introduce members to various habitats and to provide opportunity for novices of all ages to learn more about birds under field conditions.

Indoor meetings (usually illustrated lectures) are arranged in the winter months. Most of these will be of interest to all members, but some may cater particularly for specialist groups, e.g. survey workers' meetings.

There are two sub-committees:

(a) The Scientific sub-committee is responsible for organising surveys, for the collection of records and for the production of the Bird Report. It also handles liaison with the BTO, ringing groups, conservation bodies and similar organisations.

(b) The Membership sub-committee is responsible for the Newsletter, the Members' Day, conferences, indoor and outdoor meetings and the library."

Officers of all the committees are listed on page 230 of this report.

HAMPSHIRE BIRD REPORT

BACK NUMBERS

The following issues are still available

Year		Price	P&P
1991	-	£1.00	£0.75
1992	-	£1.00	£0.75
1993	-	£1.00	£0.75
1994	-	£1.00	£0.75
1995	-	£2.00	£0.75
1996	-	£3.00	£0.75
1997	-	£3.00	£0.75
1998	-	£3.00	£0.75
1999	-	£5.00	£0.75
2000	-	£5.00	£0.75
2001	-	£8.00	£0.75
2002	-	£8.00	£0.75
2003	-	£8.00	£0.75
2004	-	£8.00	£0.75

Please add postage and packing as indicated for each report order.

BIRDS OF HAMPSHIRE

Produced by John Eyre, the checklist gives a full list of all species ever reliably recorded in Hampshire with a month by month guide to their relative abundance.

Price £1.50 including P&P.

Cheques should be made payable to *HOS*

Orders to: Mrs. M. Boswell,
5 Clarence Road,
Lyndhurst
SO43 7AL